Imagining the Atacama Desert

Imagining the
ATACAMA DESERT

A FIVE-HUNDRED-YEAR
JOURNEY OF DISCOVERY

RICHARD V. FRANCAVIGLIA

THE UNIVERSITY OF UTAH PRESS
Salt Lake City

The Defiance House Man colophon is a registered trademark of
The University of Utah Press. It is based on a four-foot-tall Ancient
Puebloan pictograph (late PIII) near Glen Canyon, Utah.

LIBRARY OF CONGRESS CATALOGING-IN-PUBLICATION DATA
Names: Francaviglia, Richard V., author.
Title: Imagining the Atacama Desert : a five hundred year journey of
    discovery / Richard V. Francaviglia.
Description: Salt Lake City : The University of Utah Press, [2018] | Includes
    bibliographical references and index. |
Identifiers: LCCN 2017052578 (print) | LCCN 2017056146 (ebook) | ISBN
    9781607816119 0 | ISBN 9781607816102 (cloth)
Subjects: LCSH: Atacama Desert (Chile)--Discovery and exploration. | Atacama
    Desert (Chile)--Civilization. | Physical geography--Chile--Atacama Desert.
Classification: LCC F3131 (ebook) | LCC F3131 .F73 2018 (print) | DDC
    983/.14--dc23
LC record available at https://lccn.loc.gov/2017052578

Printed and bound in the United States of America.

Dedicated to:
The people of the Atacama Desert—
past, present, and future

Dedicación por:
Los Habitantes del Desierto de Atacama—
pasado, presente y futuro

CONTENTS

PREFACE

"The desert calls, and its voice is always heard."
—from *The Garden of Allah* (film), 1936

Of the many places that enchanted me as a child, none had more power on my imagination than the desert. This enchantment seemed peculiar to me because my formative years were spent in verdant New York state, and I had never experienced the desert firsthand. What, I wondered, could explain my fascination with the world's deserts? I believe that I was first introduced to deserts in stories such as *The Arabian Nights* and other tales that my parents, and especially my older brother, Bob, read to me as a child. I also became aware of the desert in film at a relatively early age, and I think this helped me further visualize what magical places such as the Sahara or Tibet might look like. Although these stories and films had exciting plots, to me the desert itself was as interesting as any character in a story that was set or filmed there. In these stories, though nominally set in desert locales, reference was often made to the challenges faced by those who would venture into their driest and most desolate parts; at the forbidding interior, the real "wastelands" needed to be crossed in order for people to reach exotic destinations or to escape the wrath of desperadoes. In such places, storytellers emphasize nature at its cruelest: the sun hangs overhead like a burning disk, and the land shimmers in heat waves that torture the air into producing deceptive mirages. Under this atmospheric alchemy, the air becomes a false sheet of water transposed onto the desert floor, beckoning but ultimately deceiving the traveler. And yet, to me there is something oddly appealing about a place that can produce such mind-bending effects.

By the time I was about ten I knew that every continent except Europe had such deserts, and South America was no exception; my incredibly supportive fifth grade science teacher Jerry Keshan had told us so. Mr. Keshan also encouraged my interest in mineralogy and geology, and I began a collection of rocks and minerals at that time. One of the minerals I read about and wanted to add to my collections was atacamite, which was named for the locale where it was first found—the mysterious-sounding Atacama Desert. I would later learn that deserts like the Atacama are not arranged randomly, but tend to lie well north or south of the equator, very rarely along it. Here, along those artificial lines we call the Tropics (of Cancer, in the north, and

Capricorn, in the south), drought can be seasonal or perennial. The Atacama Desert, which straddles the Tropic of Capricorn at 23 1/2 degrees south of the equator, is such a place. Over the span of several centuries, this desert came to be regarded as the driest place on earth. Many sources that I consulted stated that rainfall has *never* been recorded here—a claim that I shall examine in considerable detail in subsequent chapters. The very word "never" suggests that this desert is essentially timeless, adding to its reputation and allure.

In addition to deserts, maps also had me under their spell early on; they have, in fact, never let go of my imagination. Of all the maps I perused as a child, most were in encyclopedias and old books. However, I also took notice of them in movies, where they were used to show where the action was taking place. Naturally, those maps depicting deserts were especially intriguing. When I was twelve, my family spread out a huge Triple A map of the United States to show me where we'd be moving. The location— Tucson, Arizona—seemed romantic and exotic, but our route's marking on such an impressive map made the move seem more imminent. The American Automobile Association (AAA) had wisely marked the most direct route in red, and because it was still winter, they selected a southern route that took us through Washington, D.C., Nashville, Little Rock, Dallas–Fort Worth, El Paso, and thence to our destination. It was a memorable six-day drive in a brand-new 1956 Lincoln Premiere.

With each passing day after we arrived in Tucson in early April of 1956, the sun rose higher in the sky than I'd ever seen, and it beat down more and more fiercely as spring progressed. By late May, my parents concluded that Arizona was more like hell than heaven; in fact, one of them unapologetically branded it a "God-forsaken place." Unlike my parents and siblings, however, I simply loved the abundant sunshine, furnace-like heat, and cactus-clad hills of the Sonoran Desert that surrounded Tucson's forty-thousand souls on every side. By age fifteen, after a move to supposedly greener pastures (suburban Los Angeles), I found myself drawn to the Mojave Desert, which lay on the other side of the towering San Gabriel mountains. The Mojave was more sparsely vegetated than the Sonoran, and that made it even more attractive. As a teenager with slightly older friends who could drive, I'd visit the desert every chance I got, often with the goal of collecting mineral specimens and sketching rock formations and plants. My folks found this incomprehensible, but to me nothing was more exhilarating than hiking in the barren hills of the Mojave, or traversing the desert's backroads in what would later be called a dune buggy—an old 1940 Plymouth that had been modified by being stripped down to the chassis. I also enjoyed going out to the desert with Mr. Woodward, the father of a girlfriend who had some property he hoped to develop into a copper mine outside the desert town named Lucerne Valley. My interest in mineralogy, coupled with his firsthand prospecting experience, made every trip a new journey of discovery. Through high school my folks considered me a "desert rat," while my friends nicknamed me "Desert Rick."

By the time I was eighteen, I had graduated from high school and was living in San Francisco, where I found my first full-time job. Luckily, it was not just any job, but one at Rand McNally Map Company. At that time Rand McNally was a household name synonymous with maps of all kinds. I began as a clerk in their retail store on Market Street selling maps, globes, and travel books to those destined for exotic places; within a year, I moved into the map production part of the operation, framing and preparing maps on rollers for distribution to homes, offices, and schools. Maps were now not only my passion, but also my way of earning a living. I enjoyed working at Rand McNally, and after about two years my fellow workers realized that I was a bit different than the typical twenty-year-old. They noted that I was not only intensely interested in the maps but also avidly reading about the world around me in my spare time. Moreover, they noted that I drew sketches of natural subjects such as geological formations, plants, and entire landscapes, especially those in desert places on the other side of the Sierra Nevada mountains and beyond.

In 1963 my colleagues at Rand McNally concluded that a young man this interested in the world should go to college, a scenario that seemed out of reach. To me college was not only a scary place but also something that seemed restrictive. When they informed me that it might prove just the opposite—open up even broader horizons, as one of them put it—I agreed it was worth a shot. In retrospect, they were correct; I am forever grateful to those colleagues at Rand McNally, not only for their faith in me but also for their material support. They actually raised some money to help me enroll at Foothill College in Los Altos Hills, where I quickly discovered that my professors were genuinely excited about their subjects and my prospects. I soon confirmed that my associates at Rand McNally were right about college, for one of the first readings assigned in English class by Professor Carolyn Keene was Joseph Conrad's *Heart of Darkness*. The entire book was spellbinding, but a three-line quote by the story's narrator Marlow took my breath away: "Now when I was a little chap I had a passion for maps. I would look for hours at South America, Africa, or Australia, and lose myself in the glories of exploration. At that time there were many blank spaces on earth, and when I saw one that looked particularly inviting (but they all look that) I would put my finger on it and say, 'When I grow up I will go there.'" That quote became my mantra, and its placement in the college curriculum was proof to me that higher education was liberating to both the mind and the spirit.

Although I began college with a combined major in geology and art, an introductory geography course taught by Donald Graham convinced me that the discipline of geography was more germane. At Foothill College a visiting lecture on the Andes Mountains and adjacent deserts of Atacama and Patagonia enthralled me. The stunning color slides that accompanied this lecture revealed a landscape even more spectacular than the deserts of the American West. To complete my undergraduate studies, I enrolled in the geography department at the University of California at Riverside. UCR was known for its academic orientation, and the department was getting a reputation for applied research in several areas, including "arid lands geography" that covered places closer to home such as the Imperial and Coachella Valleys and also places much farther afield. Summer jobs at the California Map Centre in Los Angeles, as well as a research project on the remote sensing of southern California deserts for NASA, kept me involved with maps. I learned a lot about deserts from professors Harry Bailey (a climatologist) and Len (Leonard) Bowden (who supervised my NASA studies). Visiting professor Neil Salisbury helped me better understand arid lands' geomorphology by statistically evaluating the channels of arroyos. One class in particular, Economic Geography as taught by Homer Aschmann, exposed me to a place that especially burned itself into my subconscious—the fabled Atacama Desert of South America. For much of his career Aschmann had travelled throughout, and extensively studied, deserts in Mexico and California. However, he was actively preparing for a research trip to the Atacama Desert—or the Atacama, as he simply called it—and regaled the class with stories and facts. This South American desert's reputation as the driest place on earth only whetted my appetite. It was a first-class mining area with a fascinating geological history. Aschmann's vision as a historically savvy economic geographer helped me put mining in a temporal and spatial context. His class on the geography of Latin America also inspired me.

My interest in the Atacama Desert was far more than scientific. Being somewhat of a romanticist at heart, I must admit that the name "Atacama" itself was simply wonderful; both exotic and melodic, its four syllables tumbled off the tongue as methodically and liltingly as sand spilling through an hourglass. Subliminally, the name Atacama reminded me of the Amargosa Desert, which was named after the bitter river in the Mojave Desert that ends in the formidable Death Valley. The Atacama Desert, though, was even more mysterious. Aside from Aschmann's encyclopedic mind, there was relatively little information about the Atacama Desert that could supplement passages in my geography textbooks. At that time the only "modern" book specifically about it was the American Geographical Society's *Desert Trails of Atacama*. However, that informative book by geographer Isaiah Bowman had been published about half a century earlier (1924), and most of the field work had been done about a decade before that. As I eagerly read Bowman's wonderful book, savoring its superbly written text and studying its beautiful photos and maps, I hoped that someone, somewhere, was busy writing a new book about this desert. I was overjoyed when Professor Aschmann informed me that our library at UCR had recently acquired a copy of William E. Rudolph's *Vanishing Trails of Atacama*, which had also been published by the American Geographical Society. Much slimmer than Bowman's, this book updated that classic to the early 1960s.

My fascination with the Atacama Desert simmered, as obsessions sometimes do, while I attended graduate school at the University of Oregon from 1967 to 1970. My two foreign languages—Spanish and German—helped me in researching papers for my Geography of Latin America class with professor Carl Johannessen. Other, closer arid lands were my focus at that time, namely the deserts of northern Mexico and southern California. In 1968 I wrote my MA thesis on the settlement and perception of the southern Mojave Desert in the vicinity of Twentynine Palms, California. My doctoral dissertation on the role of the Mormons in transforming the landscape of the interior American West into a religious homeland (1970) also helped me better understand the arid and semiarid Intermountain West. While spending much of my academic career focused on North America, though, I put the Atacama Desert on the back burner. Nevertheless, part of my varied career was spent as an environmental and community development planning manager for SEAGO—the Southeastern Arizona Governments Organization. This stint from 1979 to 1983 enabled me to intimately understand the peoples and ecosystems of a portion of my beloved Arizona. In 1983 I accepted the position of director at Arizona's Bisbee Mining and Historical Museum, which helped me better understand the history of copper mining in the Southwestern borderlands.

Over the years, some of my geography colleagues wondered why I found myself working in positions that related more to history than geography. The unvarnished truth is that the discipline of history seemed more welcoming to my ideas, and much the same could be said about anthropology. There was an irony here, for when I would propose subjects for inquiry to my geography colleagues, I often got the reaction "interesting, but is it geography?" On the other hand, historians and anthropologists would respond "great, I like the way you use geography to pose and answer interesting questions." For my part, I was uncomfortable separating the disciplines of history and geography. As the British explorer John Smith reportedly observed about four hundred years ago, "Geography without history seems a carcass without motion, so history without geography wandereth as a vagrant without a certain habitation."

Acting on that premise, in 1991 I accepted a dream job—the combined position of history professor and director of the Center for Southwestern Studies and the History of Cartography at the University of Texas at Arlington (UTA). Although many of the maps in the Virginia Garrett Map Collection at UTA's Special Collections were of the Southwest, some covered South America. The more I studied these, the more the Atacama Desert beckoned me to decipher how it had been mapped through time. Moreover, my research into mining history, albeit mostly in the American West, had well prepared me to better understand the fabulous mining history of northern Chile. As any mining historian knows, the Atacama Desert is a treasure trove of unique mineral deposits. Naturally, as my mineral collection grew, I acquired several specimens of the rare copper mineral atacamite over the years; someday, though, I hoped to personally find specimens in the mines and ore dumps of the Atacama Desert.

My first real glimpse of arid western South America came in January of 1992, about thirty years after I learned about the Atacama Desert in my geography classes and lectures at Foothill College. On a trip with friends in the high Andes, I looked southward into coastal Peru, with Isaiah Bowman's fascinating Atacama Desert just out of reach. Like Marlow in Conrad's novel, I now vowed that "I will go there." I was, after all, fairly well prepared for a trip to northern Chile. Over the years I had collected information about the Atacama Desert, much of it in Spanish, some in German, and some in French. By the time I wound down my long academic and administrative career at the University of Texas at Arlington in 2008, I was now ready to begin serious writing on this desert. Because forty-five years had elapsed since I first studied this desert, I was now in a position to apply lessons learned from several disciplines to my own interpretation. Methodologically speaking, the focus I elected to use in telling the story of the Atacama Desert was both old and new. I called upon one of the truly venerable traditions based on sequential occupancy of a place. This recognition that different people occupy a place through time, giving it a character that differentiates it from other periods, dates from the pioneering studies of Harvard geographer Derwent Whittlesey (1890–1956). However, I was not only concerned with the actual settlement through time, but actually even more interested in the differing *perceptions* that people developed of this desert over time. Accordingly, this study owes much to both the geography of perception(s) about places and an especial debt to the history of travel narratives. By deconstructing these narratives and their associated images such as maps and sketches, it will become apparent that deciphering the impact of place on observer is my main goal. This was heady prospect indeed, for it meant I could not only study the Atacama Desert that Bowman (and others) had studied, but I could also study how it affected people who experienced it, including Bowman himself. The Atacama Desert, Bowman confessed, "attracted me more than any other part of South America." I wholeheartedly agreed with his more general sentiment that "to my mind, the desert is the most interesting place in the world for exploration and geographical study."[1]

Some may call this type of inquiry perceptual historical geography, but I have no interest in labeling it. To me, it is just as much historical anthropology, or environmental psychology, or the history of discoveries as it is geography. The process of encounter, which is variously called exploration and/or discovery, is what I seek to address in this book. At times this will mean that the actual environment, while important, is in fact less important than what the observer thinks he or she has encountered. The latter—the premise that places are as much in the mind as real—is what might be called postmodern in that it recognizes that perceptions are as important as, and perhaps even more important than, reality. Thus, my use of the term "imagining" in the title is quite deliberate and should serve as notice that deserts like the Atacama are part of a popular mindset about places, a mindset that owes as much to romantic travel literature as it does to hard science. Another premise of this book is that the separation of art and science, or for that matter literature and science, is arbitrary; these separations are artifacts of the recent past that can obscure the essence of what we study through time. My approach is, in a word, *phenomenological*, that is, based on the premise that something can be studied and appreciated from several perspectives simultaneously. Understanding a place as intriguing as the Atacama Desert requires a diverse set of perspectives: artistic expression, poetic interpretation, and scientific observation.

My intense six-year research project into this fascinating desert (2008–2014) culminated in an article for *Terrae Incognitae* (the Journal of the Society for the History of Discoveries) in 2016.[2] That article demonstrated that the Atacama Desert is a land both illuminated and obscured in folklore. Because there were so many things I had to omit from the article, the book you are about to read used the article as a basic template but expanded it into many new and exciting directions. Whereas that article offered the barest of outlines, this book provides a more complete picture. Like that article, this book analyzes maps and written narratives to demonstrate how the Atacama Desert was discovered, and then rediscovered, over nearly five centuries. However, because more space is available for both text and images, this book contains many more illustrations, far more original narratives, and a much deeper interpretation of what I encountered in both the historical literature and in the field.

As I demonstrate in this book, and provided the barest outlines of in the earlier article, the discovery of the Atacama Desert occurred in a series of four sequential phases, each of which I will later discuss in separate chapters. Here, though, is a synopsis. From about 1530 to 1700, the name Atacama represented a remote but strategic political province whose lack of population rather than desert climate was emphasized. At that time, the Atacama was simply a place to cross on the way to somewhere else, its meager ports on the Pacific Ocean funneling the riches of the Andes into the holds of waiting ships destined for Europe. After about 1700, however, the Atacama began to be identified as a desert, the result of exploration and mapping that was increasingly scientific in nature. In this transitional second phase, the Atacama was part of a broader pattern where the political mapping of empires was gradually supplemented by thematic physical or scientific mapmaking. In the third stage, which was fully evident by the mid-nineteenth century, the Atacama became linked to increasingly nationalist impulses and the rapidly growing power of international corporate developments in transportation and mineral extraction. In the fourth and current stage, starting in the mid-twentieth century, the Atacama began to be promoted and marketed as the quintessential desert worth seeing for its uniqueness—something that the earliest explorers would have found incomprehensible. For Chileans and outsiders alike, the Atacama Desert now serves as a repository for relics of the past, including ancient mummies and forlorn mining ghost towns.

However, this desert is also recognized as a graveyard of another kind: in the collective memory of Chile's recent tumultuous political past, it has been popularized as a place haunted by the memories of "the disappeared," as unaccounted victims of the 1973 political turmoil are known. In the early twenty-first century, the Atacama also began to play yet another role—providing an ecosystem similar to what may be encountered by future space exploration to Mars and beyond. My premise, worth restating here, is that time and place—which is to say history and geography—are inseparable. Moreover, as I shall show, this desert's past, present, and future reveal much about how places are discovered, and then rediscovered, through time. Thus, as the word "imagining" in the book title suggests, discovery involves far more than the initial historical physical encounter between early explorers and place. It can also be profitably expanded to address the ongoing ways places are inevitably reimagined—and hence rediscovered—by subsequent generations.

As hinted at, writing this book has been an exciting journey itself, and many people helped me along the way. An example of how long it took to finally write this book is that in the 1960s when I first became aware of the scholarship about this desert, my son Damien had not yet been born. Recently, as a man in his midforties, Damien accompanied me on a trip to the Atacama Desert where his considerable skills as an anthropologist and photographer proved enormously helpful as we explored the area's cities and immense open spaces. Writing this book also helped me connect with geographers such as Dan Arreola and Kent Mathewson, who put me in touch with others who have conducted research in this desert. Their studies confirm that geographers work at the intersection between many other disciplines, especially anthropology, political science, economics, history, and sociology.

In this book I pay special attention to the role that maps play in such interdisciplinary endeavors. Cartography is indeed one of the keys to understanding not only how places evolve, but how attitudes about places take shape. The premise is that a map includes both narrative(s) and image(s) that reflect the values of mapmaker and map user alike. In that light, map historian and artist David Rumsey gladly shared many of his maps and encouraged all to access them through his masterful digitization program. Recalling that fellow mining historians William (Bill) Culver and Bob Spude had visited the Atacama Desert, I called upon them to share insights and photographs. At the University of Texas at Arlington, former colleagues anthropologist Karl Petruso and historian David Buisseret were strong supporters of this project; both read initial drafts of the article and made numerous suggestions. Also at UTA, Ben Huseman, map curator in Special Collections, was enthusiastic about this project as the Atacama has long fascinated him, too. In Chile, Professor José Antonio González Pizarro of the Universidad Católica del Norte in Antofagasta kindly shared his wealth of knowledge about the Atacama's history and geography. At the American Geographical

Society Library in Milwaukee, Susan Peschel, Jovanka Ristic, and Robert Michael Jaeger provided information about the publications of geographer Isaiah Bowman. In Baltimore, Jim Stimpert of the Johns Hopkins University Library also kindly assisted me in accessing materials from the Isaiah Bowman collection. At Willamette University (WU), where I teach occasional courses and conduct research as an associated scholar, several colleagues were especially helpful. They include administrative assistant Leslie Cutler and librarian Shanel Parette, who were particularly helpful in providing guidance on the finer points of word processing and locating interlibrary loan resources, respectively. So, too, were WU's print services professionals Mark Bernt and Craig Wheeler, who helped me digitize images of historical maps and other illustrations in my personal collection. Professor Anna Cox of the Spanish and Film Studies departments helped me translate some of the arcane historical passages. To these colleagues and many others, I owe a lasting debt of gratitude.

Lastly, special thanks is due to two others. First, to my wife Ellen's dearest friend Ana Julia Rugel Hollis of Guayaquil, Ecuador, who helped me polish the Spanish translations of the six poems that I wrote for this book. At eighty years of age, Ana's tireless energy and enthusiasm for South America is an inspiration. Also, editor John Alley of the University of Utah Press urged me to write this book and deserves special mention. John recognized the widespread interest in the Atacama Desert, and he aptly noted that it is one of those rare "last" places on earth about which little is still known. Despite the advances in science, the Atacama Desert is a place of considerable mystery, and how that came to be is an integral part of this story.

One

"A Desert Picture of Great Beauty"

An Introduction to the Atacama Desert

Although the title of this introductory chapter suggests something visually compelling, I would like to begin it not with an actual picture, but with words. By adopting this approach, I follow the pattern set by the Atacama Desert itself, which first became known to the outside world not through visual images but rather through verbal and written descriptions. Historically speaking, narratives are a fundamental aspect of place; people often heard stories about places long before they actually saw those places on maps or in paintings or experienced them firsthand. When people describe a desert to a person who has never seen one, they may use many words to convey both its barrenness and the sensations that the place triggers. Thirst is one such sensation, as is burning heat, neither of which can actually be seen. Even when we hear the term "Atacama" and automatically think "desert," the word may trigger a complex mix of sensations such as the glare of intense sunlight, the feeling of dry heat, or the sound of shifting sands. Some of these sensations are visual but others are sensed through the skin, ears, and nose. Although narratives work hand in hand with pictures—the shifting sand dunes being something we can see, hear, and feel, for example—I shall operate on the premise that, as one of the earliest written statements in western culture goes, "in the beginning was the word."[1]

The word *imagining* in this book's title suggests that imagination was, and actually still is, central to understanding the desert. Imagination is defined as the ability of the mind to form new ideas, or images, or concepts of external objects. The inclusion of both ideas and concepts that are entirely dependent on words, as well as images that assume a picture of something, is noteworthy. And yet, because the verb *to imagine* shares a root with *image*, I would also like to begin by defining that evocative word. In the first definition in *Webster's New Collegiate Dictionary*, an image is "a reproduction or imitation of the form of a person or thing," and more particularly, "the optical counterpart of an object produced by a lens, mirror, or other optical system." That may seem to validate the visual and exclude words, but in the verb transitive, image means "to describe or portray in language, esp. vividly," or "to call up a mental picture of... someone or something." This confirms that a person can form an image of something without having actually seen it. I could use the controversial issue of how Muslims are said to have an image of their prophet Muhammad when in fact pictures of him are forbidden, but much the same thing happens when I describe the Atacama Desert to someone without using pictures of any kind. Muslims know how their prophet looks without seeing him because of the rich body of written and verbal descriptions of the life and times of Muhammad; when written down these are called hadiths, and they serve in the place of visual images, even to the point of describing how Muhammad looked, including his complexion, facial hair, posture, etc. Similarly, most people learned about the Atacama Desert in times past through such verbal and written description, not because images were forbidden, but because they were so expensive to reproduce until recently. The point to recall here is that a person's imagination can indeed take words and shape them into pictures of someone or something; similarly, pictures are readily put into words, though as the old proverb goes, it may take a thousand of them to do the job.

With the Atacama Desert in mind, I would like to now show how words and images work hand in hand to create places in the imagination. I begin by presenting a list of nine words in the box below.

Arica

Iquique

Antofagasta

La Serena

Valparaíso

Concepción

Valdivia

Puerto Montt

Punta Arenas

As readers ponder this list, some words may seem familiar and some may seem exotic, but they are in fact the names of places in present-day Chile. The places that I have selected happen to be port cites, and they huddle along the coast of the Pacific Ocean. Moreover, these nine places are arranged along that coast much as I have listed them here. In other words, this list, which may seem to be just words, is actually a diagram of how these places appear to a person traveling along the coast. Moreover, because they are arranged much as they are in the way we tend to conceptualize the world cartographically—with the northernmost at the top and the southernmost at the bottom—they are arranged on the list much as they are on a map or globe. In fact, had I drawn a simple vertical line along the left-hand margin of those names, I could claim that this list of nine words is indeed a map, albeit a crude one, with Arica in the far north, and Punta Arenas in the far south, of Chile.

My reason for beginning with place names (toponyms) is that they are key to understanding the places they represent. We often take place names for granted, but many of them are far more revealing than they first appear. Some, for example, are narrative descriptions about the character of a particular place or event in its history. The first name on the list, Arica, is one of them. Although the origin of Arica's name is clouded in the mists of antiquity, it is widely regarded to be an Aymara word (Arikka) that has long been applied to this geographic location. The Aymara are Indigenous peoples in this part of the Atacama and locally characterized in Spanish as *gente originaria* (original people). As an Aymara place name, Arikka combines two words, one meaning rocky and the other a high point or prominence. Like most explorers, the Spaniards adopted some names directly from the Indigenous people, and Arica was one of them. However, they modified it slightly. Recognizing the importance of this geographic feature as a landmark, they called it Morro de Arica (in Spanish, *morro* signifies a large, prominent rock). This distinctive topographic feature frames the city and has long been a landmark for locals and visitors alike. Mariners in particular used it as a point of reference, as confirmed by two mid-nineteenth-century illustrations. In the first, a nineteenth-century watercolor by Hamilton Williams, the Morro de Arica dwarfs the vessels harbored beneath it and the port city that huddles just north (left) of it (Figure 1.1). The second illustration is a map showing much the same scene, but looking straight down on it (Figure 1.2). On this map, which effectively shows the topography using a series of lines called hachures (from the French word to chop), the Morro of Arica is prominently indicated, as are the depths of the water offshore. Useful to mariners and land travelers alike, this map also indicates the actual layout of the town of Arica, which lies just north of the Morro de Arica and consists of a rectangular grid containing about two-dozen blocks that are positioned at right angles to each other. As hinted at, the name Morro of Arica is a reminder that names can be redundant; in fact, when a person uses the name Morro de Arica (or Morro of Arica), he or she is essentially saying "the prominent rock at the place that Indians call the rocky prominence." By learning the meaning of the name Arica, the conscientious traveler has not only added two new words—one Indian, the other Spanish—to his or her vocabulary but has also learned something important about the character of the place and its impact on the perceptions of people who occupy it.

At this point, I would like to state for the record that place names should be considered short stories; perhaps the shortest of short stories is even more apt. Yet in shorthand, so to speak, they are often full of meaning: some of it lost, some still palpable, and some of it dependent on only the limits of

Figure 1.1. *Watercolor painting of Arica, Peru, ca. 1867*, by Hamilton Williams, showing the prominent Morro de Arica. (Author's digital collection; original in the National Maritime Museum, Greenwich, London.)

imagination. As we shall see, that rocky prominence at Arica is steeped in meaning for the role it played in major events, including battles that turned the tide of history. I would also like to note that a word or name such as Arica can mean even more than what it meant in the original Indian language(s). Subliminally, for later arrivals, Arica recalls two words from their own European language(s). The first is ari, which is the root of the word ari-dus, meaning dry or parched in Latin (and by extension English and Spanish); the second is rica, which means rich in Spanish. Coincidentally, Arica did in fact become rich in commerce, for considerable wealth passed through its port. The richness of Arica also pertains to its rich history, as it is one of the oldest cities on the Pacific coast of South America. Regarding any subliminal connections to aridity, Arica is certainly a dry place, some sources claiming that it is the driest city on earth. Through such subliminal or subconscious associations, Arica and arid are wedded phonetically and psychologically to those whose origins are European.

Regarding the origins of another name on the list—Valparaíso—we can be more certain, though it too triggers visions. Valparaíso consists of two Spanish words, Val for valle (valley) and paraíso (paradise). As the name came

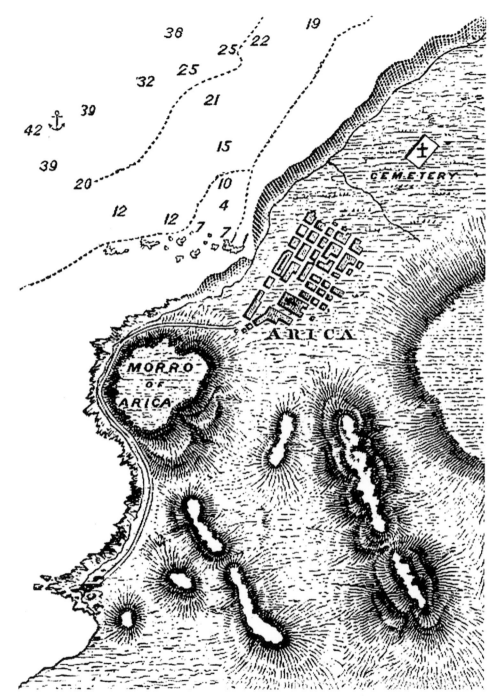

Figure 1.2. Plan of the Bay of Arica, ca. 1868, showing the prominent Morro of Arica and community lying immediately to the north, as illustrated in "The Great South American Earthquakes of 1868," by Ephraim G. Squire, *Harpers New Monthly Magazine*, Vol. 38, No. 227, New York, 1869. (Author's digital collection)

to signify, this was a beautiful location indeed, and it would soon become Chile's major port. It should be recalled, though, that the Spanish name Valparaíso eclipsed an earlier indigenous name for the same locale, a subtle (or not so subtle) way of silencing the native narrative about the place. In other words, just as explorers might perpetuate indigenous place-naming (as in Arica), they could just as easily erase it. Continuing with the theme of place names as short stories, the name of a place farther down the coast—Valdivia—honors a sixteenth-century Spanish explorer and colonizer, Pedro de Valdivia. Although the origin of the surname Valdivia has deeper roots as a geographic descriptor in Spain, it is now shorthand for this man's somewhat heroic deeds (at least to Spaniards) in Chile, much as the name Austin (Texas) honors the founding father of the Lone Star Republic.

Quite aside from the origins and meanings of the place names on this list, though, I want to again make it clear that this list is both a narrative and a graphic device. Much like an actual map, it serves the purpose of orienting the reader by indicating the actual positions of places in present-day Chile. Remarkably, the list can work this way even though not a single line or arrow has been used—just the specific placement of those words themselves. And yet, that orientation of north as up, which is to say at the top of the map, is not as "natural" as it may seem to us today. It began as a cartographic technique developed by the Egyptian/Roman astronomer and geographer Claudius Ptolemy (ca. 100–170 AD), who used it to depict zones of climate, with the torrid zone at the bottom and the frigid zones at the top. Long before Ptolemy, the philosopher Aristotle (385–323 BC) and others in the Greco-Roman world conceptualized these climate zones as horizontal belts, but Ptolemy appears to be the first to place them on a map. Even in Ptolemy's time, maps usually used other directions—especially east (likely in deference to Jerusalem as the holy city). However, Ptolemy's technique became common in Europe after about 1450 as the West (Europe) began to turn to classical scholars rather than religious authorities for inspiration. In time, mapmakers adopted the technique, and it ultimately came to dominate Western maps. However, as I shall demonstrate in the next chapter, even mapmakers in later times could place east at the top of their maps. Nor have we escaped it even in our language: although we use the noun "to orient" a map today as we position it with north at the top, we should recall that this term once literally meant to place the map so that up pointed to the east (Orient).

Before leaving the list of place names, I should note that it can serve as a map because those places are arranged linearly, which is to say along an axis. Some people aptly characterize Chile's shape as a string bean, for it is generally long and narrow. Chile stretches about 2,600 miles (ca. 4,200 km) in length but averages only about 110 miles (ca. 175 km) in width. The distance from northern to southern Chile equals about the distance from New York to Seattle, while the distance across Chile equals that from Los Angeles to San Diego, or from New York to Philadelphia. Chile's latitudinal (north–south) stretch is exceptional; the only other country that comes close is the United States, provided we include noncontiguous states such as Alaska and Hawaii. The former Soviet Union also had an impressive latitudinal reach, though that was whittled down somewhat following the disintegration of the USSR in 1989.

Had I drawn a line to indicate the coast, and hence turned the list into a graphic we would call a map, I could also have drawn a parallel line to the right of it to indicate the eastern border of Chile. However, because Chile is such a narrow country, I would have to draw that second vertical line (its eastern border) to include only the first two letters of each name! The remaining letters of those names would have extended into Chile's neighbors to the east, notably Bolivia and Argentina. In reality, these two borders are not perfectly straight lines, for they follow natural features. Chile's eastern border with Bolivia and Argentina follows the sinuous spine of the Andes mountains. Moreover, although Chile's west coast may appear nearly straight as an arrow on some maps, it is occasionally marked by small bays and frequently by prominent headlands, such as the impressive rocky prominence that looms above the desert port city of Arica. The string bean analogy fits Chile well in another regard, for that lanky South American country actually curves a bit at both the north end and especially at the southern end. I should also note something else geographical about this list of nine place names in relation to Chile's long and narrow shape. The fact that Chile extends so far north to south means that it stretches through several major zones of climate, from rather cold and wet in the south to warm and very dry in the north; being located in the southern hemisphere, Chile's climate zones are arranged opposite of Ptolemy's, that is, with cold at the bottom and warmer at the top.

Because only about the northern third of Chile has a desert climate, and may broadly be considered part of the Atacama Desert, only four of the nine places on the diagrammatic list/map—Arica, Iquique, Antofagasta, and La Serena—will be discussed, along with many others in this book. To some readers, it may seem surprising that I begin this section of a book about the desert by highlighting cities; however, there is a good reason for this rationale. As in most deserts, the Atacama is highly urbanized, with about 99 percent of its inhabitants living in settled places. Cities are not only jumping-off places into the desert but the places in which records are kept that shed light on the surrounding area. And regarding the Atacama Desert only occupying the northern third or so of the country, I should add one more note about distance. Regardless of how things look on a map, it still takes at least three full days to drive this desert part of Chile from one end to the other (Arica to La Serena) at the speed limit. Moreover, as I shall show, while the Atacama Desert is as narrow as slender Chile itself, it might take a full day to travel across it by car as the topography can be formidable, and the roads suited only to slow-speed driving.

One of the cities listed and mentioned earlier—Antofagasta—is remarkable for several reasons. In addition to being the Atacama Desert's largest and most economically vibrant city, it includes within its municipal limits a narrow strip of land marking the location of an imaginary line called the Tropic of Capricorn (Trópico de Capricornio). Geographically, this line indicates the place(s) on earth where the sun shines directly overhead on the December 21 solstice, which is summer in the southern hemisphere and winter in the northern hemisphere. Its counterpart in the northern hemisphere is the Tropic of Cancer. Girdling the earth at about 23 1/2 degrees south latitude from the equator (actually closer to 23 degrees, 26 minutes, and 14 seconds), the Tropic of Capricorn is not truly stationary as it is slowly, albeit infinitesimally, moving northward. The Tropic of Capricorn has long intrigued scientists and mapmakers. Although intangible on the landscape, this line is fixed enough in the imagination to be commemorated in several locales in the Atacama Desert. For example, on the northern outskirts of Antofagasta, it is immortalized in an abstract monument consisting of three uprights and a horizontal lintel-like crossbeam that suggests a mason's or carpenter's square. The monument, which was conceived by architect Eleonora Román and built with the assistance of the Antofagasta Rotary Club, is located near the city's airport and gives bold, tangible presence to an imaginary tool of geographers and mapmakers. To further suggest the curvature of the earth's surface, the monument ingeniously curves slightly.

Although Eleonora Román's monument near the coast is spectacular, there are others in the interior Atacama Desert where the traveler encounters this imaginary east–west line on well-traveled roads. Some, such as the standard official highway marker signs for "Línea Trópico de Capricornio," are more prosaic but nevertheless reaffirm the importance of this cartographically inspired feature (Figure 1.3). Out in the unpopulated areas of the Atacama Desert, for example, along the Pan-American highway northeast of Antofagasta and southwest of Calama, one encounters an official road sign as a reminder that even places this remote are fixed on a larger system of

Figure 1.3. Official sign on the Pan-American Highway indicating the location of the Tropic of Capricorn at an isolated spot near the Mantos Blancos mine, northeast of Antofagasta, Chile. (Photograph by Damien Francaviglia)

Figure 1.4. This large stone monument identifying the location of the Tropic of Capricorn along the Pan-American Highway was the joint effort of a civic group and mining company interests. (Photograph by author)

geographic coordinates. In addition to official signs posted by Chile's Ministry of Public Works, other markers to the Tropic of Capricorn may be regional and local efforts. For example, located just yards from one official sign is the impressive masonry monument (Figure 1.4) that represents the combined efforts of scientists, a civic organization, and copper mining company interests—an enduring partnership reflecting the importance of minerals and mineral extraction that I will describe in later chapters. The main takeaway here is that the Atacama Desert is bisected by the Tropic of Capricorn, its northern portion lying toward the equator and its southern portion reaching into the temperate midlatitudes. However, a second point worth remembering is that people have long felt the need to take the edge off such vastness by marking it with something familiar that anchors it in space.

On modern-day maps, the Atacama Desert appears as a long (ca. 1200 km), narrow (ca. 160 km), hyperarid portion of western South America. One of the simplest but most effective maps of this region was created about a century ago by geographer Isaiah Bowman (Figure 1.5). I deliberately selected this illustration, which appeared in his book *Desert Trails of Atacama*, because it does two things simultaneously: 1) it shows the locations of important

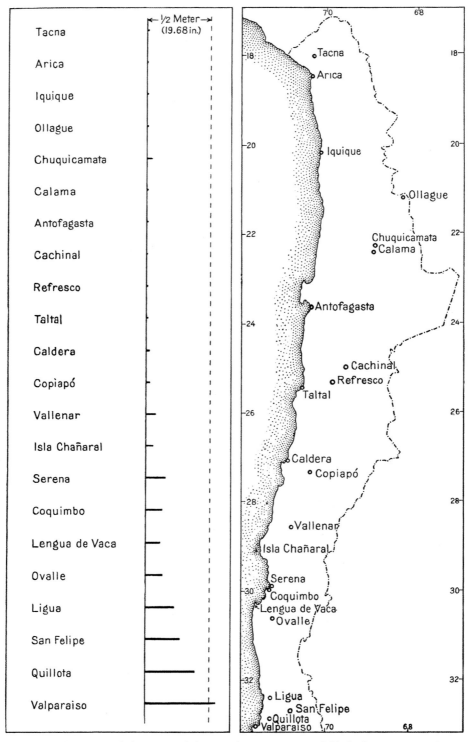

Figure 1.5. A map and diagram of the northern portion of Chile from geographer Isaiah Bowman's *Desert Trails of Atacama* indicates annual average precipitation at various communities. (Author's digital collection)

communities, and 2) it reveals something important about the climate at each place. Bowman called this two-part illustration a "diagram (and location map)" because it is a composite of two separate but closely related graphics. Bowman had good reason for placing this map fairly early in his book. Given his interest in an aspect of what makes the Atacama a desert—climate—he conveniently listed its place names in a column at the left. This column is a bit like the list of places I discussed earlier, but it is far more detailed in that it includes many more places north of Valparaíso, which Bowman included as a reference point. Next to each of these place names on Bowman's map is a black line showing the average annual rainfall there. Some of these lines are longer than others, longer indicating a greater amount of precipitation. Some places receive so little rainfall on the average that their totals only show as barely discernible dots rather than actual lines. In comparing this listing of precipitation totals to the map that appears to the right of those totals, readers can see at a glance that aridity increases northward, for each horizontal black line may be thought of as a rain gauge. In one fell swoop, then, Bowman not only shows us where important places in the Atacama Desert are located but also tells us something about their character or condition. Bowman provided his readers with a remarkably simple yet effective diagram, really a thematic map, the theme being moisture, or lack of it, in relation to the individual places shown on it.

Bowman's ingenious graphic is based on scientific data, but it confirms something that travelers have long known: rainfall decreases from south to north in this part of South America. Whereas Valparaíso is relatively well watered, receiving a little over half a meter (about twenty inches) annually, precipitation decreases northward. Places such as Serena (La Serena) and Coquimbo at the southern edge of the Atacama receive only a third of what falls in Valparaíso. Continuing northward to places like Arica and Antofagasta, the precipitation only averages a few millimeters (about half an inch) per year. By comparison, no place in the United States averages so little precipitation as the cities in this coastal desert. Moreover, regardless of where they are located in this region, either on the coast (Antofagasta) or inland (Calama), their climate is arid. Bowman's rather austere map, like Bowman himself, is scientific and disciplined. Just as each of the sentences in his landmark book seems both economical and informative, his map informs us about those places in regard to a hallmark of this region—its startling aridity. The map's sparseness, though, does something else. Because it conveys the sense of scarcity, it excites the imagination to ponder the consequences of that scarcity on the character of this place and its inhabitants. Those cities may be dry indeed, and there is considerable evidence, some of it scientific but most of it anecdotal, that the areas adjacent may be just as dry, and perhaps even drier.

Astute map readers will note that the name "Atacama" appears nowhere on Bowman's map. Among the many places on it Bowman could have shown San Pedro de Atacama, which lies in the blank space southeast of Chuquicamata and Calama where the dashed line (here the border between Chile and two neighboring countries, Argentina and Bolivia) jogs eastward. After all, San Pedro de Atacama is the namesake community of the region; however, in keeping with Bowman's restraint and focus on one major aspect of the region—rainfall—his map only shows places that had meteorological data. At that time, San Pedro de Atacama was a cluster of small farming villages connected to the outside world only by primitive roads, the classic trails that Bowman loved to travel, often on pack train but sometimes aboard cantankerous, shuddering motor vehicles that braved the poor roads. The absence of the name Atacama on Bowman's map brings up an important issue about the name itself. As I shall demonstrate, the land area shown on Bowman's map has been known by various names through time. Although the name Atacama Desert is commonly used today for much of this area, in the past many observers and residents thought of it as one part of a region with several separate deserts *within* it. Bowman in fact called the area inland from Arica and Tacna the "Desert of Tarapacá" in honor of a venerable community there. Moreover, with increasing elevation toward the Andes, one encounters a somewhat better-watered area called the Puna. Today, a century later, the Atacama Desert popularly includes this elevated area, but Bowman considered it separate and somewhat distinct from the desert.

A bit earlier than his "diagram (and location map)" in *Desert Trails of Atacama*, to prepare his readers for the many places he was about to introduce, Bowman presented a far more detailed map, entitled "General Location map of the Desert and Puna of Atacama" (Figure 1.6). Like most mapmakers, Bowman based his on a preexisting map, namely the American Geographical Society's six-sheet series of maps comprising Hispanic America. At first glance this map may seem complicated, but it reflects the massive amount of information Bowman had acquired in the field during several visits from about 1907 to 1913. Moreover, it covers a larger area than the first Bowman map I discussed, as revealed by the inset map in the lower-right corner, which includes large portions of Bolivia and Argentina. On Bowman's map readers can spot places that I've already mentioned, including the coastal cities Arica and Antofagasta and interior cities such as Calama. These places were, and still are, reached by a web of transportation routes on land. As he put it, "The heavy dotted lines represent the principal trails that supplement the railway network," which is shown as solid black lines. Bowman also identifies numerous physical features like streams and rivers (for example, the Loa River north of Calama) and salars, or dry lakes (such as the Salar de Atacama).

Careful study of this map reveals that the name Atacama appears no fewer than three times: once in capital letters just below the line of 26 degrees south, where it identifies the political district or province of that name; once for the name of a physical feature, Salar de Atacama, about 140 miles (ca. 200 km) east of Antofagasta; and lastly as the name of a cluster of villages comprising San Pedro de Atacama, which lies southeast of Calama. To place this desert region in perspective, Bowman included that inset map in the lower right-hand corner. Although the inset map clearly shows the borders of Chile, Argentina, and Bolivia, readers with a sharp eye will note that he neglected to indicate Peru, a small part of which should be shown in the northwestern portion. Given Bowman's interest in defining national borders correctly, this omission is surprising. To Bowman, this lapse must have been especially disconcerting, for it likely offended Peruvians who consulted the map. As I shall discuss in chapter four, Peru's southern border had already been pushed far northward by Chile after a costly war in the early 1880s; hence, the omission added insult to injury. Looked at philosophically, though, this exclusion serves as a reminder that all maps, no matter how well executed, are liable to include errors.

Surprisingly, nowhere on Bowman's information-packed map do we find the name Atacama Desert (or Desert of Atacama, as it is often called), though he finally clarified that omission in another more physically based map toward the end of his book. On that rather easier-to-read map (Figure 1.7), which is entitled *Desert and Puna of Atacama*, Bowman kept things relatively simple. In addition to indicating prominent towns and cities for reference, his main focus here is to show the mountains and salars. The capital letter C before some names, such as C. Copiapó and C. Guanaca, indicates a peak of that name, the word cerro in Spanish meaning hill or backbone. Some popular culture writers emphasize the openness or flatness of the Atacama Desert, but in reality much of it is hilly to mountainous in nature. On Bowman's map, the mountain ranges are indicated by names that run along their ridges, as in the Cordillera De Los Andes. Bowman is concerned with watersheds here, as evident in his showing the divides. In jet black Bowman shows the salars, which in actuality are likely to be white, but the effect is stunning, as they are the lowest parts of basins and appear almost like holes burned into the paper of the map. Where he places a white name over the black, as in "S. de Atacama," the effect is even more startling.

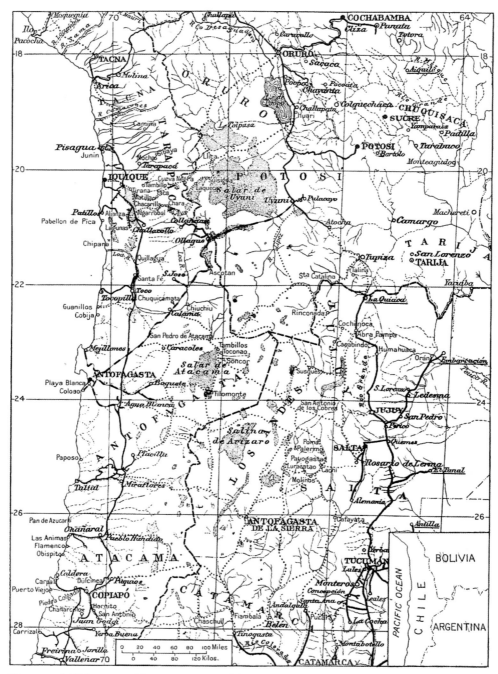

Figure 1.6. This general location map of the Atacama region in *Desert Trails of Atacama* by geographer Isaiah Bowman was based in part on the six-sheet map of South America by the American Geographical Society. (Author's digital collection)

As the map's title suggests, there is a difference between desert and puna. Generally, desert lies at a lower elevation than puna and may be devoid of vegetation in large areas. In the age of space exploration, that absence of vegetation would lead to the desert's comparison with the landscape of Mars, and some Atacama Desert landscapes do indeed resemble the red planet's rocky and sandy surface (Figure 1.8). The fairly flat interior lands of this desert are called pampas, a term that may convey grassland, as it does in Argentina; however, this pampa is usually nearly devoid of vegetation and its surface consists of soil. Although this pampa soil may be buff colored, and from a distance appear to be very closely cropped dry grass, upon closer inspection it turns out to be covered by tawny-colored dust, sand, or gravel. In other places, the pampa may be as white as snow and covered with salt of nitrate.

By contrast, the puna of the Atacama region lies at a higher elevation and is often somewhat less barren in appearance. Bowman uses the term puna to mean high desert, wording that a Westerner might better understand, that is, land that tends to be somewhat lusher—though still sparse—in vegetation (Figure 1.9). Tufts of bunch grass and clumps of bushes are indicators of the puna, which is grazing country in many areas. Given its higher altitude, the puna is also colder than the desert. The difference between pampa and puna is important locally, but to many outsiders, especially those unfamiliar with arid lands, both landscapes may appear so similar that it is impossible to tell them apart. This confusion is common today, when tourist literature brands much of the entire region as the Desert of Atacama. Despite recognizing the differences in vegetation patterns across this region, I shall respect the term Atacama Desert, as it is well entrenched in the popular literature and the result of centuries of scrutiny, and conjecture, by travelers.

One thing about these two maps by Bowman is remarkable, especially given their titles. Bowman mentions that both show desert and puna, but he does not actually delineate these subregions, that is, draw lines around them showing their exact location within the broader framework of the Atacama. Instead, he relies on his accompanying text to inform the reader. Bowman's book, in essence, is as much a sophisticated travel narrative as it is a scientific monograph. He uses many scientific measurements to position himself and the reader in time and place, but the ultimate effect of his book is to awe the reader with the uniqueness of this region. One more thing about Bowman's map of the desert and puna of the Atacama is remarkable: on it, he now boldly locates the name DESERT OF ATACAMA using tall letters that arch from just north of Caldera to just southeast of Iquique. My characterization

Figure 1.7. Isaiah Bowman's map depicting the desert and puna of the Atacama region originally appeared in *Desert Trails of Atacama*, 1924. (Author's digital collection)

Figure 1.8. Devoid of vegetation, this rock-strewn, Martian-like landscape lies north of Cerro Paranal, Chile. (Photograph by author)

Figure 1.9. A typical puna landscape in the higher reaches of the Atacama Desert southeast of Calama, Chile, sustains meager plant growth of importance to grazing animals. (Photograph by author)

as boldly, though, should not be confused with bodily. The positioning of that lettering is deliberate but should not mislead readers into thinking that this desert begins and ends here. As Bowman noted, one could use the term Atacama Desert for portions of southern Peru (as he did in his informative 1915 book *South America: A Geography Reader*, almost ten years earlier). Similarly, sections of this same desert that stretch along the South American coast to well south of Copiapó and Caldera were also considered part of the Atacama Desert by Bowman, though he conceded that they were not as arid as places farther north. To make matters even more interesting, people were prone to give parts of the Atacama Desert region their own names. Bowman's aforementioned use of the Desert of Tarapacá for its namesake city and province in the borderlands of Chile and Peru, which lies well beyond the northern border of this map, is a case in point. The Desert of Tarapacá is widely recognized by many observers as being within the Atacama Desert, and this overlapping nomenclature again confirms something that Indigenous people have long understood: this is not one desert at all, but rather many separate dry places that have their own unique character—provided we are astute enough to read the landscape in that depth.

As I shall repeatedly demonstrate throughout this book, reading someone's published descriptions of places reveals very important information about both the place and the writer (and by extension the anticipated readers). But what about unpublished and hence less-censored and more candid writings? Such musings, which may never be intended for the public to read, can be eye-opening indeed. Bowman's published writing is usually straightforward and descriptive and only occasionally offers a glimpse into the aesthetic soul of the man. For a deeper look, one must consult Bowman's handwritten field notes, which are located in the library of the American Geographical Society at the University of Wisconsin at Milwaukee. Luckily, Bowman made detailed field observations that reveal his fascination for the desert and absolute awe of the high-desert puna. In mid-May of 1907, with the storms of early winter gathering, Bowman traversed the puna in the vicinity of the border between Chile and Bolivia. His notes record that "the wind from the west has continually operated and blown great quantities of quartz sand very clean and white in small mounds over the surface against which is thrown the dark green foliage of the tola bush." This passage is descriptive and suggests that Bowman had a more aesthetic side that was tuned to this austere and fascinating place. That suspicion is confirmed in his next sentence: "At sunset as we crossed this strip it was lighted up in the best possible way and made a desert picture of great beauty."

That Bowman had a deep appreciation for this landscape becomes even more clear in subsequent passages that reveal how he could even overlook physical discomfort to discover the sublime. The next day, his group trudged onward despite the wind blowing "with a force and coldness beyond anything that I had yet experienced." The wind was so strong that the group was unable to pitch their tents that evening, opting to push on to an Indian village. Regardless of the punishing weather, the sights were amazing. Bowman noted, "As we reached the summit of the pass we could see off to the east the snow-capped peaks of the eastern range of the Andes across the great wide pampa between." Having set the scene geographically, he continued, "The low crimson and purple sunset clouds of the trades [i.e., trade winds] blew in long streamers to eastward from the crest we were passing." As Bowman succinctly put it, this was "the experience of a lifetime."[2]

When he made that observation, Bowman was just beginning his extensive South American fieldwork. Although he was cognizant of the different physiographic zones that he traversed, he considered all the country under the broad rubric of "Atacama." Bowman had numerous maps in hand, and he consulted others in archives, in order to describe the region. My goal, though, is to go beyond describing the Atacama Desert and to determine how travelers such as Bowman and others perceived the region. Accordingly, throughout this book I shall use maps to show how the name Atacama Desert—and hence the desert itself—took form. This happened fitfully, at first meaning something very different than it does today. It was a process that involved equal parts of historical narrative and geographic imagination. That process was cumulative. It involved the coalescing of several isolated, individual geographical descriptions into a broader mindset that ultimately gelled as the larger Atacama Desert in the popular mind.

Deciphering when, where, and how this happened is an exciting endeavor, for it involves considerable mapping of what historian Katherine Morrissey calls "mental territories."[3] There are, however, a few caveats. First, I should note that even today there is not complete agreement on what the term Atacama Desert signifies; some people reserve the term for only the most austere and climatically hyperarid portion of arid northern Chile, which some romantically call the "heart of the Atacama." Still others have used the term Atacama Desert for just about everything north of La Serena all the way toward Lima, Peru, and even beyond. Nevertheless, both of these perceptions, one restrictive and the other very inclusive, can and will be respected as representing poles on a continuum. Between these poles lies a rich variety of ways that the Atacama Desert has been constructed. In other words, there is no right and wrong way to determine where this desert lies, just an array of perceptions that have varied through time.

While on the subject of perceptions in this introductory chapter, which I began by emphasizing the importance of words, and hence names, I should forewarn readers that the noun Atacama may refer to two very different kinds of places, namely: 1) the political jurisdiction of Atacama in Chile, so designated officially, and 2) the name of an arid portion of South America, now most often conceptualized as being entirely within Chile but extending into adjacent countries, notably portions of Peru, Bolivia, and Argentina. Of the latter, Atacama as a desert place rather than a political jurisdiction, I and many others may simply use the term "the Atacama" as synonymous with the Atacama Desert. In other words, I may mean "the Atacama Desert" when I say "the Atacama." Hopefully this flexibility will not confuse readers, but in my defense I am simply following the lead of many Chileans who do the same. In their language, they often use "el Atacama" as shorthand for "el Desierto de Atacama," even though they may then say the full name— "el Desierto de Atacama"—shortly thereafter. There may seem to be no rhyme or reason to this, but I believe there is, for I have noticed that Chileans often use the full name with some deliberation, perhaps a few sentences later, when they want to emphasize the challenging desert character of this region.

To our aesthetic preferences today, Bowman's maps may appear austere or stark. We now expect color and perhaps a bit more drama in our maps than black lines, dots, and splotches on white paper. For that matter, we can easily go online, locate a Google map of South America, and scroll along the coast in the vicinity of about 22 degrees south latitude. In that satellite-generated view, this desert region appears as a slender band of buff and ocher running along the western edge of South America. This roughly north–south trending desert lies between the Pacific Ocean on the west and the Andes Mountains to the east. Its proximity to the Pacific Ocean, whose Humboldt (or Peru) current brings cold water northward, is a key factor in the prevailing stable atmospheric conditions that cause prolonged drought. The high pressure in this zone near the Tropic of Capricorn tends to be anticyclonic, which in the southern hemisphere is counterclockwise. The center of the high-pressure zone lies over the ocean to the west of this coastal desert; accordingly, the winds are normally southerly (i.e., flowing from south to north). Another factor in the region's aridity is the position of the spectacular Andes mountain chain that, except on rare occasions, effectively blocks any moisture-laden air that may move westward from the Atlantic Ocean.

Although this coastline appears fixed on maps such as Bowman's and even satellite images, in reality it is not only constantly being eroded by the force of the Pacific Ocean's waves, but also subject to tremendous forces that are lifting it relentlessly as the floor of the ocean west of it is shoved beneath the continent. This subduction of the Nazca plate accounts for the many earthquakes experienced in Chile, some of them the strongest ever recorded on earth. With jolts that shake the landscape and rattle nerves periodically, the Chilean coast is rising, as is the land east of it.

Although any particular earthquake may raise the land only a few inches or feet, the cumulative result over millions of years is the region's towering mountains, including the Andes. That term "only" in the last sentence is relative, for the damage resulting from such earthquakes can be catastrophic. For example, an 1869 report on the Arica and Tacna railroad company mentions the "disastrous earthquake which, on the 13th of August [1868]...was felt with fatal effect at Arica, where the sea swept away three-fourths of the town, including the company's station, buildings, machinery, stores of fuel and other necessaries, besides three locomotives and more than half their rolling stock." The company promised that "due economy would be used in every department to restore the line and works and to equip it for traffic as early as possible,"[4] but the disruption lasted months. That earthquake, or *terremoto*, which jolted the area from Arequipa (Peru) southward to Arica (also then in Peru), is estimated to have been between an 8.5 and 9 magnitude and may have claimed upwards of twenty-five thousand people. In Arica the earthquake was widely covered in news journals worldwide, as it also destroyed many ships from various countries, including two U.S. Naval vessels that had anchored there to escape a yellow fever epidemic in the tropics. Illustrations from the time of the earthquake show catastrophic damage being wrought on Arica (Figure 1.10).

As the description by the railroad and the lurid illustrations from the press suggest, earthquakes in coastal towns such as Arica typically deal a punishing one-two punch. First, a city is severely damaged structurally and sometimes swept by fires that break out. Shortly thereafter, sometimes within minutes, a huge wave measuring sixty or more feet (twenty or more meters) high may rush ashore, destroying remaining buildings and drowning stunned inhabitants. To make matters worse, the wave then sweeps people and debris out to sea, making it difficult to determine the full extent of loss of life. Erroneously called "tidal waves," these were subsequently understood to be caused by the earthquake, hence the German name Erdebenwelle (earthquake wave) was coined; in the twentieth century the Japanese term

EARTHQUAKE IN PERU—SUBMERSION OF ARICA.—[SEE FIRST PAGE.]

Figure 1.10. As seen in this illustration from *Harper's Weekly* (November 21, 1868), the massive earthquake and tsunami at Arica, Peru, on August 13, 1868, washed ships ashore and resulted in the submersion of the city, which claimed thousands of lives. (Author's digital collection)

tsunami (long wave in a harbor) was adopted worldwide.[5] In the Atacama Desert region, other major earthquakes of about 8 to 9 magnitude occurred at Iquique in 1877 and 2014, but thanks to better construction design, the latter claimed far fewer lives.

By the late 1890s, the Russian/German climatologist Wladimir Köppen designated the Atacama as true desert (BW) according to the now-famous climatic classification system[6] that bears his name. Furthermore, Köppen recognized that deserts could be hot or cold, the Atacama being an example of the former, and Antarctica representing the latter. Later editions of Köppen's world map often depicted the hotter arid and semiarid areas using yellow and orange, respectively, perhaps as a subliminal suggestion that they are warm and devoid of a green vegetation cover. A map showing the distributions of warmer desert climates worldwide based on Köppen's system, which is today better known as the Köppen-Geiger climatic classification

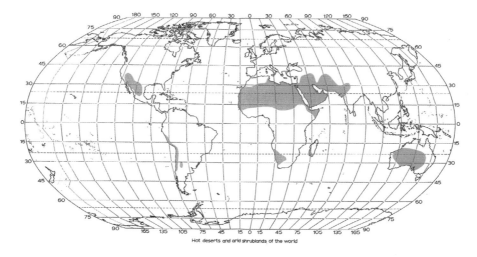

Figure 1.11. World map from *Ecosystems of the World, Volume 12A: Hot Deserts and Arid Shrublands* (Evenari, Noy-Meir, and Goodall, eds, 1985), confirms the great latitudinal (north–south) reach of the desert region containing the Atacama, which straddles the Tropic of Capricorn, shown as a dashed line. (Authors digital collection)

system in reference to Köppen's German protégé, revealed that deserts are not randomly distributed; rather, they are positioned into two zones in the lower midlatitudes, one in the northern hemisphere and one in the southern hemisphere (Figure 1.11).

This distribution is partly a result of atmospheric circulation systems, notably the subtropical high-pressure zones where the air subsides and is hence more stable. As implied in the term subtropical, these high-pressure zones lie poleward on either side of the equator. These gyres, or large masses of stable but slowly rotating air, are related to and complicated by the cooler temperatures of adjacent oceans off the west coasts of continents; the net result is the seeming paradox that these deserts march right up to the coast-line. The fact that deserts in the northern hemisphere are much larger is a result of larger land masses there than in the southern hemisphere. Of all the continents, South America has the smallest percentage of land in zones of desert climate, which is another seeming paradox because it is home to the Atacama, which is widely considered the driest desert on the face of the earth. But as the map of South America reveals, this continent is widest just below the equator—where the magnificent, lush rainforests of the Amazon are found—and narrows markedly toward the south. The closest counterpart to the Atacama Desert is the Namib Desert of South Africa, which likewise lies along the west coast. The Atacama Desert, though, is even more arid than the Namib.

In South America the arid strip of land along the west coast represents the world's longest latitudinal stretch of desert. Remarkably, it reaches from just south of the equator to about 30 degrees south. In part of this strip lies the quintessential coastal desert, the Atacama. Although the name Atacama is often followed by the designation of driest place on earth, that also demands some clarification. The term desert suggests perennially clear skies, but the Atacama is known for its occasional overcast and foggy conditions, especially near the coast. Most often occurring in the low-sun months of June through September, these fogs usually form a layer about a thousand feet thick. This layer does not usually hug the ground but more commonly hovers about 1,500 feet (500 m) above sea level. When a traveler is in such a camanchaca, the skies are gloomy, but about 2,000 feet in elevation can bring him or her into startling bright sunshine. Viewed from above, the camanchaca lies like a fluffy blanket filling the valleys. In 1875, while perched on a rugged hilltop at Mina Guanaca, watercolor artist John Marx beautifully captured the contrast between ruddy, arid hills and white fog below (Figure 1.12). This persistent fogginess has led some to classify the climate of the desert along portions of the South American coast as BWn, the "n" derived from the Latin nebulous signifying cloudy.

Despite this cloudiness, little actual rain falls from the camanchaca, but when it does, locals call this light drizzle a garúa. Although the Atacama is often portrayed as a land of no rain, in this chapter and elsewhere I will point to notable exceptions. For example, in November 1899 portions of western South America were hammered by a storm system that caused much damage. Even though the coast south of Valparaíso was especially hard hit, the damage extended well into the Atacama Desert region, including the normally dry and serene port town of Taltal. Here, as we learn from another railroad's annual report back to England, nature could be capricious and destructive; the Taltal Railway's property, including its docks and loading pier, were destroyed in that same storm.[7] Usually, though, measurable rainstorms are rare, and the entire Atacama Desert is what climatologists call hyperarid. Climatologists use this term to differentiate the very driest of climates from those that are simply arid.

With increasing rainfall toward the perimeters of such deserts, the traveler encounters grass- or brush-covered landscapes that developed under

Figure 1.12. In 1876, watercolor artist John Marx captured the arid, rust-colored mountains in the Atacama with a characteristic *camanchaca* (fog) obscuring the landscape below. (Author's digital collection; original in the Colección Iconográfica, Archivo Central Andrés Bello, Universidad de Chile)

semiarid climates. Farther beyond those margins, climates transition into subhumid, and beyond that, humid climates are found. All of these designations assume that better-watered places are the norm; this, of course, reflects the position of well-watered locales such as Europe and the eastern United States in the development of the atmospheric sciences.

These classification systems ultimately address what American geographer Charles Warren Thornthwaite (1899–1963) identified as the root cause of deserts, namely, the fact that precipitation is exceeded by evaporation from the earth's surface as well as the moisture lost by plants. Thornthwaite called this loss "potential evaportranspiration," or PET. These two losses of moisture (from plants and the ground surface) are dependent on temperature, for it had long been understood that the higher the temperature, the greater the amount of moisture would be lost to the atmosphere. Further refinements to this system by meteorologists factored in aspects such as the actual variation of the ground surface and effects of atmospheric conditions such as cloud cover. Although far from perfect, they do provide a reasonable index for decision makers. However, it is deceiving to simply compare rainfall totals to temperatures, especially in coastal deserts where fog and dew may provide a considerable amount of the water that plants receive. This, as we shall see, is extremely important in the Atacama Desert.

Given its geographic location, the surface air moving into or over the Atacama Desert tends to blow from southwest to northeast for much of the time. Moving in from off the cool ocean, this air is normally fairly mild in temperature; one becomes aware of it as a constant breeze that may stiffen to a strong wind on occasion, often in afternoon. So predictable is the flow of this air here that one of the railroads in the pampa actually used sail as motive power, though heavier trains still required steam power. Isaiah Bowman was impressed enough by a wind-powered railcar in the vicinity of the inland desert city of Calama that he featured a photograph of it in his book (Figure 1.13). Sail-powered railroads were almost unheard of; the only few successful examples were in scattered locations in northwestern Europe and the south sea islands. In the Atacama and those other locales, winds were actually strong enough to blow these light-rail cars uphill slightly. To return to their starting point, the driver simply lowered the sail and the car rolled back downhill by gravity—the ultimate in sustainable ground transportation.

In the Atacama, these prevailing breezes bring slightly moist air with them. Therefore, it can be cloudy at times, especially when this air passes over mountain ranges and condenses high above the ground. These clouds rarely bring rain, and although the skies in this region may be deep blue for days at a time, high, thin, wispy cirrus clouds are also common. The sail-powered railcar operation near Calama reflected the high cost of fuel in the Atacama. Firewood for steam locomotives was expensive as wood is so scarce. Moreover, the fossil fuels usually burned by those steam locomotives (coal early on, and oil later) had to be imported because the geology of the Atacama, while rich in some minerals, is completely devoid of coal and oil deposits. This continues in the diesel age for the railroads and the era of gasoline-powered or diesel automobiles and trucks on the region's roads. Although Chile has some coal deposits in the south, most of these were mined out more than a century ago, and it is now dependent on other countries for the fossil fuels that still power much of modern life.

In the Atacama, topography and geology also play an important role in affecting climate. Here, as in other deserts, the surface of the earth may vary widely in color from nearly snow-white sands to veritable mountains of dark-colored rock. The latter absorb heat while the former reflect it. The density and porosity of the ground is also another related factor that affects the availability of rainwater to plants. The lay of the land is yet another important factor influencing habitats. As noted, although the term desert

Figure 1.13. In this early twentieth-century photograph by Isaiah Bowman, the fairly strong and predictable winds of the Atacama Desert were harnessed to propel a sail-powered railcar near Calama, Chile. (American Geographical Society Library, University of Wisconsin at Milwaukee)

may convey flatness, the topography of the Atacama is highly varied. To help readers better understand the lay of the land here, I created a physiographic block diagram (Figure 1.14). This graphic represents a very generalized composite of the landscape along a line of about 22 degrees south latitude, but

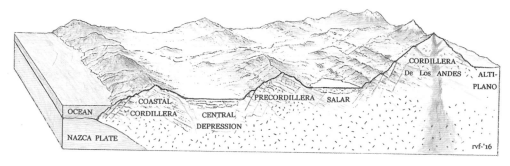

Figure 1.14. This physiographic block diagram by the author presents a generalized and vertically exaggerated cross-section of the Atacama Desert region looking generally northward in the vicinity of the Tropic of Capricorn.

much of the entire Atacama Desert region is similar in that it generally consists of three topographic/geological zones running roughly north–south. This diagram is conjectural, that is, it does not depict specific geographical features but provides a broad glimpse of how both surface and subsurface features are configured.

I should also note that the vertical relief is greatly exaggerated, for the mountains are not nearly as high as they are indicated relative to the horizontal distances. Moreover, the subsurface geology is merely suggested to emphasize the fact that different types of rocks—igneous, sedimentary, and metamorphic—underlie portions of this region. Underground fault zones fracture these units, and in many places above ground the general lack of vegetation enables travelers to see these fault lines as telltale scarps. There is a reason why I colored the landscape of the area above ground a bit more intensely than what lies below the surface. In reality, the coloration of the Atacama Desert is a result of its bedrock geology and the weathering that results from exposure to the elements. The rusty reddish or yellowish ocher-like colors in places are in part the result of oxidization of iron-rich minerals. Other metals such as copper and manganese also oxidize green and purplish, respectively. Additionally, as a result of the dry climate, the surface of the rocks may be coated by a thin, slightly glossy veneer of oxidized metals and organic material called desert varnish. By chipping away at this satin-sheened, dark-purplish veneer, one can easily expose the true color of bedrock, which is often much lighter.[8]

For readers unfamiliar with interpreting physiographic block diagrams such as this one, which represent a slice or section of the earth viewed obliquely and looking northward, I will now briefly narrate a journey across it from west to east (left to right), which is to say from the Pacific coast inland. On this journey, one first encounters the coastal mountain ranges, often from a small port tucked into one of the numerous small harbors behind which loom tall mountains. As the diagram suggests, the coast is a zone where the underlying Nazca Plate and the rocks above it come in contact as the westward-moving rocks of the continent are forced above it. I have tried to suggest the powerful forces that are released here by showing how the lands adjacent to the coast are vertically fractured as the ground rises. This mountainous zone is locally called the Cordillera de la Costa, its North American counterpart being the Coast Ranges of California and Oregon. Like its North American counterpart(s), the Cordillera Costal is not a single range but rather several different mountain ranges running in the same direction. And like the coastal ranges of the North American west coast, the topography here is not the result of any particular kind of rock, but instead the effect of continued uplift breaking the bedrock into huge blocks (called horsts) along the fault lines that parallel the coast.

Although the types of bedrock vary, with granite-like igneous rocks in places, these mountains generally consist of sedimentary rocks of varied ages. Sandstone and siltstone are common here, as are fossils of animals such as giant ammonites that lived millions of years ago when the area was submerged by the sea. In places, the rocks contain fissures rich in copper and silver, although much of this range is not especially productive mining country. This coastal mountain range province is generally only about thirty miles (45 km) wide but runs the length of the desert as tall, corrugated parallel ridges. The steep rise of the Coastal Cordillera from the Pacific Ocean is noteworthy, one source claiming that these are "some of the highest coastal cliffs in the world." That same source quantifies this by noting that some of their slopes run about 2,000 meters (ca. 6,500 ft) in length while "dropping abruptly into the Pacific Ocean."[9] The Cordillera de la Costa is a lofty mountain chain, especially in the south, but generally does not exceed about 8,500 feet (2,600 m) in elevation. The highest of these coastal ranges is the Cordillera or Sierra Vicuña MacKenna, which reaches almost 10,000 feet (ca. 3,000 m).

Moving inland from the coastal ranges, one encounters a broad north–south trending valley, which is sometimes called the Central Depression because it is bordered on the east and west by mountains; more to the point, most of the precipitation that happens to fall here flows into broad basins and evaporates. The term depression may suggest that it lies below sea level like Death Valley but that is not the case, for this part of the Atacama is actually an elevated plateau-like area that lies about one thousand meters (ca. 3,200 ft) above sea level. To the observer on the ground, the Central Depression appears to be level, but the land's surface slopes slightly to the west. Its counterparts in North America are the great Central Valley of California and the Willamette Valley of Oregon, though they are much closer to sea level. Toward the western side of this central depression in northern Chile, which is one of the driest places on earth in the latitude from about 21 to 25 degrees south, large deposits of nitre (sodium nitrate) are found; hence natives call these areas the "nitrate pampas." In this regard, the physiographic block diagram best characterizes the central portion of this desert. In the southern part of Chile's desert north, especially in the vicinity of Vallenar and La Serena, the central depression gives way to a series of transverse (east–west trending) mountain ranges and there are no nitrate deposits.

Returning to the block diagram, the traveler moving eastward from the Central Depression will note that the land rises, sometimes fairly steeply, into rather tall mountains that can reach to about 13,000 feet (4,000 m) above sea level. Chileans use the term "Precordillera" for this elongated mountainous zone, which consists largely of stratified rocks of various types and ages whose beds may be fairly horizontal but in places distorted and fractured. Here, as in much of the region, earthquakes are frequent; sometimes the faulting is horizontal, but vertical faulting is especially common. Igneous rocks abound in these mountains, sometimes as intrusions and sometimes as ash flows from the volcanic country that lies farther to the east. The prominent hills or small mountainous areas that rise along the eastern edges of the Central Depression region of the Atacama are part of this zone. In many places these mountain ranges may be fragmented and highly mineralized, containing deposits of gold, silver, and copper. The impressive Sierra de Domeyko, which has long been the site of metals mining, is the quintessential Precordilleran landscape. On the eastern side of the Precordillera lie valleys where salars are found. Although the term salar signifies a dry, salty lake bed, in wet periods water may stand for an extended time, turning lake beds into salty marshes.

To the east of the Precordillera and the Salars de Atacama, the impressive Andes Mountains reach skyward, their peaks often shrouded in snow and distantly glimpsed from sweeping valleys. Their North American counterparts are found in the upper reaches of the Cascade Mountains of Oregon and Washington, where volcanoes punctuate the horizon. By comparison, though, the volcanic peaks in the Cascades are far lower—about a vertical mile—than those found in the Andes. As in western North America, some of these volcanoes are active. The block diagram reveals the source of this red-hot magma deep underground. In places to the west of the crest of the Andes, lakes are found, and in locations where heat is conducted from deep underground, geysers are encountered, the famous El Tatio geyser field being the best example. Also in common with their North American counterparts, these volcanic mountains stand on the shoulders of other older mountains. For example, in the vicinity of Chuquicamata, the western margin of the Andes range consists of a rugged zone of steep-sided mountains of medium elevation (about 10,000 ft or 3,200 m) that comprise a treasure house—a rich porphyry ore body consisting of minute crystals of copper sulfides dispersed throughout igneous rocks. Here, a grayish colored intrusive igneous rock called diorite often forms the bedrock. Because it cooled fairly slowly from its molten state deep underground, diorite consists of visible grains and crystals of minerals such as white feldspar, black biotite mica, greenish black hornblende, and pyroxene. The elevation at Chuquicamata is 9,350 feet (2,850 m) above sea level, but a look eastward across the Loa River valley reveals even taller mountains, the Andes, where the topography becomes considerably more rugged. In this labyrinth of steep-sided slopes and deep canyons, travel slows considerably. Geologists classify this wide area as the CVZ (Central Volcanic Zone) of the Andes mountain range, which is often regarded as the spine of the entire South American continent.

In one last caveat about this physiographic block diagram, I should also note that at the far north end of the Atacama Desert region, the Andes Mountains curve to the west at around the Arica bend, and are hence closer to the Pacific Ocean as the landscape is more compressed horizontally. In this area, the Central Depression is far less pronounced, actually petering out into a series of smaller bowl-shaped pampas. The Precordillera, too, is often less pronounced and consists of smaller ranges that appear to be foothills of the Andes. That explains why many travelers considered a trip inland from the coast in the vicinity of Arica to be a more or less direct ascent into the Andes.

In the rugged country at the base of the Andes and up to its towering heights, one commonly finds a fine-grained grayish to purplish colored extrusive rock called andesite, which comprises outcrops and litters the landscape as rocky talus. Because it cooled somewhat faster than diorite, andesite consists of small barely visible crystals of hornblende, black mica, and a light-colored plagioclase feldspar. Andesite is so ubiquitous here that the Andes mountains became the type-locality for this rock, no matter where on earth it is found; thus, whenever geologists use the name andesite, it evokes its namesake mountains of South America. It is humbling to realize how interconnected this region is, geologically speaking, with the rest of the

Figure 1.15. In this scene at Bahía de Nuestra Señora, Chile, arborescent multi-stemmed cacti such as species of *Eulychnia* are found close to the shoreline in parts of the Atacama Desert. (Photograph by author)

world. Geologists note that the Andes are part of "an 'andesite line' formed by thousands of volcanoes (consisting of some andesite)" that "circles the entire Pacific Ocean, corresponding to the subduction of an oceanic plate with an essentially basaltic composition beneath the continental plate." They also confirm that "the major andesitic volcanoes are in the Andes (hence the derivation of the name)."[10]

In my geological collection are a number of specimens from places named after them; for example, basalt from Basalt, Nevada; and granite from Granite, Montana; but none is more prized than a specimen of andesite that I personally collected from its type locality in 2015. Wielding the same Estwing rock pick that my parents gave me for Christmas in 1955, I took considerable pleasure knowing that my oldest (and most prized) of possessions was still helping me make my most recent of discoveries, this time in the Andes at the eastern edge of the Atacama Desert. With that blocky, fist-sized, rectangular specimen of andesite now stuffed into my backpack, I paused to take in the view of this breathtaking landscape. Although the bulk of the Andes range is high country, much of it above 15,000 feet, its crest or spine is punctuated in places by gently sloping volcanic cones that reach even higher, to nearly 23,000 feet (7,500 m), so high and consequently cold that they are perpetually snow covered. Among its numerous volcanic peaks are Tacora (an active volcano), Parinacota, Irrupuncu, Guallatiri, and Lázcar. Most of these average about 19,700 feet (ca. 6,000 m) in the vicinity of the Atacama. This geological complex and its surrounding countryside is not only high but deep, comprising what geologists describe as "the thickest crust on Earth: 50–65 km thick below the Altiplano and Puna plateaux and > 70 km thick below the bounding mountain belts."[11] In terms of physiography and hydrology, the Andes represent the continental divide of South America; westward from its crest the precipitation that falls here, either as rain or snow, heads toward the Pacific, though it usually does not make it all the way, collecting instead in the desert basins and evaporating. Eastward, the waters flow to the Atlantic Ocean, though here too some of it may be collected in desert basins, notably in Argentina.

In this part of South America, which is by most accounts desolate, vegetation takes on a new importance by virtue of its near absence. A world map by A. W. Kuchler (1907–1999) authoritatively declares "vegetation largely or entirely absent" in this arid portion of South America.[12] However, that term *largely* is extremely important, for it suggests exceptions. The biggest distinction is the difference between the lower-lying portions of this region, which, as noted, some locals call the Atacama Desert, and the higher areas which may have patches of vegetation such as bunchgrass and may be called the puna(s) of Atacama. Closer to the coast, where moisture condenses from frequent camanchacas, the vegetation may consist of scrub plants or cactus communities. Northeast of the small port of Huasco, the lush Llanos de Challe represent an area high in biodiversity. Farther north, in the Pan de Azúcar area, columnar cacti thrive right up to the shores of the Pacific (Figure 1.15). A bit farther inland, in the vicinity of Paposo, barrel cacti abound in the sometimes scrub-covered hills (Figure 1.16). In what some observers call "the fog zone of the Atacama Desert," the small cactus *Copiapoa dealbata* "forms impressive mounds that dominate the landscape in places in a virtual monoculture." The larger clumps may be more than a meter (ca. three feet) across and "may be hundreds of years old, so slowly do they grow, watered only by fogs, as actual rains can be nonexistent for years at a time."[13] These areas sustained by camanchaca fogs represent the lushest places in the Atacama Desert region outside the river valleys. Just a few miles east of these brushy locations, the landscape farther inland is often totally

Figure 1.16. Sustained in part by moisture from camanchacas (fogs), scrub plant communities near the coast also feature barrel-shaped cacti of the genera *Copiapoa* (pictured above) and *Eriosyce*. (Photograph by author)

devoid of vegetation, or nearly so. As I shall demonstrate, any sprig of vegetation in this entire region was (and is) of value to local Indigenous people and would later be of increasing interest to science.

Whereas the characterization of vegetation being nearly absent is true enough for large areas that receive very little rainfall, phreatophytes (plants with deep roots that can reach ground water) are found in many places. In the Atacama, even the most barren of places may surprise the traveler with greenery that at first seems a mirage. Such scenes are encountered where water from far-distant sources reaches the desert, as in the verdant riparian oases found along river valleys where runoff from the neighboring Andes Mountains flows into the desert. There are many such locations in the Atacama where a seemingly out-of-place river sustains a veritable corridor of trees in what is otherwise a hyperarid landscape. It is here that trees such as Chilean mesquite (*Prosopis chilensis*) and willows (such as *Salix humboldtiana*) are found. Although most of these rivers never reach the ocean, the aforementioned Loa River is a major exception. Rising in the Andes, it

Figure 1.17. Flanked by bare hills and low mountains, the Copiapó River Valley features riparian shrubs and scattered trees along the normally dry watercourses. (Photograph by author)

flows generally westward into Chile. Like a serpent, the Loa sashays lazily along the desert floor, helping sustain the fields and mines near Calama and Chuquicamata, respectively, on its circuitous course to the Pacific at Iquique. Farther south, the Copiapó River also reaches the Pacific, though its bed may be dry for years at a time (Figure 1.17). In the 1910s, Isaiah Bowman took a stunning panoramic photograph of this valley, with its tall Lombardy poplar -like willow trees standing like sentinels and often planted in rows (Figure 1.18). The way those trees are leaning is the result of both the topography and the meteorology: the breezes that sweep through the region from off the Pacific are in this location funneled through the narrow Copiapó River valley, effectively increasing their velocity through a phenomenon known as the Venturi effect.

At this point, I should note that the idea that a river must reach the sea is largely of European origin. In popular culture we may assume that the romantic sentiment "like a river flows, surely to the sea, darling so it goes, some things are meant to be"[14] is a truism, but it reveals that those who write such songs are from areas where precipitation, and hence runoff, is plentiful. This is especially the case with much of Europe and the eastern portion

Figure 1.18. In Isaiah Bowman's early-twentieth-century panoramic photograph of the Copiapó River Valley, tall Lombardy poplar-like willow trees lean as they are blown by the strong prevailing southerly winds. (American Geographical Society Library, University of Wisconsin at Milwaukee)

of North America. By contrast, in most of the world's deserts it is more or less a given that any watercourse will *not* make it very far, and certainly not to the sea, unless sustained by huge amounts of water. Thus, when another popular song claimed "lonely rivers flow, to the sea, to the sea, to the open arms of the sea,"[15] it assumed a setting in a well-watered land where all rivers do so obediently. The exceptions to this are both legendary and important for sustaining urban centers in the desert, as is found in the cases of the Nile of North Africa, the Tigris and Euphrates in Iraq and vicinity, the Indus of Pakistan, and the Colorado River in the American Southwest. These rivers help sustain, respectively, "desert" cities such as Cairo, Baghdad, Karachi, and Las Vegas/Los Angeles.

Despite its proximity to the coast, the Atacama is for the most part what hydrologists call an endorheic region, that is, one having interior drainage that does not reach the sea. Nevertheless, in places in the Atacama, dry river channels do run all the way to the coast. These, though, only convey water when very heavy rains or rapid snowmelt from the Andes provide a large amount of water, which may then flood communities nestled along their normally dry channels. The Rio Loa in the northern Atacama is the most apparent of these desert rivers that reach the sea, but other smaller ones include the Rio Copiapó, which enters the Pacific at Caldera and Bahia Inglesa in the central portion, and the Elqui River, which reaches the coast at La Serena and Coquimbo in the southern part of the Atacama.

Like all deserts, the Atacama is difficult to depict with precision on maps. As noted earlier in this chapter, geographer Isaiah Bowman knew this and wisely refrained from drawing lines around it. In places the boundary may seem clear, but such boundaries are often more imaginary than real. For example, in many places toward and beyond the eastern edges of the desert valleys of the Atacama, the term "puna" replaces "desierto," especially in the area adjacent to the Andes Mountains. This puna, as noted, is also rather desert-like but often features distinctive clumps of bunchgrass. The higher elevation of the *puna* (much of it well above 3,000 m or ca. 10,000 ft) makes it considerably colder, and some claim even less hospitable, than the Atacama Desert. That may be something of an exaggeration, though, as the grasses found in the puna sustain grazing animals.

Adjacent to the peaks of the Andes in Bolivia lies the Altiplano, which appears to be perfectly named for its root words in Spanish/Latin, meaning a high, flat plain. However, the Altiplano is actually large and bowl-shaped, a portion of which is occupied by the huge freshwater Lake Titicaca. In the higher elevations, with their colder temperatures and fairly predictable run-off from the mountains, are lakes that appear to be deep blue slabs of lapis lazuli set into the brownish and purplish landscape. Some of these lakes, though, have a startling red color, the result of algae suspended in the water. These and their adjacent marshy areas are home to huge flocks of flamingoes.

South of that high country, the deserts of Argentina are also characterized by sparse vegetation, but they normally have different names, all the result of history and local traditions. Here one encounters deserts that are punctuated by tall cardón cactuses reminiscent of the giant saguaros of the American Southwest. This desert on the other side of the Andes from Chile is remarkable in its own right, though not as dry as the Atacama of Chile.

However, as noted by a Cambridge University team that studied the Atacama Desert in the late 1950s, "the name Atacama is also used to describe this part of northwestern Argentina…[and] furthermore the Puna de Atacama lies in Argentina, although different writers use the word Puna to refer to different parts of the high Andes." Recognizing that the desert and puna generally varied by color of the landscape as well as the colors indicating varied elevations on the hypsometric maps they were using, they thankfully added, "For the sake of peace of mind and [to avoid] confusion, we shall use the word Atacama to mean the parts of Chile and Bolivia we explored, whatever their height or colour on the map."[16] For my part, I shall introduce and discuss many diverse cartographic representations of the Atacama Desert, as well as written passages about it, my purpose being to show how delightfully varied this desert has been—and can still be—in the popular imagination.

Returning to the Atacama Desert as widely accepted in most of the literature, I shall now direct the reader's gaze to the western side of the Andes, where the lake beds in the higher puna country may be full of water or marshy, and temperatures chilly indeed. In this high country, especially in the months of May, June, July, and August (winter), nighttime temperatures well below freezing are common, though daytime temperatures are usually above freezing, and—provided that the wind is not blowing—comfortable. These conditions favor hardy cacti, and they abound here. Along the base of the Andes in Chile, where daytime highs are well above freezing despite the cold nights, the tall cactus *Echinopsis atacamensis* (Figure 1.19) thrives, reaching as tall as seven meters (ca. 20 ft). These giants are covered with tawny spines that almost make the entire bristly plant seem to glow a pinkish color in certain light. Sometimes they branch but typically stand as singular columns. Well adapted to large swings in temperature, they often grow on steep, rocky outcrops that seem to have little or no soil.

Continuing west, the climate becomes warmer and drier. The lakebeds here tend to be seasonally or even perennially dry. With every mile westward from the puna, then, the land drops into a series of hills through which streams sometimes rampage, leaving deeply scarred arroyos that are locally called quebradas. When the traveler continues westward and reaches the wide, lower desert valleys in the central part of the Atacama, the temperature rises noticeably. However, even here cold winds can sometimes sweep down into the interior portions of this desert from the Andes. Generally, though, given its relatively lower altitude and latitude, it is not surprising that the Atacama Desert is considered a hot desert (BWh) not unlike the Sonoran Desert of southern Arizona and northern Mexico, as opposed to

Figure 1.19. The tall columnar cactus *Echinopsis atacamensis* is common in higher elevations of the Precordillera and in the foothills of the Andes in Chile, as well as in adjacent portions of Bolivia and Argentina. (Photograph by Damien Francaviglia)

a cold desert (BWk) such as those found in much of northern Nevada and Utah, or, for that matter, Tibet. This is ironic, because the Atacama may be classified as a hot desert, but few portions of it could be considered hot. Moreover, parts of it can be as cold as many cold deserts, at least on occasion. In other words, the climate of the Atacama may approach extremes, but it is largely mild, with relatively few temperature extremes. This is especially true compared to deserts of the northern midlatitudes, such as portions of the Middle East and the Great Basin of western North America.

Still, one of the things that characterizes this desert, as it does many deserts, is the wide range of temperature between day and night, especially of the elevated areas away from the coast, where diurnal temperature ranges of 40 degrees Fahrenheit are common. The Andes Mountains adjacent to the Atacama offer a clue to the climatic differences on either side of the range: on the moister eastern side, the level of permanent snow at

27 degrees latitude is about 17,000 feet (ca. 5,180 m), while on the western side the snow line is at about 21,000 feet (6,400 m). Farther north along the Chilean–Bolivian border at between 21 to 25 degrees south latitude, the line of perpetual snow is at about 18,000 feet (5,486 m) to the east, and virtually nonexistent on the west side. The difference is explained by the fact that none of the peaks, excepting Llullaillaco, is high enough to hold permanent snow.[17] Where snow remains in broad patches, it may assume strange shapes. In the case of the imaginatively named penitentes, the high rate of evaporation causes the snow to sublimate, that is, change from the solid state (ice) into the gaseous (air) without melting into water. By this unique process, the remaining snow forms vertical razor-sharp spires that, to some early devout travelers, resembled praying figures.

Although the Atacama Desert has an alpine countenance as it reaches into the Andes, much of it westward of that great range experiences far milder temperatures. The proximity of the Atacama Desert to the cool waters of the Pacific Ocean is a major factor in keeping its temperatures relatively moderate. Throughout most of this desert, daytime temperatures rarely exceed 90 degrees Fahrenheit (32°C). Therefore, despite its reputation for aridity and heat, the weather in this desert is generally fairly temperate. A glimpse at the weather records for the interior desert city of Calama (elevation ca. 7,410 ft or 2,260 m) is instructive, for the highest recorded temperature is only 90 degrees (32°C), while the lowest-ever recorded is 15 degrees above zero. By comparison, at Santa Fe, New Mexico, which lies at almost the same elevation (7,199 ft or 2,194 m), the highest temperature ever recorded is 101 degrees (38.3°C), and the lowest is 17 degrees *below* zero (-27.2°C)! Small wonder that early Spaniards in highland New Mexico regarded it as the coldest place they had ever encountered.

Farther south, in Mexico, the city of Durango (elevation 6,170 ft or 1,880 m) is perhaps a more comparable location, as it lies nearer the Tropic of Cancer, much as Calama lies nearer the Tropic of Capricorn. However, Durango also has recorded far more extreme temperatures than Calama—a record high of 103 degrees (39.5°C) and a low of 10.4 degrees (-12°C). In the Atacama, as elsewhere, weather stations at lower elevations are warmer, the exception being the coastline where the Pacific Ocean keeps temperatures more moderate. In keeping with this truism, the lower-lying places toward the western edge of the Atacama's Central Depression region experience the highest temperatures in the region. Copiapó (1,293 ft or 391 m) is perhaps typical of an interior location: most days year-round are pleasant, with highs normally in the 70s (20s C). Record high temperatures for every month reached about

90 degrees (32°C), with the all-time high at 94 degrees (ca. 34°C). Similarly, low temperatures are moderate, usually averaging well above freezing, though the lowest ever recorded at Copiapó was 28 degrees (-2°C), a reflection of its distance inland (about 60 miles or 100 km) from the coast.

Closer to the coast, snow may fall in the higher coastal mountain ranges, but frost and subfreezing temperatures are virtually unknown along the coastline, which some observers have equated with eternal springtime. The coastal city of Antofagasta is perhaps typical: the highest temperature ever recorded is 86 degrees (30°C) and the lowest 35 degrees (1.7°C). By comparison, the resort city of Cabo San Lucas in Baja California, which is not far from the Tropic of Cancer, has recorded a record high temperature of 111 degrees (44°C) and a record low of 31 degrees (0°C). Farther north along the Chilean coast, the temperatures in Arica are slightly more moderate than Antofagasta. Arica, in fact, is known for a climate that is not only moderate but positively sublime. Arica's lowest recorded temperature is 42.4 degrees (5.8°C) and its highest 88.7 degrees (31.5°C). Most of the time, the temperature remains in the 60s and 70s (10 and 20s C), and the city has earned its perennially vernal characterization.

As suggested, the differences in temperature extremes between locales in North America and South America likely reflect the greater continental conditions in the northern hemisphere (that is, larger land masses). Thus, although travelers often comment on the heat in the Atacama Desert, they may be judging that in terms of the more temperate places they know. In reality, this South American desert is relatively milder in temperature than its North American counterparts. One of the best examples of this is the warning that many plant dealers selling Atacama Desert cactus provide, that their plants are not able to take the punishing heat in places such as Phoenix, Arizona, where they must be grown indoors during summer. The main factor to recall about the climate of the Atacama Desert is that it varies less from north to south than it does from west to east. As a rule, the isotherms (lines of equal temperature) in this region run remarkably parallel to the coast. This is a reflection of at least three factors: the relative uniformity of ocean temperatures; the sustained, fairly strong southerly winds; and the increasing elevation from the coast to the crest of the Andes. In fact, it is likely that no other place on earth has such a long stretch of longitudinally (east–west) determined temperature zonation. On any given day, the temperatures in the coastal towns of Coquimbo and Arica, which are separated by about 800 miles (ca. 1,200 km) or about 12 degrees of latitude, can be identical, or close to it.

The impact of the coast is sensed palpably, if subtly, by traveling westward from the interior portions of the Atacama. With each mile westward into the coastal ranges, travelers can tell when they are approaching the ocean well before it comes into view. Degree by degree, the temperature becomes ever so much more moderate, the sky takes on a slightly milky haze, and the air feels "closer," that is, becomes more humid. Confirming these observations, the character of the landscape changes as one leaves behind the completely barren land of the interior and encounters small cactus plants and shrubs. At first these plants appear to be a mirage, as they may be much the same color as the brownish hills. Continuing west ones sees them turn into veritable groves of spiny erect cacti. Somewhat optimistically, these are called *bosquetas* (little forests) of cactus, and they make a striking sight indeed. The condition and color of these plants may make one wonder whether they are dead or alive, but they are a reminder that many living plants in the desert are not necessarily bright green; when seen on a sunny day, they might best be described as a greenish bronze color, in part because they have a slightly metallic luster, perhaps from a light coating of dust. Under cloudy skies, however, they appear dull and nearly lifeless. This portion of the desert is still arid, but, as noted, many of these scrubby plants get their moisture not from actual rainstorms, but rather from the low clouds that hang over the land like a damp blanket at night and sometimes well into the morning. The traveler engrossed by this plant community may be surprised to hear, in muffled pulses to the west, breakers hammering the shore. This desert plant habitat is literally clinging to the western edge of the continent. With one more twist along a ravine, or glimpse from a hill slope facing west, the Pacific Ocean comes into view.

In the Atacama it is the aridity, rather than the temperatures, that visitors are much more likely to write home about. Although the region has a gradient of increasing rainfall southward, it also varies east to west, with the coast being drier than the adjacent Andes. I will discuss this subject in more detail throughout this book, but I should note at the outset that animal life here has adapted surprisingly well to the arid conditions. For a hyperarid region, the Atacama has a surprising diversity of animals, including insects and reptiles, many of which hunker down during the day time and move about at night. Birds are also encountered here and vary in size from vultures to smaller Peruvian sparrows as well as hummingbirds. Although many descriptions of deserts imply they are devoid of mammals, especially large ones, the Atacama Desert is home to several camel-like creatures, such as vicuñas and guanacos. These animals are related to the llama, which lives in

Figure 1.20. Nearly blending in with the tawny and ruddy colors of the desert and puna landscape, guanacos may appear when least expected, as seen in this hastily taken photo from the interior of a moving car northwest of San Pedro de Atacama. (Photograph by author)

the higher mountains. The sight of an animal as large as a guanaco, which is the size of a deer, may be startling (Figure 1.20). The fact that they seem to come out of nowhere and have little upon which to subsist is one of the marvels of the Atacama Desert. Guanacos are found throughout the Atacama from the Andes to the coast and are adept at finding similar feed in the hills. Given their migration patterns, they can go long distances between drinking and feeding. The vicuña has a somewhat more restricted range, but it is common in the higher elevations of the desert at the base and slopes of the Andes. It too can survive with little food and drink at times. Guanacos and vicuñas are, after all, the forerunners of camels that do so well in similar conditions in the Old World.

The Atacama Desert may seem inhospitable to humans but it has been occupied for a long time, perhaps even longer than the current conservative estimates of 11,700 BP. It was once commonly believed that the desert posed a barrier to the earliest peoples moving southward, but "recent research has detected evidence of freshwater plants and animals buried beneath the arid surface." This same research "suggests that between 9,000 and 25,000 years ago, the Atacama may have contained wetlands that could have sustained and even aided early human colonization."[18] This corresponds to the cooler and wetter conditions that were found in the late Pleistocene worldwide, but was likely somewhat more intense in the northern hemisphere, with its larger land masses. As in the American West and elsewhere in the northern hemisphere, the early hunters in the Atacama are associated with Pleistocene mega fauna. At one site in the Tuina Hills in the vicinity of the Calama Basin, projectile points were found with remains of a horse, a species that became extinct in the New World at about this time. The bones of a camelid were also found in what is today the Atacama Desert, leading an archaeologist to speculate that "this find probably represents the last remnants of a Pleistocene fauna."[19]

The presence of horses suggests available grass and forage, but the camelid would likewise suggest that such forage might have been scarcer at times during this period.[20] In other words, although climatic fluctuations likely occurred at the end of the Pleistocene, which is the time when people first arrived here, the climate was slowly becoming warmer and drier. An important point to recall here is that the appearance of the landscape of the Atacama region has changed through time. As will become evident in the chapters that follow, even what observers see in one year or decade may vary from what other observers had recently encountered. The expectations of those observers also play a role, a subject I will discuss in considerable detail in subsequent chapters.

Fast-forwarding to the present, the Atacama Desert currently occupies a noteworthy position politically as well as geographically. On today's map much of this desert lies in Chile, but it is part of an arid region that continues northward along the coast of Peru. A century ago, geographer Isaiah Bowman was amazed by how quickly the transition occurred at the northern end of this desert region. As he put it, "Of all the interesting features of the west coast of South America none is perhaps more lasting in the mind of the traveler than the startling suddenness with which he comes upon the coastal desert in sailing south from Ecuador."[21] Bowman had seen the verdant coast south of Guayaquil (Ecuador) from his steamship in the evening before turning in, and he awoke the next morning surprised to experience the desert of northern Peru. As noted, this desert region continues along most of the coast of Peru, intensifying as it reaches the Arica bend at the modern border between Chile and Peru. This desert region thus stretches to include portions of Chile and Peru—a reminder that the political border between these two countries is a cultural artifact rather than a natural demarcation.

Although the landscape does not noticeably change at the border, a number of factors have conspired to enforce the perception that the Atacama Desert "belongs" to Chile. I will say much more about this later when I discuss the forces that led to this border change, namely mining activity in the later nineteenth century that triggered disputes and then outright war.

For their part, present-day Chileans view their arid north with a mixture of awe and pride, and they now divide it into two zones: 1) the extremely arid Norte Grande, which begins at the Peruvian border and extends to about 27 degrees south, and 2) the Norte Chico, which extends southward to around 33 degrees latitude. These two divisions reflect both history and geography. Whereas much of the Norte Grande is extremely arid, the Norte Chico benefits from occasional rains characteristic of midlatitude west coast locations. At its southern edge a few hundred kilometers north of Santiago, this Norte Chico portion of the Atacama Desert grades almost imperceptibly into a transitional zone of classic Mediterranean (CSb) climate with its warm, dry summers and cool, moist winters. Those adjectives "grande" and "chico" for the two divisions of the Atacama have some basis in vernacular perception as well as scientific observation. The term chico implies familiarity and close proximity, while grande suggests the desert's increasing size and increasing intensity of drought, with distance from the nation's most populous city: the capital, Santiago.

In terms of broad-brush geographic analogs, the latitudinal climatic zonation in coastal South America is similar to, but directionally the reverse of, that found in western North America, which grades from Mediterranean climate in central coastal California and increases in aridity southward to the coastal desert of Baja, California. In other words, traveling southward from Los Angeles through San Diego and thence to La Paz, aridity increases noticeably, much as what happens when a traveler drives northward from Santiago through La Serena and thence to Copiapó, Antofagasta, and Arica. However, topography complicates things considerably, for mountains tend to receive more precipitation than lowlands. In regions where rainstorms come from one particular direction, the area downwind is usually drier, hence these mountains are said to create "rain shadows." This pattern is especially noticeable in California, where prevailing winds and storms are from the west, and the land lying east of them is noticeably drier. In South America, mountains also have a considerable influence on the climate, but the intensity of aridity is even greater due to other factors. Travelers crossing the Andes from Chile into Argentina may be surprised to discover that the areas on *both* sides of this tall mountain range are rain-shadowed. True, the area eastward of the Andes receives more precipitation than the Atacama Desert of Chile, but it may be as desert-like in appearance as parts of the Sonoran Desert of Arizona. Even though the Andes are taller than the Sierra Nevada of California, and are snowcapped in many places, the atmosphere on both sides of the Andes is so dry that it does not sustain tall coniferous tree growth like it does in the Sierra Nevada. For those who love the prospect of desert mountains in the literal sense of the word—impressive ranges unencumbered by tree growth, where every detail of the geology is starkly revealed—this part of the Atacama is paradise.

While on the subject of physical geography, I should note that deserts, at least in the western mind, call for superlatives. Hence, a particularly arid portion of North America—Death Valley—is often cited as the place where the highest temperature on earth (134 degrees, or 56.6°C) was recorded more than a century ago. However, as noted, being a coastal desert rather than a below-sea-level depression like Death Valley (or the area adjacent to the Dead Sea), the Atacama cannot boast of such high temperatures but instead is widely known for its aridity. As a quick search of websites reveals, the Atacama Desert is popularly regarded to be the driest place on earth, and I will discuss these claims, some of them grandiose, later. Unlike Death Valley, which normally experiences some measurable precipitation annually (though a year or two may witness none), weather stations in the Atacama may not record any precipitation for decades at a stretch. This, however, has misled some people into believing that it *never* rains in the Atacama—hyperbole that has been voiced for centuries and has now become part of the folklore, if not a trademark, of this desert.

As we shall see, it is one thing to say "it never rains in the Atacama," and quite another thing to say that it has never rained here. The former may be true for the personal experiences of someone here, but the latter implies an immense time period indeed. Then again, as Albert Hammond, the British-born songwriter whose popular 1972 song "It Never Rains in Southern California" sarcastically observes in a meteorological metaphor, it may instead pour in that usually sunny, semiarid area. Hammond uses that oft-quoted metaphor to point out how tough it might prove to break into the California entertainment scene, but Chileans optimistically, and sometimes opportunistically, tout the Atacama Desert as a place where it never rains, and they urge those interested in such phenomena to experience it firsthand.

A place where it rarely or never rains may suggest bleakness, but there is another side to the picture. Although it is true that some parts of the Atacama such as the nitrate pampas may be a monotonous grayish white, one of the biggest surprises awaiting travelers in much of the region is the variety of colors that they will encounter. William E. Rudolph, a mining engineer who lived in the northeastern portion of this desert for forty years, observed that the geology helps determine the color of the landscape. As Rudolph noted in his 1963 book *Vanishing Trails of Atacama*, "the rock varies in all the shades of brown and red, with light cream color in places, and purple and even shades of green." The position of the sun plays a role in enhancing this. Rudolph concluded, "for a few minutes at sunset these colors and the shadows form picturesque and fantastic combinations."[22]

In their book *The Useless Land*, John Aarons and Claudio Vita-Finzi note that even the austere puna in the vicinity of Laguna Colorada has startling color variations determined for the most part by the geology. "The basalt is black, the volcanoes purple, and their exposed interiors yellow and red. The beach is grey and the lake pink, topped with the icing of iceberg-like masses of salts." [23] The yellow, it should be noted, is from sulfur, with some of the world's purest deposits of this mineral found in the volcanic fissures of the Andes of Chile and Bolivia. In some places on the coast range, the tawny colored mountains are banded with black-striped strata much like a zebra. One's first reaction is that these might be volcanic basalts, but in places here, surprisingly black limestone strata also laces the landscape like bunting. Indicating contortions of the earth from ages past, these strata sometimes run diagonally and in some cases even appear to crisscross each other (Figure 1.21). As a whole, the kaleidoscopic variation in the Atacama Desert's palette is striking—so distinctive, in fact, that it was a major factor in my publisher's decision to print this book in full color.

Despite the color variations, one portion of the spectrum that is seemingly omnipresent is a rusty reddish orange. The normally deep blue sky accentuates this ruddy color, and the contrast between both and the jet-black asphalt of a recently paved highway is striking. Along the well-maintained Pan-American Highway in the vicinity of Las Bombas, my son Damien and I stopped to soak in the solitude, which was broken by an occasional vehicle. Selecting a super wide angle spanning nearly 180 degrees, Damien captured the quintessential Atacama Desert landscape panorama that seemed to swallow up both us and our Citroën rental car (Fig. 1.22).

At this point I would also like to add a few comments about the photographs that my son and I took for this book, including the one just mentioned, which features a seemingly timeless landscape in which a modern automobile is prominent. As a historian, I admit to being fascinated by images made by early travelers such as Bowman, who experienced this desert

Figure 1.21. Bare multicolored mountains in the background expose the complexity of Atacama geology in the coastal cordillera between Las Bombas and Caleta Pan de Azúcar. (Photograph by author)

Figure 1.22. A quintessential Atacama Desert landscape. In this wide-angle panorama scene near Las Bombas, Chile, the ruddy tones of the land and blue of the sky contrast with the trappings of modernity, including a white rental car and jet-black asphalt highway. (Photograph by Damien Francaviglia)

a century or more ago. Even in a book in full color, his black and white photographs still have much to convey. I also confess to being especially intrigued
when a scene includes a horse or mule nearby, and sometimes even a wagon.
After all, in these scenes Bowman freely discusses how he got around this
vast desert region. For most of my career, though, the thousands of slides I
took in varied locales frequently found me taking some pains to hide modern
conveyances such as my then-new 1963 Volkswagen or my now-aging 2008
Subaru Outback. I'd find something I wanted to photograph, perhaps some
sedimentary strata or an extinct volcano, and make sure I included only the
"natural" scene by avoiding capturing the roads, automobiles, traffic signs,
and the like. I still tend to do this almost instinctively, but on our recent trip
through the Atacama (2015), I noted that my son had no compunctions about
including the rental cars, paved highways, and modern road signs. These not
only provided some chronological context but also helped underscore how
we got around on a truly memorable road trip.

Although some readers may find these modern elements distracting or
intrusive, I now find them just the opposite—reminders that modernity
makes getting around the Atacama Desert far easier than ever. Secretly, I take
some pleasure in noting that readers of this book far into the future will find
a 2015 Citroën to be a positively quaint fossil-fuel-powered conveyance with
few safety features such as anti–lane-departure warning buzzers and even the
driverless option that will someday be *de rigueur*. On this most recent trip,

we experienced a problem with the Citroën we had rented in Antofagasta, and we swapped it for a Renault diesel Picasso in Copiapó. Both rental cars had five-speed manual transmissions, something only five percent of my students would know how to drive (based on a few informal in-class surveys). I can imagine some readers in the not-too-distant future saying wow, those Francaviglias were sure pioneers, driving stick-shift cars and experiencing the hands-on, old-school way of getting around. For the record, we were so intrepid that we didn't even have GPS in those primitive rental vehicles.

To return to the subject of the Atacama Desert's geology, especially in light of the climate, I should again mention sand, especially sand dunes. Although vast areas covered by towering sand dunes are part of the popular image of deserts (no doubt influenced by the sandy Sahara and Arabian Deserts), they are relatively rare in the Atacama. There are, however, some notable exceptions. In the vicinity of Copiapó, for example, the mountains are often veneered with fields of sand that reach halfway or more to their summits. Few places in the world offer such a sight, which seems almost surreal. The accumulation of such a huge amount of sand is a dead giveaway that the sparse rainfall here is unable to remove it. With more frequent rainfall, this sand would be washed down into the valley below and ultimately out to sea. To my knowledge, there is no similar mountain-covering landscape, or rather sandscape, at this impressive scale anywhere in North America; typically, areas of sand dunes (e.g., White Sands, NM, or the dunes at Kelso and Death Valley, CA) lie between mountain ranges. Elsewhere on earth such sand-veneered mountains are rare, though found in the Sahara Desert's Acacus mountain range of southern Algeria, and in rugged portions of southern Libya, as well as—on a much smaller scale—the sand-swept volcanic hills of west-central Arabia.

The method by which these sandy wastes are formed and sustained was on my mind as Damien and I drove toward Copiapó with the large mountain called *El Bramador* looming high on the horizon. I instantly recognized this mountain from Bowman's photograph in *Desert Trails of Atacama* (Figure 1.23). Even from a distance, one can tell that El Bramador is barely able to keep its head above the sand, for it stands covered to its shoulders in it. Whereas most mountains in deserts, even the Atacama, tend to rise respectably above fairly compact alluvial fans, this one and many of its neighbors are clothed in dun-colored dunes. The contrast between their dark, steeply inclined masses of rock and their sweeping off-white aprons of sand is striking. As Bowman discovered when he hiked up this mountain more than a century ago, that sand was clearly not derived from the mountain itself, but rather blown in

Figure 1.23. As described and photographed by Isaiah Bowman about a century ago in *Desert Trails of Atacama*, the mountain called El Bramador stands above the Copiapó River Valley and is covered nearly to its summit by sand dunes that emit roaring sounds when disturbed by hikers or earthquakes. (American Geographical Society Library, University of Wisconsin at Milwaukee)

by winds that lift the fine particles from the valley floor. Whereas the other mountains in the vicinity of El Bramador also seem to be struggling to keep themselves above the sand from the nearby Copiapó River Valley, this one is well known for the roaring sound that it emits, hence its name El Bramador, in Spanish, "the one who roars." More prosaically, English speakers call it the Roaring Mountain of Toledo.

As recounted by Bowman, the sound emanating from this mountain appears to be the result of sand particles descending the dunes at certain times, especially when they are disturbed by hikers or when an earthquake hits the area. Ever curious and scientifically inclined, Bowman experimented to see if he could get things started. To his delight, by shuffling along in the dunes he was able to generate the sound by starting the sand moving; the more sand, the louder the roar, which struck him as being more like a humming sound. Now highly engaged in getting answers from this mountain, Bowman noted that he "experimented with this with watch in hand, and found that the sand required stroking 15 times in 20 seconds to produce the effect, and there could be little variation from this period without the noise ceasing."

Despite this demystification, Bowman was truly impressed by the phenomenon, which varied depending on how much the sand was disturbed. He noted that the condition of the sand is critical: "The volume of the sound clearly depends on the state of the sand, whether it is piled ready to slide down at a touch on the steeper slopes of the dune or whether it lies on slightly flatter grades." Bowman visited the mountain when the dunes were fairly quiet, but it could indeed roar. "At the time of an earthquake great masses slide down over each other as the trembling of the earth dislodges sand that the wind has piled up for days or months beforehand." On such occasions, Bowman noted, El Bramador can be heard to roar for up to a mile away.

In other deserts worldwide these rare, sound-emitting sandscapes are called "singing dunes." Sand Mountain in west-central Nevada is North America's best example, but it is adjacent to and not hugging the side of a taller mountain. Given their peculiar sound-producing qualities, such places are normally associated with considerable folklore. El Bramador is no exception. Confident that he had found the answer to why this mountain roars, Bowman nevertheless added "naturally this has given rise to superstitions of one sort or another, and popular explanations given in the valley are that the noises emanate from caves in the mountain or from some concealed volcanic crater from which an eruption may someday rise."[24]

With such superstitions in mind, I wrote a short poem entitled "Escuchando al Desierto (Listening to the Desert)."

ESCUCHANDO AL DESIERTO (Listening to the Desert)

Breaking the desert's silence
you sometimes hear a low roar
like an airliner streaking far above,
or the imaginary sound of the wind
rolling around in your ears
when trapped by a large seashell.

Instinctively looking upward, though,
you see only the blue of the heavens
and conclude that nothing is moving
on this air as motionless as stone.

Perhaps it is something far deeper
like the mysterious humming

of the slumbering earth itself,
fitfully awakening to primal forces
you've only read about in books
but never expected to experience yourself.

Like Bowman, I make a point of stopping to check out places that seem out of the ordinary. On these stops I talk with people about things that seem interesting or are part of older folklore. I also consult as many maps as I can. In addition to buying the most modern maps, I always look for old ones that may contain interesting information that has been discarded by the more recent mapmakers. In this introductory chapter I have relied on information from the last century or so, but I shall now prepare readers to step back much further in time in the following four chapters. This time travel will also be based on narratives and on historical maps. The latter often feature a remarkable combination of verbal and graphic information, reminding us that maps are texts in and of themselves, for they contain a wealth of information depicted in varied form and style. Many of these maps are as colorful as the places they depict, another reason for printing them in full color. They are telling documents indeed, for just as history and geography are difficult to separate, maps too are inseparable from the time in which they were prepared.

In other words, maps portray places—or rather attitudes about places—at particular points in time, and thus serve as historical documentation. That said, it should be understood that maps as historical evidence need to be, like all types of evidence, corroborated with other sources such as the written record. Although hundreds of maps were consulted in my research for this book, for convenience I will showcase only about forty that represent or characterize this desert in particular periods or epochs. Similarly, although the Atacama has been written (and talked) about by outsiders for nearly five hundred years, I will highlight a few hundred key passages to drive home the point that although this desert may seem timeless, it has indeed changed through time in the popular imagination. Studied simultaneously, these maps and narratives suggest that this desert has been perceived differently in different time periods, each explorer or observer selecting things that he or she emphasized; for convenience, these differences in perception from the 1500s to the present will be treated as four distinct time periods or phases—each comprising a separate chapter.

One final caveat needs to be introduced at this time. What I said about maps assumes that they are still in existence or were well enough described in words early on to give us an idea of how they looked and what information they contained. That is a noteworthy concern because maps are often either separated from their original reports or sometimes even deliberately disposed of when superseded by newer, and hence seemingly better, maps. How many of us have kept all of the many road maps we've ever owned? Certainly not I, even though I disposed of each with some regret. In truth, discarding an outdated map is more or less understandable, for we likely will never again use it anyway. However, to the cartographic historian, and indeed all historians, it represents a significant loss in our understanding of how places were perceived in the past. Luckily, though, some maps as well as other documentation have survived and form a key part of reconstructing past places. In the following chapters, I will use maps that have escaped destruction to help tell the story of how the Atacama Desert has been represented over the last five centuries. I believe that these surviving maps are representative of others that may no longer exist, and they hence constitute a valuable record of place making and place recording in the Atacama Desert. Like all maps, they tell a story using two types of information—visual (or graphic), and narrative (verbal or written)—another reminder that images and words work hand in hand.

Two

"El Gran Despoblado"

Early Narratives and Images of Empire Building,
ca. 1530–1700

Before I make good on my promise to introduce historical maps of the Atacama, I would like to pause long enough to introduce, and then deconstruct, the word "desert." It is humbling to comprehend how powerful and complex such a simple word can be. To most people reading this book, the word desert might be associated with a place that is visibly dry, with little or no natural vegetation. However, what is clearly a desert to one person may be something (or someplace) quite different for others. As anthropologists have long noted, Native peoples dwelling in the desert may not use the term desert for places that outsiders would consider as such. The aboriginal peoples of Australia, for example, do not consider themselves to be living in a desert, but rather a complex land that offers resources in abundance, provided one knows where to find them.

At a recent conference on Death Valley natural history that I attended, a representative of the Shoshone people reminded attendees that her people actually resent the name Death Valley, and for that matter even the name Mojave *Desert*, for a place that has sustained them for so long. Moreover, even in Western or European culture, although the term desert may convey a parched and barren land today, such was not always the case. In centuries past the word desert did not necessarily refer to an arid area, but rather one with few or no people. It was, in a word, wilderness. We know this use in the term "desert island," which may be a lush tropical place but devoid of fellow humans.

The island in the Tom Hanks film *Cast Away* (2000) depicted such a place, but its literary prototype is likely Juan Fernando Island off the coast of central Chile, which is covered by considerable tree growth and was likely the setting for the island experienced by *Robinson Crusoe*. That wildly popular 1719 novel by Daniel Defoe is loosely based on the actual accidental abandoning of an English sailor on Juan Fernando Island in the late 1600s. More to the point, though, a place like the Atacama Desert, which lies farther north along the west coast of South America, fits another use of the word desert, namely, a place that is barren or devoid of vegetation. This characterization, as we shall see, has interesting origins and significant consequences.

To better understand how the word desert came to be shorthand for places with little or no vegetation, we need to briefly transport ourselves to fifteenth-century Europe. One source in particular, the writings of Felix Fabri (ca. 1441–1502), sheds light on the articulation of desert in the modern sense of the word. Fabri, a Dominican theologian from Zurich, experienced the Holy Land on two separate pilgrimages in the 1480s. At that time, the Crusades were winding down after exposing Europeans to the Holy Land over several centuries. Although geographical aspects of the Holy Land are hinted at in the Bible, and were well known firsthand to Crusaders, *The Book of the Wanderings of Felix Fabri* offered an insightful section called "A description of the wilderness, the solitary place or desert, setting forth its length, breadth, and barrenness."

According to Fabri, who apparently had an analytical as well as spiritual side, there are four types of wilderness or desert in the world. The first three types are solely defined by the absence of people, namely: 1) places in which people might dwell but do not; 2) places that may have fields, pastures, and gardens but which people do not occupy; and 3) places that may have considerable vegetation but which are occupied only by beasts such as bears and wolves.

Lastly, and most importantly for our purposes here, Fabri noted that the fourth type of desert was demonstrably different. As he put it, "that part of the world is called the wilderness wherein nothing grows for man or beast to eat, neither trees nor herbs, and wherein neither men, beasts, nor birds can live, both because of the want of water and the intolerable heat of the sun, the barrenness of the ground, and, in short, because of the lack of all things appertaining to the support of life." Realizing that lands described in his last definition might be foreign to most of his readers, Fabri was specific about where one could encounter them. As he noted, "such a wilderness is that which reaches from Gazara to Mount Sinai; not indeed everywhere, but in the greater part thereof." Fabri further noted that although Europe possessed the first three types of wilderness or desert, the fourth kind was only to be found elsewhere.[1]

Fabri's observations were prescient. He was among the first to define and differentiate deserts in the modern sense of the word; he also gave these desert places a spatial position as being distant from Europe. Above all, he noted that deserts may vary somewhat even within their own boundaries, which were evidently hard to place with precision. Fabri was also as much a dramatist as a scholar, though, for he added one of the most evocative descriptions of the desert ever written. He began by placing the desert in a cosmographical perspective: "Firstly, this country is called the desert because it seems to be, so to speak, deserted by God, by the heavens, and by the world." As a theologian, Fabri quickly noted that "it is deserted by God, because it is empty and void, as though God had used it to improve or adorn the rest of the universe." That was evocative enough, but he then proceeded to further note that the desert was a "solitary" and "lonesome" place where no man dwells or even visits because it is inhabited by "serpents, scorpions, dragons, fauns, and satyrs." In a description that would resonate down through the centuries, however, Fabri presented the desert not only as a bleak land of "devils and temptations," but—as Michael Welland astutely observes in his 2015 book *The Desert: Lands of Lost Borders*—one also known as "a place of meditation, devotion, and contemplation, 'a place where great merit is acquired.'"[2]

This contrast between the desert as essentially godforsaken, yet the very place in which one could find deep spirituality, dates back to early Judaic history and continued into the formative years of Christianity. As historian William Harmless noted in his groundbreaking book *Desert Christians: An Introduction to the Literature of Early Monasticism* (2004), the desert-mountain sanctuary of St. Anthony in Egypt became a hallmark of the soul-searching Christian. As opposed to monks who left their villages and went to desolate places close by for short periods in order to sort things out spiritually, St. Anthony made the radical move to commune with God in "the great desert" of extreme eastern Egypt. After twenty years in his isolated desert fortress, and finally able to triumph over its demons, Anthony emerged as "a mystic initiate, mystery-taught, and God-inspired." To Christians, that experience placed Anthony among the earliest of monastic monks. However, as Harmless correctly noted, "the most significant biblical type for Anthony is Elijah," an Old Testament prophet who likewise triumphed over evil in the desert.[3] These experiences helped configure the desert as a major player in the drama of spiritual conversion, ultimately laying the groundwork for the close association between the words "desert" and "religion."[4] Although Fabri informed his European readers that this truest of deserts was geographically distant, his positioning it in terms of familiar places in the Holy Land also reminded them that it was as close as the Bible.

Fabri's comparison of the desert as a place that God created through subtraction, as it were, in order to build a better, wetter land for humanity is reminiscent of the desert as depicted in the Qur'an. In that holy book of the Muslims, heaven and hell are contrasted using a number of metaphors. Whereas paradise/heaven is associated with running water and verdant gardens, the fire of hell is akin to the blazing sun that tortures the desert much like the fire that will consume the unheeding and unrepentant sinner. Of signs that should be heeded, the Qur'an asks, "Have you not considered that God sends water down from the sky, guides it along to form springs in the earth, and then, with it, brings forth vegetation in various colours, which later withers, turns yellow before your eyes, and is crumbled to dust at His command?" In the same passage, the Qur'an immediately adds words to the wise: "There is truly a reminder in this for those who have understanding" (39:21). The Qur'an is ever cautionary that people can go astray in the desert, making numerous references to the Bedouins, who are considered obstinate desert dwellers who ignore such signs. Rich in metaphor and irony configured as environmental parables, the Qur'an warns, "But the deeds of disbelievers are like a mirage in a desert: the thirsty person thinks there will be water but, when he gets there, he finds it is nothing" (24:39).

Although the Qur'an may seem far distant to Christianity, it is part of the same Abrahamic tradition. I cite it here because Christians in the fifteenth century were more familiar with Islam's holy book than they are today, even though they may have used it to cast aspersions on its followers, as Spaniards commonly did. Islam also likely influenced Dante Alighieri, the Islamophobic Italian writer whose classic *Inferno* features a fiery hell that is more likely derived from the Qur'an than the Bible. The Muslim connection was especially relevant in Spain in the fifteenth century. The year 1492 is widely recognized as when Columbus left Spain on the first of his four voyages of discovery to the New World. Not so coincidentally, however, 1492 also marks the year in which Muslims and Jews were driven from Spain. In an indirect but palpable way, then, the same forces that fueled the *Reconquista* wherein Spain freed itself from the infidels were also in full force when Columbus set sail for the Indies, failing to find that fabled land but instead stumbling upon the Americas. A few of Columbus's best sailors were Muslims who hailed from the same intellectual tradition that had informed Spain for centuries. For the record, though, as cultural anthropologist Carol Delaney recently documented, Columbus's ultimate goal was to reap enough wealth from the Indies to assemble an army that could ultimately retake Jerusalem from the Muslims.[5]

Now that I have set the scene for the so-called discovery of the Americas, I think it wise to discuss how early travel narratives and maps worked together to depict the new land wherein the Atacama Desert would be encountered. These were linked to the impulse called exploration, which is both timeless and modern. As historian Stewart A. Weaver notes, the noun exploration derives from the verb to explore, and both words are relatively new to the English language, just under five hundred years old. The word explore was originally used metaphorically—as a way of better understanding something—but quickly became linked to place. Weaver notes that the modern *Encarta Webster's Dictionary* defines exploration as "travel for discovery," or, more specifically, "travel undertaken to discover what a place is like or where it is." Although the English first used the words explore and exploration in the mid-1500s, they were building on accomplishments begun fifty years earlier by Spain. Christopher Columbus's 1492 voyage launched Spain on its trajectory as a transatlantic power and the world into the modern age. Even though Columbus was not the first European to reach the Americas (the Vikings had done that five hundred years earlier), he was the first whose exploration would be recorded in print and hence reach a huge audience. As Weaver astutely notes in his recent pocket-sized book *Exploration: A Short Introduction*, "to qualify as an explorer, one must not only find something new but write about it, publicize it, draw the attention of others to it."[6]

Exploration led to discovery, but exactly what a particular explorer had discovered was sometimes unclear. The archetypical example is the prototypical explorer Columbus himself, who did not comprehend the location of the land mass that he encountered. In fact, despite making four separate voyages to the New World and back, Columbus still thought that he had encountered the Indies, hence his use of the term *Indios* for the people living there. I should also note that Columbus's misunderstanding of exactly what he had encountered ultimately led to the permanent confusion between two similarly named places—the West Indies and East Indies. By extension, the naming also included peoples of these two widely separated locales, the West Indians and East Indians. We know the former peoples as Native American Indians today, and they consisted of many diverse tribes, but their counterparts ten thousand miles to the west are in fact not only the "real" Indians who live in India but also many others, such as Indonesians, Malaysians, and the like.

Despite being geographically as well as culturally challenged, Columbus gained some idea of how the islands he reached were positioned with regard to one another as well as what appeared to be much larger land masses. However, he had no idea of the full extent of the land in these places. Thus, although exploration led Columbus to "discover" the Americas, he never knew that he had stumbled upon an area containing two unknown (to people in the Old World, at least) continents. The closest he came to those continents was a glimpse of the southern edge of one (North America, in the vicinity of Florida) and a brief landfall on the northern shores of the other (northern South America, in what is today Venezuela).

In the late fifteenth and early sixteenth centuries, Europeans who made and used maps were also confused as to just what had been discovered and what was already known from earlier voyages. Not long after Columbus's voyages, some who studied earlier maps claimed that these maps already showed the lands that he supposedly discovered. Such claims surround the World Map by Henricus Martellus Germanus (Heinrich Hammer), which was made in the period ca. 1480–1496. To some observers, this so-called Hammer map appears to show South America.[7] However, there is no substantiated record of any voyage this early that reached "la cola del Dragón" (the tail of the dragon), which extends southward from the continent depicted on the map. More likely, it is not southern South America at all, but rather a fanciful rendition of the Malay Peninsula. The imagination can confuse two widely separated places, in this case, the narrowing land mass of southeast Asia and the southern part of South America, both of which also have a tapering shape and jut southward into an ocean. This funnel-shaped part of South America includes Chile and Argentina and with good reason later came to be dubbed "the southern cone" for its distinctive tapering shape on maps. Although developments in exploration and mapmaking during the past five centuries now confirm that the shape of this part of South America bears far less of a resemblance to southeastern Asia than originally thought, few can deny the power of maps to suggest connections between peoples and places, some real and some imagined.

Despite keeping detailed journals, Columbus left no known maps showing the overall area that he explored. From the text, we can deduce that it represented a considerable part of the vast Caribbean Sea, parts of coastal Central America, and the extreme northern portion of South America. At this point, I would like to introduce and illustrate the first published map of the Americas that shows its full extent, or a rough approximation of it. On this 1507 map by the German cartographer Martin Waldseemüller (1470–1520), one can imagine the embryonic conceptualization of South America taking shape (Figure 2.1). Waldseemüller's seminal map, published in Saint Dié, France, was part of a coordinated effort by a group of scholars there to depict the New World according to the latest information as well as traditional accepted sources. The map's full title—*Universalis cosmographia secundum Ptolomaei traditionem et Americi Vespucci Alioru[m]que illustrationes*—alludes

Figure 2.1. The 1507 world map by Waldseemüller presents a cartographic mystery: although western South America appears recognizable to us today, this part of the continent was not visited by Europeans until about two decades later. (Author's digital collection)

to its sources, and it singles out the Italian explorer Amerigo Vespucci (1454–1512), who traveled to the New World on several voyages on behalf of Spain between 1499 and 1502 and left detailed correspondence about it in 1504. The fact that we use the term America for this part of the world is a result of Waldseemüller using Vespucci's given name on this map. The main reason I elected to use the Waldseemüller map, though, is because it embodies a mystery about the depiction of the region that cradled the Atacama Desert, namely western South America.

On Waldseemüller's map, the west coast of that continent appears lined up along a roughly north–south axis, as it does on modern-day maps. It is natural to assume that Waldseemüller's map resulted from actual exploration, but that is very problematical because no known European voyages reached this part of South America this early. Some historians suggest that the area shown on Waldseemüller's map is actually a portion of Southeast Asia.[8] Others propose that the information was conveyed by Native American informants based on their own earlier voyages and travels. According to cartographic historian John Hébert, the latter scenario is "a possibility that present scholarship considers unlikely." We have no evidence that Indigenous peoples provided the information to Spanish or other mariners who may have in turn conveyed it to Waldseemüller, and to claim otherwise amounts to speculation.

Still, Hébert concedes that the land masses on this map, though rudimentary, are so configured that "as we view the map, we can recognize the world as we know it today."[9] I think it both ironic and telling that the first known map depicting the area where the Atacama Desert lies is shrouded in mystery, for as we shall see later in this book, so would the desert itself be enigmatic enough to delineate with precision. During the same time period in the early sixteenth century, other maps would offer a more sobering, or perhaps more accurate, assessment of the geographic knowledge of the time. For example, on the Piri Reis map of 1513, which was also based on knowledge imparted by Columbus, only the eastern (Atlantic) portion of South America is delineated. This makes perfect sense. Columbus, after all, had never even seen the Pacific Ocean. That distinction was claimed by Vasco Núñez de Balboa in 1513. On the Piri Reis map, the absence of the western coast of South America is a reminder of how remote this part of the continent was from European centers of power, namely Spain, in the early 1500s.

Claims about Native American informants are intriguing, and sometimes plausible. By the time the Spaniards eyed the arid portion of western South America, it had been settled for thousands of years. These Indigenous peoples knew the environment intimately and were adept at using it effectively. On the coast north of Antofagasta, the Las Conchas dated from about 7,500 BC and were well connected to Andean culture. They were dependent on marine resources such as shellfish, while inland into the desert hunting and gathering were practiced, later followed by irrigated agriculture. In her 1994 book *Ancient South America*, archaeologist Karen Olsen Bruhns suggested that this mountainous desert coastal part of South America was a region of major cultural development.[10]

By around the turn of the sixteenth century, maps were only beginning to delineate the coastlines of the Americas. That would soon change as the Crown became aware of the riches that existed in the Aztec Empire in Mesoamerica (1519–1521) and shortly thereafter in the Inca Empire in Peru (1528). The latter empire concerns us here, for it stretched along the axis of the Andes for a considerable distance, perhaps 3,500 kilometers (ca. 2,100 m), from Ecuador into Chile. For Spain, determining the extent of the Inca Empire became a preoccupation. This quest would serve as the springboard into the Atacama region, which lay just west of the lofty, volcano-crowned spine of the Andes. As historian Edward L. Goodman observed in his book *The Explorers of South America*, "with the exception of the extreme southern portion of Chile, most of the early exploration of the west coast of South America was closely tied up with the conquest of Peru and Chile."[11]

The early Spanish exploration in the northern part of what is today Chile was nominally under the command of Francisco Pizarro (ca. 1471–1541), well known as the conqueror of Peru. In 1534, under orders from Pizarro, the mercurial and personally ambitious Diego de Almagro (1475–1538) began preparations to explore southward from Peru. Almagro's goal was to discover, conquer, and settle the lands and provinces along the coast of the South Sea, as the Pacific Ocean was then called. Despite Almagro's headstrong disregard for the advice of the Natives, which cost his expedition dearly, he became the first European explorer to reach the Atacama Desert, which he encountered in 1536 following a grueling trek along the Andes Mountains.

Almagro apparently decided to forego the better-travelled Inca road that ran along the eastern side of the Andes into present-day northwestern Argentina. Still, as he moved southwestward into present-day Chile, he likely traveled the more rugged tributary trails established by the Inca that reached down into the Atacama, where he was sorely disappointed by what he found and entrusted his loyal captain, Rodrigo Orgóñez (in some sources, Ordóñez), to continue southward. Luckily, for the Spaniards at least, the route the captain followed led to populous places in the Atacama such as the oasis community of Copiapó.[12] Therefore, although Almagro is widely credited with blazing new trails, the land his expedition encountered was not entirely trackless, nor had it been unoccupied. Even at that time the landscape of this desert region was littered with ruins of fortresses (called pukarás or pucarás). These had been constructed by warlords after the collapse of the Tiwanaku Empire (ca. 1000 to 1450 AD) and subsequently destroyed by the Incas, who were expanding southward from Peru. Now the Spaniards (under Pizarro, and by extension Almagro) were superseding the Incas as the major power players in this part of South America.

In a manner of speaking, then, the Spaniards exploring this area not only found ruins that were testimonials to earlier hostilities, but they were in the process of supplanting the settlement pattern as they dominated various places and their inhabitants (collectively called pueblos) here. As historian Stewart A. Weaver also observed, the Age of Exploration quickly turned into the Age of the Conquistador. Weaver notes that separating these two stages of Spanish expansion is not always easy and is even somewhat artificial, yet he effectively shows that in the first stage explorers head somewhere unsure of what they will encounter, while in the second they aim to subdue and control what they encountered.[13]

In 1536, after arriving at the lush oasis of Copiapó which huddles along the river of the same name, forces under the direction of the explorer-turned-conquistador Almagro forced the Native peoples into submission before continuing southward, encountering stiffer resistance by Indians who halted their progress. In the process of conquering this area, Almagro's expedition encountered a diversity of Indigenous peoples but likely had only a vague understanding of their territories. In retrospect, Spain was attempting to replace the Inca Empire with their own, a task that began forcefully enough but never fully succeeded, as evident in the survival of independent Indigenous people today in the Andean highlands, especially in Bolivia. In the mid-1530s, Almagro personified Spain's use of force in extending its empire. Almagro cared little for the niceties of diplomacy, his main mission being domination rather than assimilation. This, it should be recalled, was a time when Spain was uncertain about how to deal with Indigenous peoples. Ideally, they could be brought under control peacefully, but another school of thought was to show and use force first while the element of surprise was in the Spaniards' favor. It should be recalled that a few short years after Almagro entered the Atacama Desert with a heavy hand, the nominally pacifistic explorer Alvar Núñez Cabeza de Vaca faced much the same choice on the other side of the Andes, and he too elected to use force.[14]

At this point, I would like to set the scene for the intense cultural interaction and conflict that was about to take place in the region we today call the Atacama Desert. This region had much in common with other remote parts of New Spain but differed in other regards. In her masterful study *Wandering Peoples: Colonialism, Ethnic Spaces, and Ecological Frontiers in Northwestern Mexico, 1700–1850*, Cynthia Radding reminds readers that cultural contact is a very complex subject. On the one hand, colonialism "implies political domination by the Spanish Crown over its American territories and peoples, economic exploitation and the transfer of wealth from colonies…and diverse responses from colonized indigenous nations and enslaved workers." However, as Radding quickly adds, "Latin American colonialism does not reflect a binary distinction between dominant and subordinate strata, or between colonizer and colonized, owing to the multiracial hybridity of American societies and the complex networks of exploitation, complicity, negotiation, and resistance that developed over three centuries of Iberian rule."[15] Radding's work was among the seminal studies in the

Southwestern borderlands that questioned a simple dominant (Spaniards) and subordinate (Native peoples) scenario.

As historian Claudio Esteva-Fabregat notes, "In many cases, the Indians' facility of movement over vast and hostile territories made it practically impossible to subject and control them." Although Spaniards might muster sufficient forces in one place, they had considerable difficulty maintaining permanent control as reinforcements were often unavailable, especially in lightly populated frontier areas. As Esteva-Fabregat concludes, "these conditions lasted a long time, particularly in the northern regions: northern Mexico, and in the Atacama [Desert] in South America" as well as other remote areas. Several factors governed how Spaniards and Native peoples interacted, foremost the scarcity of Spanish women, who were subject to capture by Indians. One solution to this, at least for Spain, was that although the male Spanish explorers initially overpowered the Native population, they almost immediately mated with Indian women; one result of this was very few families of purely Spanish origin but rather a largely *mestizo* population. In the Atacama Desert, as in many parts of Latin America, *mestizaje*—the mixing of Native American and Spanish cultures—became a fact of life. This pattern was established early and had long-lasting consequences. By 1583, according to Esteva-Fabregat, the ratio of Spanish-born women to Spanish-born men in Chile was only 50 to 1,150—or less than five percent.[16] I introduce this concept of *mestizaje* here because it will be a recurring theme in the way that cultural identity in the Atacama Desert developed. That cultural development in turn affected how the identity of the desert itself took form in the popular imagination.

The landscape of the Atacama Desert even today reflects the long-lasting power of Indigenous culture despite Spanish intrusion dating to 1536. This is in part a result of the desert climate, in which vegetation rarely obscures anything, and the lack of demand on land for cultivation or other uses, at least in areas where water is scarce. The Indigenous people knew where to find water, though, and created many oases in an otherwise barren region. A noteworthy archaeological site near Turi, Chile, offers an example. This site consists of a series of clearly visible but ancient fields that had been developed as early as ca. 1000 AD. Initial excavation of this site suggests that a group of Indigenous people had carefully carved the parcels out of the desert, creating terraces and using runoff from the Andes to irrigate crops. University of New Mexico archaeologist Frances Hayashida notes that the fields were used for hundreds of years, perhaps in close contact with the Incas. This intensive farming lasted until the Spaniards arrived, and there is some suggestion that these newcomers found themselves under the forced control of Spain briefly. In any event, the fields appear to have been abandoned so quickly that, as Hayashida notes, "it looks as though the farmers laid down their stone tools and just walked away."[17]

In another example, anthropologists from the University of Chile in Santiago are working in consultation with the local Aymara community of Nama to conserve and protect stone pukarás and chullpas (adobe funerary towers), create paths within the site, and address strategies for protecting these features. At this impressive site in the northern Atacama Desert, stone remains of residential structures as well as agricultural terraces serve as reminders of Nama's Indigenous past. Moreover, the Aymara's active involvement in the preservation efforts confirm that the Indigenous past is also part of the region's future.[18] The early Spaniards frequently commented upon these impressive stone features that were testimonials to earlier peoples, some of whom had vanished, leaving few other traces.

Prior to the arrival of the Spaniards, the Indigenous peoples of the Atacama Desert lived in a complex pattern that is just coming to light through investigations by Chilean and international archaeologists. Historical sources suggest a general pattern involving at least four major groups of Indigenous peoples who occupied the area at about the time the Spaniards arrived: 1) in the northern part of the region adjacent to the Bolivian Andes, the Aimara (Aymara) had a well-developed agricultural and herding economy and were in communication with the Incas; 2) farther south and along the current border between Chile and Argentina, the Atacameños cultivated crops in the numerous lush oases that huddled along the rivers flowing into the desert, and they also herded llamas on the flanks of the adjacent Andes Mountains; 3) in the coastal desert area on and near the shores of the Pacific Ocean, the Changos built watercraft and subsisted mainly on fish, but they could also eke out a living from the coastal scrub plant communities on the western slopes of the coast ranges; and 4) in the southern portion of the desert and well into the Norte Chico, the Diaguita Indians lived in oases where they cultivated crops such as corn, potatoes, and squash.[19] Depending on elevation and access to water, then, a variety of native crops, including beans, quinoa, potatoes, and maize, were grown in this region. In large, irregularly shaped areas between these four groups, however, the countryside was virtually devoid of human habitation, and that barrenness made an impression on early Spanish explorers. Native people wandered across this area foraging for what little edible plants they might find, and occasionally hunting animals, but it was likely the least-populated portion of the Americas.

It is tempting to think that the Spaniards encountered a stable Indigenous population in this region, but that was not the case. In the mosaic of four Indian cultures just outlined, the Atacameño were originally much more widespread. In the early anthropological literature, the Atacameño Indians were thought to represent the prototypical people of this desert region, though their origins and culture were, and remain, somewhat nebulous. According to anthropologist Wendell C. Bennett, who wrote a seminal essay on the Atacameño in the *Handbook of South American Indians—The Andean Civilizations* (1947), they had once occupied much of the arid region we call the Atacama Desert. This includes most of northern Chile, notably today's provinces of Tacna, Arica, Tarapacá, Antofagasta, and Atacama—as well as portions of adjacent Argentina, namely the northwestern provinces of Los Andes, Salta, and Jujuy.

Like the Incas and a number of other Andean peoples, the Atacameños had mined and smelted copper, even alloying it with tin to create bronze. They also mummified their dead and buried them in cemeteries. Why the Atacameño Indians had lost so much territory by the time the Spaniards arrived remained an enigma, exacerbated by the fact that, as Bennett glibly observed, "the Spanish historical sources furnish little information about these people." The Atacameño's former range, if not omnipresence, was deduced by the distinctive place names they left across a large area, despite the fact that the Atacameño language, or Kunza, had itself all but disappeared. Bennett was in agreement with historians and anthropologists who speculated that the name Atacameño for these people likely derived from the town that Spaniards came to call San Pedro de Atacama, about which I shall say much more in subsequent chapters. In trying to answer the first of many vexing questions associated with the Atacama Desert, especially concerning the identity of its earliest people, it is wise to recall the malleability of culture. As Bennett concluded of the Atacameños, "culturally and linguistically they have been absorbed by *Aymara* or Spanish."[20]

Although Bennett lamented the fact that only limited ethnological studies of the Atacameños had been conducted, their numerous archeological sites confirmed that they had effectively adapted to the varied physical environment. Some Atacameños lived along the coast, comprising small fishing groups that supplemented their marine subsistence with hunting. In the interior regions of the Atacama, herding llamas and alpacas was widespread; moreover, many Atacameños also lived in productive farming villages such as the oases lining the Rio Loa near Calama. Bennett noted that well before the Spaniards had arrived, "the Atacameños had been badly reduced; first by the *Diaguitas*, and later by the Inca, who assumed total political control."

Despite these challenges, Bennett was able to reconstruct how the Atacameños used the land and built their settlements. "The typical village," he noted, "consisted of a number of stone houses arranged along irregular streets and surrounded by an enclosure wall." The basic building blocks of the community, so to speak, were the houses, which were "rectangular with flat roofs made of perishable materials." The fact that the Atacameños lost ground to other Indigenous peoples is noteworthy, even though they were armed. As Bennett recounted, "certain weapons were developed for warfare, including the bow and arrow, the sling, and the wooden knuckle-duster. Cloth or leather armor was used, and villages were often fortified."

However, the Atacameños may have been assimilated into other groups for another reason having little to do with failed defense. In Bennett's assertion that "the *Atacameño* were great traders" lies a clue. As he concluded, "It appears at present as if a basic *Atacameño* culture persisted over a long period of time, and that significant changes were due largely to trade with neighboring cultures."[21] Even though archaeologists consider the Atacameños' influence on the cultures with whom they traded to be minimal, the reverse may not have been the case. They may have quietly adopted traits from the more influential cultures and ultimately ceased to be a recognizable or distinctive people, assimilating into Diaguita and/or Inca. Given this dynamic, it is not surprising that further assimilation would occur under the Spaniards.

The Spaniards relied on Inca technology to get them through this desert. Today, as in the 1500s, the imprint of the Incas remains palpable in many features, including their elaborate communication system, which totaled more than 14,000 miles (ca. 23,000 km) of road constituting "one of the greatest archaeological monuments in the Americas." In reality, this Inca road system throughout western South America is a composite; whereas portions of the Inca roads elsewhere can be traced to other earlier peoples—in other words, the Incas appropriated the work of these peoples—a stretch of road in the Atacama Desert explored by archaeologists John Hyslop and Mario Rivera in the early 1980s was unique. First, it was definitely built by the Incas. Moreover, the road, which ran through three oases (Peine, Tilomonte, and Puquios) south of San Pedro de Atacama, was "unlike other known Inca routes for a number of cultural and environmental reasons." For example, the roadway is narrower (a maximum of three meters or about ten feet) or about half as wide as most Inca roadways—likely a reflection of less traffic volume or flow; next, it is the only part of the Inca road whose course is consistently marked by piles of stones (wood was used elsewhere, even in the coastal desert regions of Peru, which featured some agriculture and trees). Hyslop and Rivera attributed this use of stone to the scarcity of wood and the abundance of stone in this part of the Atacama.

Additionally, the road here features *tampu* (Inca roadside lodgings or storage buildings of rectangular rooms) that are usually found on the west or northwest side of the road and are spaced farther apart along the road than elsewhere. Between these *tampu* are stone structures that may have served as posts (*chaskiwasi*) for messenger-runners (*chaski*) who traversed the roads. According to Hyslop and Rivera, the Spaniards found such roads so useful that they "maintained part of the *chaski* system for some years after they conquered the Inca Empire." The turquoise fragments found at these sites led Hyslop and Rivera to conclude that the road was part of a road system that connected the turquoise and gold mines of Chile with the Inca capital at Cuzco (Peru).[22]

Despite the mineral treasures that lay beneath its surface, the Atacama Desert appeared impoverished to the earliest Spaniards under the command of Diego de Almagro, especially when compared to the fabulous wealth that the Incas had amassed in Peru. Disillusioned that he had found no gold, making enemies among the local tribes, and characterizing much of the country as sterile, Almagro headed back across the Andes and into even more trouble and political intrigue, being executed by the Crown in 1538.[23] In that same year, Almagro's loyal captain Orgóñez died fighting against Pizarro in the ill-fated battle for power in Peru that had earned Almagro his death sentence. Such epic internecine warfare threatened to put Chilean exploration on hold. A year later, however, the southwestern portion of South America again beckoned to Spaniards, who thought Chile was worth a second, closer look. This time, Pedro de Valdivia (ca. 1500–1553) also explored southward from Peru, traversing much of the desert before reaching the relatively lush Coquimbo Valley at its southern edge. Although this area seemed more promising than the Copiapó Valley, Valdivia was also unimpressed by the surrounding barren landscape. He would later characterize this area in more detail for the Crown, but only after continuing southward and finding what seemed to be, at least in comparison to the desert he had traversed, pure paradise. Here, about 300 miles (ca. 480 km) south of Coquimbo, Valdivia discovered Chile's magnificent and well-watered Valle Central,[24] which became a major center of colonization and the location in which Santiago would develop. The peripatetic conquistador Francisco de Aguirre (ca. 1507–1581) also played a significant role in the conquest of the Atacama—first in the vicinity of Tarapacá (Peru) and then in the Coquimbo area. Aguirre served Valdivia as lieutenant governor of the area between Atacama and the Choapa River near La Serena.

Figure 2.2. This nineteenth-century lithograph showing the port of Huasco appeared in Claudio Gay's *Atlas de la historia física y política de Chile* (1854), but certain elements in the scene can be traced back to Native American and early Spanish times. (Biblioteca Nacional de Chile, Diego Barros Arana Collection, via Wiki images)

Figure 2.3. In this 1860 view by naturalist Rudolph Philippi, the same raft and its occupants are transposed onto a scene at El Cobre. (American Geographical Society Library, University of Wisconsin at Milwaukee)

Valdivia and Aguirre helped implement Spain's colonial policies in Chile's desert north, but they left very different legacies. Valdivia, who served as Aguirre's confidante, gravitated toward Santiago and helped unify Chile, serving as its governor in 1549. He was killed in the prime of his energetic life while fighting Araucaria Indians in southern Chile, where he had founded several villages. For his part, Aguirre moved about the Spanish Empire and had numerous run-ins with Spanish authorities even in Chile, despite his close former association with the late Valdivia. Many years after Valdivia's death, Aguirre ultimately settled down in La Serena, the colonial desert town on the Coquimbo River that was filled with memories for him. Aguirre had not only helped build La Serena but also rebuild it after Indians burned it down in 1549. When Aguirre died there in 1581, he had reached the ripe old age of about seventy-five. By that time he had been nearly forgotten, and remains so today, while Valdivia's name is firmly, and patriotically, enshrined in Chilean history.

Despite the turmoil that the early Spanish overland expeditions and interactions caused the Native peoples, these forays enabled Spain to get a firm foothold in Chile, though the sea had proven an easier and more direct way to get there from Spain. In fact, almost two decades before Valdivia had arrived, Ferdinand Magellan had opened up the western coast of South America to transoceanic sea travel in 1519–1521. Although the Portuguese Magellan sailed by the Chilean coast on behalf of Spain, the latter nation controlled the Pacific coast and would soon introduce larger sailing ships to this area.

For their part, the Indigenous peoples along the desert coast of Peru and northern Chile were skilled mariners, though their travel was coastwise rather than transoceanic. Their rafts and sailing vessels were smaller, often made of reeds or inflated animal skins, sometimes with sails made out of woven reeds. By contrast, the European ships were wood hulled and carried canvas sails. An illustration of an Indigenous raft from the mid-nineteenth century (1854), but of a design that was also used in much earlier times, appeared in naturalist Claudio Gay's *História Física y Política de Chile* (1844–1848). In the foreground, two men paddle their raft at the desert port of Huasco (Figure 2.2), which was then located in the far north of Chile.

At that time, as well as in the Spanish colonial era, the rafts were used as lighters to reach deeper draft vessels that were compelled to anchor farther out in the harbor. The raft pictured by Gay appears to be made of two canoe-shaped hulls tied together and is reminiscent of the catamarans used by Indigenous peoples in the Pacific. The upswept prows on both ends is a distinctive feature of these rafts, which contrasted with the larger European sailing and, much later, steam-powered vessels. Many sources claim these cylindrical hulls or pontoons were made from sea lion skins that were inflated. So typical was this kind of watercraft on the Atacama coastline that an *exact* copy of it—including the very same men paddling it! (Figure 2.3)— was later (1860) used by naturalist Rudolph Philippi in a publication that I will discuss in more detail in the next chapter.

In the scene by Gay, the port of Huasco huddles along the shore, and tall cacti (likely *Eulychnia*) crown the rocky outcrops. The numerous goats cavorting, grazing, or standing on the rocks seem quite at home here but were introduced by Spaniards along with horses and mules. Indigenous people and mestizos wearing colorful striped shawls and tall hats dominate this scene. Paralleling the shore, a dusty road serves as Huasco's main street, which is lined by small, simple gable roof buildings. Before the Spaniards' arrival, Huasco likely had a less developed appearance, but this view gives a valuable glimpse into the past. In the background, the barren Coastal Cordillera looms under sky partly filled with cumulus clouds.

Huasco began as a pre-Columbian settlement, but with the advent of the European sailing ships, larger ports began to develop. In early colonial times, the most prominent port city in Chile was Valparaíso, which served the prosperous interior hinterland around Santiago, but Concepción was also important. Farther north beyond Huasco, in the area that was then part of Peru, Arica was one of the larger ports, but others such as Pisagua were strategically positioned to serve as jumping-off places into the wilderness adjacent to the Peruvian Andes. With time, that wilderness would become a hinterland as trade developed between the interior and ports. Of the many ports on the Atacama coastline that began as Indigenous coastal fishing villages, some would flourish, while others would disappear as trade was realigned.

For the most part, however, the earliest Spanish exploration of the interior Atacama Desert region was by land from Peru. Thus, Spaniards first became familiar with the Atacama Desert through a well-worn set of trails through the bleak country lying between the fertile Valle Central and the rich highlands of Peru. Traversing this area involved caravan-like outfits of men on foot and horseback, frequently with horses and mules (and sometimes llamas) carrying cargo; often, though, the Natives themselves were pressed into servitude carrying heavy loads. For all involved with these caravans, it was often plodding, even grueling travel through rugged terrain, especially in the Atacama Desert, where finding potable water and firewood were major sources of concern. Exploration aimed at conquest was normally considered men's work. However, a notable exception was Inés Suárez, who participated in the conquest of Chile in 1540. Suárez was honored for not only fighting alongside soldiers and tending the wounded, but also for digging a well that supplied water for thirsty troops.[25]

I would now like to reflect more on what Pedro de Valdivia personally experienced in the desert north of Chile. Suspended between the wealthy Andean province of Peru and the prosperous Valle Central of the new kingdom that Valdivia founded, the Atacama region soon gained a reputation as a desolate area, thanks in large measure to his descriptions. In a letter to Emperor Carlos V dated September 4, 1545, and written from La Serena, Valdivia described his experiences in the "valle de Atacama" as "el gran despoblado que hay hasta Copoyapo" (the great unpopulated area until Copiapó is reached). It should be noted that Valdivia emphasized the concept of population, rather than aridity, in this description. The term despoblado simply referred to a paucity of people rather than any specific environmental factor. Nevertheless, Valdivia's mention of Copiapó is an important reminder that there were well-populated places in this desert region. Like the villages at [San Pedro de] Atacama, Copiapó was a veritable oasis where water was plentiful, at least during most years. So dense was the vegetation at Copiapó that the official name given the place by the Spaniards included the word "Selva," which means jungle in Spanish.

The main factor determining how these oases fared was the rainfall and snowmelt from the Andes, which found their way into the desert through a series of channels, some deeply incised into the landscape. Rarely, when this runoff did not materialize, water would become scarce and crops might wither. Generally, then, although the Atacama province was largely arid and barren, there were within it a few well-watered valleys sustained by runoff from the Andes. These river courses in the Atacama, such as the Copiapó and Loa Rivers, tended to have the highest concentration of Indians and the best agricultural prospects; therefore, they were most desirable for settlement. This is another reminder of the importance of Indigenous culture here, for the locations long ago settled by these groups were the places that lured the Spaniards. To both Indians and Spaniards, the country between these oases was less desirable. Indians knew how to traverse it but did so with care. To Spaniards, it had absolutely no merit and was literally no man's land. In another letter to Carlos V written from Santiago about four years later (July 9, 1549), Valdivia continued the theme but made his point even more explicit, calling the entire region "el gran despoblado de Atacama."[26]

References by Almagro and Valdivia appear to be the first to characterize the Atacama as a region with a distinctly barren quality, but it should be reiterated that "Atacama" was the name of a province, not a desert per se. Moreover, although the origins of the name Atacama are clouded, it appears to refer to either a particular tribe or a black duck (referred to in Spanish as pato negro) in one dialect, not the forbidding desert we associate with the word today.[27] In any case, it was a Native name and hence exotic to Spaniards. That, though, did not inhibit them from branding or fusing it with one of their own words, the versatile adjective despoblado. It is noteworthy that it took just a short time for this hybrid terminology—despoblado de Atacama—to emerge. By 1549 the use of this term for the region evolved in narratives; that should come as no surprise, as it accompanied the interbreeding of the Spanish with the Indigenous population. In other words, like the population here, the name Despoblado de Atacama itself was the result of *mestizaje*. Moreover, the "el gran" in that early terminology already conveys a scale or size. This was not only a largely unpopulated place, but a large one that would have to be reckoned with as this part of South America was colonized.

This hybrid naming confirms an important historical development: within about a decade after their arrival here, the Spaniards had nominally appropriated, actually branded, this large and largely unpopulated area for their own. Ironically, the fact that they did this with the sometimes-reluctant assistance of Indians is a reminder of how insistently Indigenous forces work to shape the discourse of colonialism. In her 2015 book *Capturing the Landscape of New Spain*, literary historian Rebecca A. Carte notes that the cost of such contact could be high for the Indigenous population and serve the conquerors in many ways: "Naming, at once a means of taking possession, Christianizing the landscape, and making sense of a seemingly vast and unfathomable territory, served explorers on the ground…by giving them some sense of direction and stability in a world not their own." Carte also claims that what she calls topophobia—a fear of desolate landscapes— appeared to be a factor.[28]

In contrast to the northwest of Mexico, where the term Jornada del Muerto (journey of death) characterized routes north of Paso del Norte (today's El Paso, Texas) that were traversed by ill-fated expeditions, there is little or no evidence that Spaniards in this part of South America saw stark landscapes as something to fear; rather, they were an impediment that had to be carefully navigated. In the Atacama, I think the term *respect* rather than *fear* is more appropriate. Nevertheless, given the care needed before conducting a successful reconnaissance here, I think that the characterization of the Spaniards' attitude as one of *healthy respect* is probably even more apt.

A study of historical or antiquarian maps supports the idea that this desert region was first named not for its lack of rainfall but rather for its Indigenous groups. That population, albeit relatively small in number, was adaptable, clustering in oases or making use of the sea and the sparse vegetation of this arid area. The scarcity of published Spanish maps from this early period is noteworthy, but understandable. As cartographic historians have long known, it was a result of the Crown's jealously guarding its hard-earned geographic information, no doubt concerned that many maps of the lands it claimed were being prepared "by French or German geographers working for rival European crowns."[29] However, it is likely that Spanish military manuscript maps reflected the frequent mention of either tribes or major landmarks in the narrative reports rather than emphasizing this region's aridity.

Similarly, on the earliest published maps, the term Atacama originally referred to the political province of that name rather than to the physical characteristics of the landscape. Thus, European maps from the 1500s and 1600s invariably used the name Atacama to identify a political jurisdiction; this name now referred to the entire province of that name, though it was likely far more limited in Native usage. As near as I can determine, historians and anthropologists are correct that the word Atacama as a place name appears to be Indigenous in origin. Moreover, it was originally geographically restricted to the area southeast of Calama, perhaps in the vicinity of the community today called San Pedro de Atacama, which has long been sustained by runoff from the Andes and was strategically connected by trails to Salta, which lies east of this cordillera in present-day Argentina.

As the region's namesake community, San Pedro de Atacama deserves some additional comment on its Indigenous past in relation to Spanish exploration and colonization. In his informative and comprehensive book *Vida y Cultura en el Oasis de San Pedro de Atacama*, Lautaro Núñez Atencio places its development in the context of a broader area comprising the puna and slopes of the adjacent Andes. He notes that people arrived in this area as hunters more than ten thousand years ago and that agricultural and pastoral activity began in the period from ca. 1,200 to 500 BC. The development of stable settlements began ca. 500 BC to 100 AD, and San Pedro can be traced to ca. 100–900 AD based on the remains of forts (for example, the Pukará de Quitor, which lies adjacent to the community) and the villages which comprise the sprawling oasis. Núñez characterizes the landscape at that time as settled villages concentrated in *ayllos* (verdant irrigated areas) surrounded by large tracts of desert land. The Inca dominated this area from 1450 to 1536, when the Spaniards arrived.

Núñez describes and illustrates in his hand-drawn sketches the arrival of Diego de Almagro, and the battle that occurred at Quitor between the Spaniards and the Atacameño under "El Conquistador de Atacama don Pedro de Valdivia." By 1540 the Spaniards described the "pueblo of Atacama" as occupying a "flat, wide and long valley," and whose houses were built differently from those elsewhere in the area. Adobe houses with wooden viga beams holding up the roofs were common here. The Spaniards were especially impressed with the large and abundant *algorrobales* (algorrobo trees) whose wood was strong and from whose seeds the Indians make bread and a delicious brew. Moreover, there were also large chañar trees, which likewise served many uses, including as beverages. Irrigation ditches (*acequias*) diverted the waters to fields in which varied crops—corn, potatoes, beans, quenos (quinoa?)—were grown.[30] To the Atacameños and Spaniards alike, San Pedro de Atacama was indeed the quintessential oasis.

Although this part of the Atacama seems to correspond to the land originally occupied by the Atacameños, there is more to the story. As historical anthropologist José Luis Martínez noted in his book on Indian and Spanish interaction in the seventeenth-century Atacama Desert, anthropologists have tended to take at face value that the naming of a province signified one major ethnic group. As paraphrased by anthropologist Anita Carrasco, "studies of these populations have tended to establish a homology between the name of a territory; the name of the group that lived there; and the assumption of the existence of a single sociopolitical unit." Homology is an apt word for this process by which similarities are attributable to a common origin, and Carrasco uses it to remind readers that care must be exercised. Thus, although the Spanish use of the name Atacama suggests that it has been adopted from the Indians, and supposedly means the same thing that Indians meant, this process involves many assumptions. Typically, it oversimplifies the original meaning(s) and can ultimately cause us to oversimplify our view of Indigenous peoples.

The name of the Calama vicinity provides a case in point. Whereas some claim that this may be the name for Indigenous people who occupied the area, in fact as many as five separate groups existed here in the seventeenth century. In other words, the situation was likely far more complex than a single name—such as Atacameño—for people here might suggest. Carrasco further notes that currently "the Chilean state also operates under that assumption," that is, employs the name Atacameño to connote a single cohesive identity. I introduce this point here as a caveat; although Indigenous identity is important, the Indigenous names involved may be more modern artifacts than we realize. Ultimately, the names may say as much or more about Spaniards' perceptions of the time than they did (and do) say about the actual distribution of Indigenous peoples.[31]

This much is certain: at the time it was first used by Spaniards in the mid-sixteenth century, the reference to Atacama on maps is political in that it identifies a place (or places) of that name claimed by the Spanish Crown in the sixteenth century. The name has something to do with the Indians, but we cannot be sure what, especially because they may have had little or no input in the process of affixing those names on the map. Of this we can also be certain: the name Atacama at that time did not refer to the Atacama Desert as we know it today (an area possessing an arid climate). Readers should grasp this irony, namely, that the Atacama Desert bears an Indigenous name that conveys aridity to us, yet the name may convey no such notions to those Indigenous peoples. This was, after all, part of New Spain where Indians were adept at living in varied physical environments and surviving in the face of outside power, be it Incan or Spanish. Moreover, they related to this locale in ways that Spaniards were scarcely able to comprehend.

For its part, the Spanish Crown was intent on colonizing this region, although the natural and cultural environment—or more accurately percep-tions of those environments—played a major role in how that was done. Throughout this entire region, Spain allocated a series of land grants to enterprising and well-connected noblemen in the better-watered valleys where agriculture might thrive. In this regard the Atacama was similar to the Sonoran Desert of northern Mexico, where "several bodies of imperial legislation informed...classifications of land and water rights in Spain and America." In particular, "Castilian law was applied to the sphere of private commercial transactions, inheritance, and similar areas of family law in the colonies," although shortcomings in this unwieldy medieval system ultimately led to the *Recopilación de leyes de los reynos de las Indias* in 1680.[32] In the Atacama, and throughout the deserts of Latin America, the contrast between lush, irri-gated lands and stark desert had existed well before the Spaniards arrived. Now, however, a host of new crops imported from the Old World would thrive here after being introduced. These included wheat, oats, olives, and grapes.

As noted, the Spaniards also brought with them horses, mules, and many other animals from the Old World; these supplemented indigenous animals such as vicuñas, llamas, guanacos, and alpacas. The small rodents called chin-chillas existed in large numbers in the foothills of the Andes, but with the arrival of the Spaniards in the sixteenth century, they began to be hunted to

Figure 2.4. Seen under repair here, the church at San Pedro de Atacama is constantly maintained as a vital element in the historic townscape. (Photograph by author)

near extinction for their soft fur; today they are listed as critically endangered. The Indigenous people here had since time immemorial collected various useful plants such as the yareta shrub (*Azorella compacta*) and also domesticated animals. Along with their now-domesticated labor and their virtual enslavement in mining, Indigenous peoples here helped make this a viable, if remote, part of Spanish South America.

The pre-Columbian Indians also helped predetermine the settlement patterns we see today. Typically, Spaniards colonizing an area sought the same well-watered places that the Indians had already settled, such as the aforementioned San Pedro de Atacama and Copiapó. About fifty years ago, geographer Khairul Rasheed noted two distinctive characteristics of communities in his study area northeast of Calama. The first was the persistence of Indian settlement morphology. As he observed, "the pre-Hispanic nucleated pattern of settlement was not altered by Spaniards, rather it was continued and the settlements were made more compact." The second characteristic was imported by the Spaniards and has also left a lasting legacy. As Rasheed put it, "the desire to spread Christianity to the New World prompted the Spaniards to build a church in each settlement." As the Spaniards settled the area, they grasped the importance of the existing village layout, which usually featured a part that was densely settled and situated near a stream or spring. At this prime location, the Spaniards elected to build the iglesia (church).

Figure 2.5. Shaded by beautiful native pepper trees, the plaza at San Pedro de Atacama is a key element in the town's character. (Photograph by author)

Moreover, in accordance with Spanish decrees dating from as early as 1573, they created a square open space called the plaza adjacent to the church.

In one fell swoop, then, the Spaniards simultaneously preserved Indigenous aspects of the community while imposing their own ecclesiastical and political rule. This hybridization of place is evident in the morphology of agricultural villages throughout the Atacama Desert today. Only those communities initially established by Spain are characterized by a rigid grid pattern mandated in the Law of the Indies. Many other communities dating from Indigenous times are far more complex and syncretic. As Rasheed noted, "the reason that many oases in the study area do not show a clear grid plan is that the modern settlements have been superimposed on the original Indian villages with winding roads and alleys can still be observed." In a statement that confirms the hybridization that had occurred, he added, "the result is a nucleated unit of settlement, which is part planned and regular, part unplanned and irregular."[33]

This pattern is apparent to travelers in the oases of San Pedro de Atacama today, where some streets run arrow straight and intersect at right angles while others are crooked and intersect at odd angles. The center of San Pedro de Atacama, which is more densely settled than the adjacent parts of the community, is typical. Here, along a major watercourse, the Spaniards created a plaza and built a church (Figures 2.4 and 2.5) as testimony to their effectiveness in colonizing the region. This pattern of plaza and church holds

throughout the region, and in fact much of Latin America, and the two elements (one spatial, the other architectural) are a dead giveaway of a powerful
Spanish presence. San Pedro de Atacama is a tourist haven today, and great
pains are taken to keep the plaza verdant and the church well-maintained. In
the photo of the church, a tarp covering part of the building protects the section under reconstruction from the elements. In the photo of the adjacent
plaza, native paper trees (*Schinus molle*) provide beauty and shade, the "molle"
in their scientific name derived from the Inca word for the tree—*mulli*. For
their part, the Spaniards called these trees "pimientos" and spread them
widely into the parts of the New World that they colonized; subsequently,
this attractive evergreen tree spread worldwide. At Toconao, where tourism is
less emphasized and a more traditional pattern of Indian life exists, one finds
a more rudimentary expression of the church and plaza, as seen in a photograph taken about fifty years ago (Figure 2.6).

This cultural hybridization of Indigenous and Spanish elements also
extends to residential architecture, which varied from one part of the desert
to another depending on the cultural group who designed and constructed
buildings. Geographer Rasheed noted that a common type of house in the
desert on the western side of the Andes was similar to that associated with
the higher mountains—a rectangular stone structure with a gabled roof and
small windows. Typically, the ridgepoles of the house were parallel to the trail
or roadway along which these homes were built (Figure 2.7). The roof beams
and rafters were often made of the ribs of columnar cacti or from the algorrobo and chañar trees that were found near the riverbeds. The roof itself was
often made of brush or thatch laid flat and sometimes covered with mud.
These simple houses typically consisted of only two rooms, a living area and
a bedroom. Animals might also be kept inside at times, though pens were
often found in a courtyard outside. The kitchen was sometimes located in a
small outbuilding separate from the house. As we shall see, different types
of Indigenous buildings prevailed in other parts of this desert region, but
all were effective adaptations to the normally dry climate. In response to
the strong prevailing south winds here, rocks were sometimes placed on the
roofs to keep them from being damaged. Wisely, Rasheed warned readers
against the temptation to think deterministically when it comes to the desert
environment. Human ingenuity always plays a role in adaptation. In fact, he
concluded, the physical environment, which was not totally prohibitive, was
at times actually conducive to innovations and improvement in Indigenous
settlement and architecture.[34]

One of the most impressive Indigenous architectural traditions in this
region is the use of stone. Given the region's scarcity of wood, stone was

Figure 2.6. In Toconao, Chile, a small formerly Indian town southeast of San Pedro de Atacama, two Spanish elements—the plaza and church—are visible. Photographer William E. Rudolph noted that the church stands separate from its tall bell tower due to concerns about earthquakes. (American Geographical Society Library, University of Wisconsin at Milwaukee)

Figure 2.7. This ca. 1960 street scene at Toconce, Chile, reveals elements of Indigenous and Spanish architecture, its photographer William E. Rudolph noting that the woman was rushing indoors to avoid being photographed while the child lagged behind. (American Geographical Society, University of Wisconsin at Milwaukee)

Figure 2.8. The stone walls of a corral in the foothills of the Andes near Guatín, Chile, preserve the masonry traditions of Indigenous peoples in the Atacama Desert. (Photograph by Damien Francaviglia)

used for building whenever possible. Although stone required heavy lifting to move and set in place, this effort was still easier than ranging great distances to find wood. The fragmented nature of natural stone rubble in many areas made quarrying unnecessary, and the Indians skillfully constructed stout walls without mortar. In other places where semiround water-worn stones were found, they could be rolled or carried to sites where they were piled atop each other to form walls. In addition to using stone to construct the walls of houses, Indigenous people used it for making terraces, bounding fields, and even for corralling animals—a venerable tradition that continues to this day (Figure 2.8). Stone was also a key ingredient in constructing defensive forts. Geographically speaking, stone was the building block of highland Andean communities, but its use extended into the desert as the cultures were connected through trade and conquest. Throughout much of the Atacama, the presence of human-made stone features, many of them relics from earlier times, serves as a reminder of early settlements, many of which vanished even before Spaniards arrived. Stone is so abundant, and was so abundantly used, in this region that it may be said to comprise the bedrock of community building since prehistoric times.

As they traversed this region in the 1500s, the Spaniards had much to learn about this rock-ribbed environment, and they quickly obtained important information by interacting with Natives. Honing their skills of observation, they then dispatched that information along to Spain in the form of official reports called *Crónicas* and *Relaciónes*. The Spaniards' interest in documenting discoveries had actually begun with Columbus, whose observations on flora and fauna were noteworthy. It was Columbus who not only recognized cactus plants as unique to the part of the world that he had discovered, but who also brought specimens back to Europe. These cacti were soon viewed as both a curiosity and an emblem of the uniqueness of the New World.

In this light, I would like to introduce a remarkable account of Valdivia's expedition by don Gerónimo de Bibar, also known as Jerónimo de Vivar. Though his birth and death dates are unknown, Vivar's writings place him firmly in the mid-sixteenth century. This account is dated 1558 and entitled *Crónica y Relación Copiosa y Verdadera de lo que vi por mis ojos, y por mis pies anduve y con la voluntad seguí en la Conquista de los Reynos de Chile.* That unwieldy title can be paraphrased as an eyewitness and true account of what Vivar personally experienced on the expedition across the Andes with Pedro de Valdivia. In this fascinating account published about eighteen years after the original expedition, Vivar described in detail an unusual "tree" that the explorers encountered. Leaving no doubt that this plant was new to them, and would likely be to his readers in Europe, Vivar penned a classic description of a "strange" plant, the likes of which they had never seen in Spain. Vivar noted that these peculiar trees had no leaves but were instead covered with dense, needlelike spines arranged in rows. The plants grew erect, and Vivar measured them much as he might a horse, noting that they were ten or more "palms" tall. Assuming a palm is close to a hand as a unit of measurement, that translates to about four inches (10 cm); thus, these cacti were about four feet (slightly more than 1 meter) tall, though some may have been as tall as six or seven feet (ca. 2 m).

Today we would call this plant a barrel cactus, a fitting name for its stoutness and cylindrical shape. Vivar likewise characterized its appearance impressionistically, noting that the plants grew as "stout as a thigh" from bottom to top, that is, they did not taper very much. These cacti were crowned with flowers; Vivar observed that "some have yellow flowers, others white and very large." Overall, this is a fairly accurate description of cacti that are found on the mountain slopes in the Atacama. However, like most observers of the time, Vivar was less concerned with science or aesthetics and more with the utility of these peculiar plants. As he added, "After flowering, there appears a fruit. These are stout like big figs, and inside are full of small black seeds in the manner of mustard seeds mixed with a kind of liquid much like honey." The words fruit, figs, and honey hinted at something reward-ing indeed, and as Vivar confirmed, "When they ripen they open a little and are delicious." Vivar also confirmed that he was not the first to discover this plant and its delicious fruit, adding "the Indians call them in their language 'neguey.'" To this description Vivar added a statement about their distribu-tion or range. Noting that these strange plants could be found in all the mountainous areas, he made one of the first direct references to the climate of this arid region in relation to plant growth: "They grow in the dry places which receive no water."[35]

The observant Jerónimo de Vivar was not alone in describing the lands that Spain was colonizing in South America, although most of their descrip-tions naturally focused on the Inca Empire. Among these chroniclers was Pedro de Cieza de León (1518–1554), who wrote the groundbreaking work *Chronicle of Peru: The First Part*. Of interest to us here is the map that accom-panied Cieza's chronicle (Figure 2.9). Posthumously published in Venice in 1560, it was ambitiously entitled *Brevis exactaque totivs novi orbis eivsque insvlarum descriptio recens a Joan[ne] Bellro edita* (A brief and exact updated description of the entire New World and its Islands, edited by Jean Beller). As the title makes clear, Cieza may have provided much of the information, but he had assistance in positioning it on the final product. Cieza's map is a reminder that cartography is usually both cumulative and collaborative. The Cieza map, as it is commonly called, is a masterpiece. On it we can immedi-ately recognize South America, even though in reality the Pacific coast of this continent is not as straight as he depicted. Like many such maps, most of the place names are on the perimeter; the Atlantic and Pacific shores of this map bristle with them. The emphasis on places at the perimeters of continents is a hallmark of Portolan charts and is typical of the times. Cieza, though, was also concerned about the interior regions. He prominently depicts the Andes as the backbone of the continent. On Cieza's map, they run nearly arrow straight north and south, while in reality this chain curves westward—as does the west coast of the continent at the so-called Arica bend. To the east of this simplified mountain chain, the sinuous Amazon River flows northeast-ward to the Atlantic Ocean.

The prominent position of the Andes on Cieza's map is a reminder of the importance of mountains to the Native peoples and Spaniards alike. To the latter, mountains were respected as topography to be reckoned with, as they might contain mineral riches or pose obstacles to travel. To Indians,

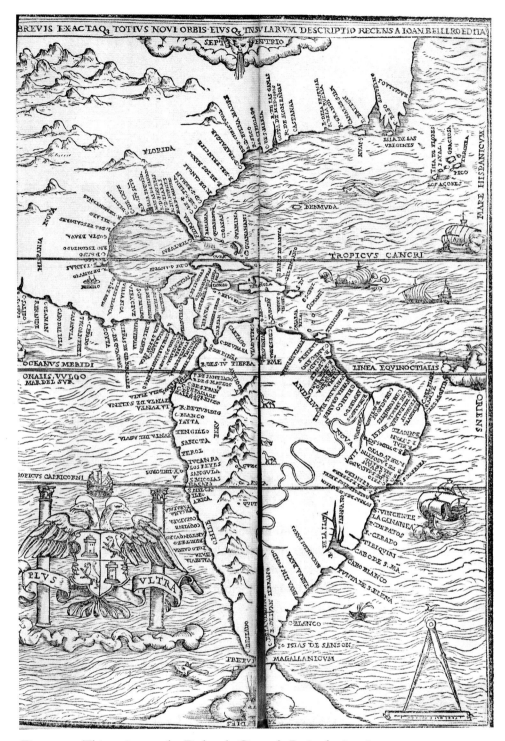

Figure 2.9. This 1560 map by Pedro de Cieza de León depicts important geographical features such as the Andes Mountains, but not the Atacama Desert. Instead, on the coast adjacent to and north of the name CHILE, it shows many ports, including Arica. (Author's digital collection; original in Beinecke Rare Book and Manuscript Library, Yale University)

however, mountains have a much deeper spiritual significance. As anthropologist Gabriel Martínez noted in an article on the Gods of the Peaks of the Andes, "mountains are a deity" that provide sustenance and to which Indigenous people give thanks.[36] To Spaniards, this reverence for a mountain would seem like so much idolatry, and someone who communicated with one might be considered crazy, but it helps explain how Indians related to the land.

As anthropologist Anita Carrasco conducted interviews with villagers in the high desert country of the Atacama Desert adjacent to the Andes, she encountered a man named Pablo, who stated, "Mountains are alive. They are like people." Leaving little doubt as to exactly what he meant, Pablo added, "Mountains are life. You can have a conversation with a mountain." The Indians here had long recognized their dependence on the mountains for their agriculture, their herding, and even their mineral extraction. Similarly, the canals channeled precious water originating from the mountains to be directed for agriculture. This was a life-giving blessing from Earth Mother, a prominent figure in the Atacama region at the base of the Andes. Maintaining these waterways was, and is, as much an act of devotion and propitiation as it is work.[37]

But to return to Cieza's map, that part of it depicting the coastal desert region adjacent to the Pacific Ocean is worth a much closer look. Among the place names here we can easily recognize Arica, which is erroneously shown as lying south of the Tropic of Capricorn. South of Arica the place names are printed upside down, but by rotating the map we can read them. With some leeway for translation, most of these names appear to depict communities still in existence today. Among these names strung down the coast are Tarapacá, Pisagua, Copiapó, Coquimbo—all of which also appear on Bowman's map, albeit in somewhat different positions. Interestingly, one name appears to be Tacama, while another is Antonqvaro. The former may mark a very early use of the name Atacama on a map, but then again it could also refer to a place such as the port of Tocopilla. As for the name Antonqvaro, which we might naturally think refers to present-day Antofagasta, it is an unlikely match for two reasons, one geographical and the other historical. Antofagasta is located far to the north, but more to the point, it came into existence about three centuries after the map was published as a mid-nineteenth century port to serve the landlocked Antofagasta region of Argentina.

Farthest down the coast on Cieza's map is Valdivia, which was then, as it is now, a port of considerable importance. When interpreted even more deeply, Cieza's map reveals something startling indeed. On this map, the forested area south of Valdivia is shown as a blank space, suggesting either a complete lack of information or few or no inhabitants; however, the desert north section of the map, in the area which we presently think of as barely populated, contains most of the place names and confirms that despite its aridity, it had many ports of call. Paradoxically, although the Atacama has indeed been a frontier to Chile in many senses of the word, Chile's real frontier has always been the verdant south, with its Mapuche Indians who put up ferocious resistance well into the nineteenth century. Lastly, it should be noted that although Cieza could have placed all of these names on the continent instead of in the waters of the Pacific, he had a good reason for not doing so. That might have detracted from the Andes, which impressed him mightily. Moreover, leaving that land adjacent to the coast unencumbered provided Cieza with the perfect opportunity to put yet another name there: in large letters, he placed the name CHILE. This not only makes the name stand out, but it also subliminally emphasizes the north–south axis, and hence the slenderness, of this remote province of New Spain.

As hinted at, Spain was not alone in its desire to dominate western South America. Throughout the sixteenth century, in fact, England had eyes on this same area. Although there were constraints toward directly attacking Spanish communities, partly a result of religious decrees by the pope, the Protestant Reformation of 1517 had set in motion the growing independence of northern Europe from such restraints. As early as the mid-1540s, in fact, English explorers began to contemplate making forays into Spanish territory in the New World. The West Indies were particularly attractive, as were portions of South America said to contain fabulous gold deposits. That anti-Spanish zeal slowed somewhat when Queen Mary I (1516–1558) embraced Catholicism and married Philip II—an act which made Mary and Philip queen and king of Chile. Alas, as fate would have it, Mary became Bloody Mary when she was deposed and beheaded by Protestants, no doubt in retaliation for her execution of their coreligionists. After Mary's demise, England again eyed Chile as a potential prize to be taken from the Spaniards. Among the noteworthy Englishmen obsessed with South America in the late 1500s were Sir Francis Drake (ca. 1540–1596) and Sir Walter Raleigh (1554–1618).

These adventurer-explorers were among the zealous Englishmen who looked westward across the Atlantic well aware that Spain, Portugal, and France had claimed portions of the Americas. Undaunted, they nevertheless sought opportunities for expansion into those places where Spain's hold was tenuous. As historian David Narrett observed in his recent book *Adventurism and Empire*, these British exploits "were inspired by the quest for personal glory and wealth as much as by the advancement of national power." More to the point: "During the Elizabethan era, merchants as well as privateers were essential players in the panorama of overseas expansion."[38] The sailing ship was their method of travel and conquest, and those numerous ports along the shores of Spanish America promised quick access to the treasures of the interior. At this time, the Pacific Ocean was widely called the South Sea and was also widely recognized not only as a key to reaching the Orient but also a vast realm of potential treasures and commerce. As it now appeared on maps of the world, the South Sea was cradled by the Americas on one side and by Asia on the other, and the focus of exploration was the continental land masses that embraced this sea. As noted by historian Glyndwr Williams, "in the 1570s, as relations with Spain worsened, plans were afoot in England for ventures into the South Sea." Going through the Strait of Magellan or around Cape Horn at the far southern tip of South America was now considered a safer and easier option.

In 1570, the types of places and treasures that England might find were portrayed in the colorful language of the times in two letters to the Crown of England. Both letters were proposals for expeditions. As the first document enthusiastically put it, "we are assured to expecte golde, Siluer, Pearle, Spice, with grayne, and such most precious marchaundize, besides countries of most excellent temperature to be inhabited." That description could characterize the bounty of many places, but South America appears to have been part of the strategy. The second document provided a geopolitical context by explicitly noting that "since the Portugall hathe atteined one part of the newefounde worlde to the Este, the Spaniard an other to the Weste, the Frenche the third to the Northe, nowe the fourthe to the Southe is by God's providence lefte for Englonde."

For historian Williams, the 1577–1580 voyages of Sir Francis Drake suggest that for England "arguably the establishment of trade and perhaps even bases along the Chilean coast south of the Spanish presence was very much to the fore." In February 1580 Drake raided Arica, the Peruvian port city that, as readers may recall, was difficult to access, as it stood on a narrow shelf of irrigated land at the base of rocky, towering hills. Drake was both an opportunist and a strategist, though at times it was difficult to separate the two. Williams aptly notes that Drake's voyage along the western coast of South America was neither accidental nor whimsical; instead, it constituted very serious and focused reconnaissance.[39] Understandably, Spain increasingly viewed such voyages with concern and by 1585 became involved in an undeclared war with England. Two years later, in May 1587, Englishmen Thomas Cavendish and Richard Hawkins likewise raided Arica. A year after that foray, Spain suffered a devastating naval loss in the north Atlantic as its armada, or prized fleet, was badly damaged by storms. This further emboldened England, which was engaged in the undeclared war until 1604.

In late sixteenth-century Europe, mapmaking kept up with these events as people began to follow international events with increasing interest. Through written chronicles and graphic mapmaking, the physical and cultural characteristics of South America had come into focus for Europeans by about 1570, when the Antwerp mapmaker Abraham Ortelius (1527–1598) created his *Theatrum Orbis Terrarum*, a monumental atlas (book of maps) featuring every part of the known world. In depicting South America, Ortelius's maps were no better than Cieza's; in fact, some feature a South America that is far more misshapen. Rather, it was the packaging of these maps into a beautiful and impressive book that made them almost irresistible. For many, the *Theatrum Orbis Terrarum* was both definitive and authoritative. Once open, it was difficult to put down, for it engrossed readers by presenting the world in far more detail—accurate or otherwise—than any single map could.

By 1575 Ortelius was appointed geographer by Phillip II, King of Spain, and Ortelius's name became a household word, at least in well-off homes. Ortelius's atlases were published in many subsequent editions, each improved and more popular than those printed earlier. It was, in fact, Ortelius who not only helped popularize mapmaking but also created a new kind of reader— the armchair traveler who, as scholar of the history of discovery Tom Conley noted, could now "find cartographic images of unknown worlds discovered by virtual voyages within the happy confines of a library or home."[40] This was ironic, for although Ortelius and many mapmakers might freely produce maps of exotic places such as South America, many of them had never travelled there, either. One of the maps in Ortelius's atlas from 1570 (Figure 2.10) shows this part of South America having a far more distorted Pacific coastline than we recognize today. It also shows a cluster of islands off the southern coast of Peru that is likely an ambitious depiction of the Islas Ballestas. On the other hand, it does show a number of port cities such as Arica and Pisagua, as well as the city of Copiapó, and hints at the existence of the Andes. However, on maps this early, including the Diego Gutiérrez map of 1562, there is not yet any indication of the Atacama Desert as such.

Professional commercial mapmaking as well as carefully compiled—and often closely guarded—sea charts suggested a world full of promise. One part of it, South America, continued to tempt interlopers. Although England

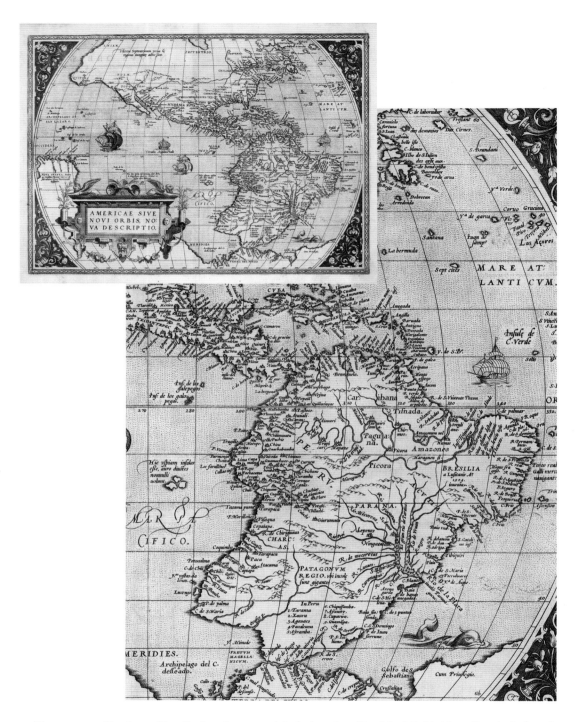

Figure 2.10. Abraham Ortelius's atlas map titled *Americae Sive Novi Orbis Nova Desriptio* (1570) greatly distorts the coastal desert region of South America and does not name nor depict the Atacama Desert, but it still indicates many places such as Copiapó, Tarapacá, and Arica. (David Rumsey Map Collection)

officially respected Spain's presence, it was clearly looking for chinks through which a South American empire could be developed. At this time, it should be recalled, boundaries were vague and maps even vaguer. By around 1600, North and South America were appearing as large land masses recognizable to us even today, but these continents had yet to be fully explored and accurately mapped. For several reasons, maps depicting South America varied wildly and erred mightily. The actual form of South America (and by extension Chile) was hard to depict on maps, in part because of conflicting information from varied sea voyages and land expeditions.

Perceptions of New World geography were also a factor. As maps of the southern hemisphere took shape, some still claimed that land masses south of the equator should occupy about as much space as those north of it; that is, they speculated that South America *should* be about the size of its North American counterpart, though in reality it is about twenty percent smaller. Speculation about the size of continents was in fact one of the premises for a huge Antarctic continent—about as large as Asia—that formed the southern boundary of the Pacific Ocean and extended on some maps from the tip of South America all the way to Asia. To some, conceptualizing South America as otherwise would be to challenge the natural order—as several observers sincerely noted, the earth might wobble and come to ruin if this were so— or run counter to deeply held religious beliefs in geographical symmetry as a reflection of God's perfection. Although Magellan had proven that South America ended near the strait that bore his name, the belief in the existence of land lying southward from there fueled visions of a huge continent (called Terra Australis Incognita). That fantastic vision of a huge Terra Australis would linger in the popular imagination, and on maps, for about two centuries. And yet South America was exceptional in many ways, including the fact that it extends to a point much farther south than Africa or Australia.

There were, however, far more mundane or scientific reasons why it took several centuries to accurately determine the shape of South America. Foremost was the problem of how to measure longitude, and thus confirm how wide the continent actually was. It was relatively easy to calculate latitude, and hence place what was discovered in its proper context in reference to the equator. Provided one knew the correct day of the year, one could tell how far south (or north) of the equator they were by calculating the position of the sun at noon; by extrapolating from that, it was easy to determine how far southward South America extended. Explorers reckoned it was about 58 degrees south of the equator and that the continent was not, as some originally imagined, linked to the Antarctic continent. By contrast, getting longitude right was much harder. It depended on knowing the exact time at two widely separated places; the first place, where one was making observations from the field, could only be determined by knowing the exact time that it was at another important place: in this case, the meridian one used as a baseline that served as the original survey point (Madrid, London, etc.).

In the meantime, the variation and discrepancies in maps encouraged adventurism and triggered international disputes. In retrospect, one thing may help explain why the west coast of South America was mapped well enough to indicate that portions of it run nearly north–south for considerable distances: the fact that some compasses—depending on how well their needles were balanced—worked well here. Sailors working their way up and down the coast of Chile were aware that they followed the same compass bearing for days on end. That magnetic pole lies in the northern portion of the North American continent, and even moves about slightly in position over time, but has long been relied upon to help one fix a position, at least approximately. As I discovered for myself in the Atacama Desert, despite its being in the southern hemisphere, standard magnetic compasses point northward with a relatively small amount of declination in Chile, so that they can be relied on somewhat more there than in many other parts of the continent. It is humbling to ponder the power of that magnetic force, coaxing a needle to point toward a source so far to the north. Then, too, it is equally humbling to think of the mineral magnetite, or lodestone, embedded in the same needle and perpetually seeking its mate 7,000 or more miles (ca. 11,000 km) distant. Using such tools, mariners and explorers figured out that the coast was remarkably straight, but they did not know exactly where it was located. At this time, the natural magnetic declination was virtually unknown this far south. The main problem to be overcome was not the compass, though, but rather devising a clock that could keep time on a moving, and often swaying, vessel or wheeled wagon— the motion of which played havoc with clock mechanisms. Between this and a failure to read compasses accurately, we can understand why—to their dismay—some mariners abruptly encountered the rugged Chilean coast hundreds of leagues from where they expected it to be.

This part of South America was not only hazardous to navigate by sea but also fraught with dangers on land. In Chapter One I briefly discussed some of Arica's environmental misfortunes such as earthquakes and tsunamis, but I will now introduce another. In a highly stylized image from 1615 that was likely made from atop that city's fabled El Morro and looking north, the artist depicts the small city as highly compact and crowned by church steeples as it stands atop the cliffs that drop vertically into the Pacific Ocean

Figure 2.11. The desert port city of Arica, seen through falling ash from the volcanic eruption of Huayna Putina in the nearby Andes, as depicted in Felipe Guamán Poma de Ayala's *El primer nueva corónica y buen gobierno*, 1615. (Author's digital collection; original in the National Library of Denmark)

(Figure 2.11). The artist's main purpose for rendering the city, however, was not to record it accurately but impressionistically—in this case, under a menacing shower of volcanic ash falling from the eruption of Huayna Putina in the nearby Andes mountains. Using only squiggly lines, the artist created a memorable image indeed, which bleeding ink subsequently made seem hazier, though one can only imagine the distress its occupants felt at experiencing sulfurous, gritty ash raining down from a darkened sky. Typically, as we have already seen, most illustrators depicted Arica and other ports more accurately, as safe navigation was a major concern along this rugged coast. Farther south during this period, ports such as Coquimbo and La Serena served the mineral-rich interior area near Copiapó. Many travelers at this time were well aware of the harsh nature of the area inland, a brooding landscape huddled behind the bleak coastal mountains, but it was not widely known at that time as a desert in the climatological sense. Rather, the name Atacama signified a very sparsely populated area nominally claimed by Spain but lying in a liminal space between Peru and Chile. Still, as any traveler who disembarked here knew, especially as informed by the local Indigenous peoples, this was a desolate land of little rain and even less vegetation.

By the early- to mid-1600s, Chile was not well known in Europe, at least according to Alonso de Ovalle (1603–1651), the Chilean-born Jesuit priest who sought to rectify his homeland's obscurity by boosting its image. While visiting Spain in 1642, Ovalle was dismayed by the prevailing ignorance about Chile and began writing an antidote. Published in Rome in 1646, Ovalle's report, *Histórica relación del Reyno de Chile* set the record straight by showing how the Jesuit missions had brought order to a formerly chaotic land—a sanguine assessment to be sure, but one in line with his loyalties. Of special interest is Ovalle's map entitled *Tabula Geographica Regni Chile*, which accompanied his report. Influenced by earlier maps such as Pedro Cieza de León's and especially the more recent *Mapa de Chile* by the Franciscan priest Gregorio de León (ca. 1625), the map by Ovalle continued the tradition of depicting geographic features such as mountains in highly stylized form. On Ovalle's map, which is oriented with east at the top, the Andes appear as an arrow-straight wall of mountains punctuated almost evenly by numerous sharp peaks, about sixteen of which are smoking, if not erupting, simultaneously. Across that barrier lies Patagonia, which Ovalle depicts as much wilder. Cartographic historian Mirela Altic characterizes Ovalle's rendering of Patagonia as "a carpet of wilderness, desert, unusual animals, and even stranger people," including "Patagonian giants, men with tails, and a woman clothed in mud."

By contrast, Ovalle's homeland Chile is "civilized" and features numerous "Catholic cities and Christian missions." On Ovalle's map there is no indication of a desert in Chile's far north, although a large mountain—perhaps suggesting mineral wealth, using the term "turquoise"—is shown in the vicinity of Copiapó. Most interesting, though, is the way Ovalle's remarkably straight rivers obediently make beelines for the Pacific from the Andes. However, although everything seems tamer here than in Patagonia, Chile is not perfectly uniform: the portion that we today consider the desert north appears more hospitable, or at least better settled, than Chile's far south. Despite apparent attempts at subtlety, Ovalle is here conceding that this is Chile's most challenging area, as it contains more blank space than the north. Placing all of Ovalle's map in the broader context of the time, Altic concludes, "the map's apparent ambiguity composed of the colonial power's benevolence and the representation of the land's native inhabitants as barbarians was by no means unique to Ovalle."[41] In fact, Ovalle's map and report became an important part of that literary and cartographic tradition. It even influenced French cartographer Nicolas Sanson, who published a similar (albeit more refined) map of Chile a decade later in Paris (1656) and in subsequent versions for two decades.

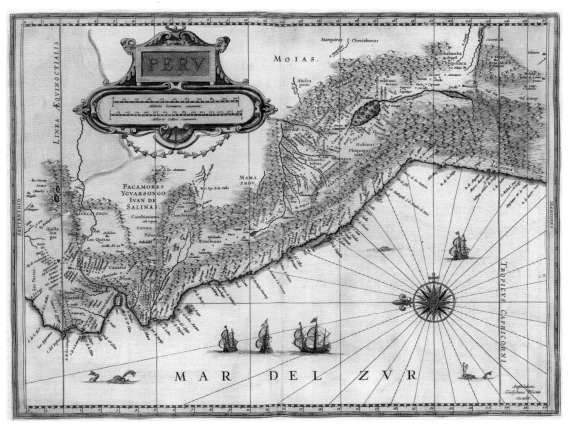

Figure 2.12. In Joan Blaeu's majestic work *The Grand Atlas* (1662), the map of Peru is oriented with east at the top, and it uses the name Atacama not as a reference to the desert, but rather for a province of that name that straddles the Tropic of Capricorn. (David Rumsey Map Collection)

The process by which the name Atacama became a fixture on European maps in the mid- to late-1600s is fascinating indeed, and brings us to the work of Dutch cartographer Joan Blaeu (1596–1673) at a time when national and provincial borders were becoming increasingly important. Maps depicting portions of western South America in Blaeu's *The Grand Atlas* (1662–1665) typify this political emphasis in later colonial times. Although one looks in vain for the name Atacama on Blaeu's map *Chili*, (which does show the river and valley of "Copayapo" but is a bit less accurate than Sanson's earlier maps), the search becomes more productive when one consults the map *Peru*, which is to say the Viceroyalty of Peru (Figure 2.12). There, toward the

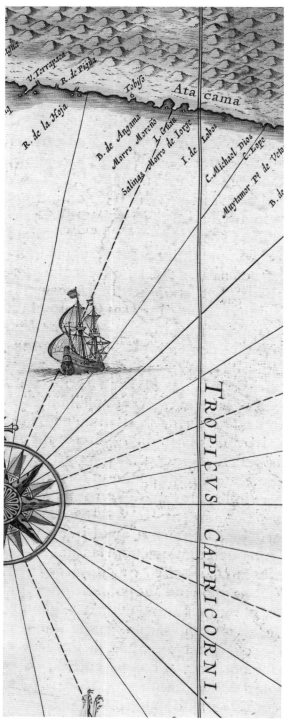

Figure 2.12. detail.

far south (i.e., right) side of the map, one finds the name "Atacama" straddling the Tropic of Capricorn, where it identifies an administrative part of New Spain rather than a desert. In other words, Atacama was, like any other provincial name, shorthand for land of interest to, and ideally governed by, the Crown. However, unlike many other names that originated in Spain and were then transplanted to the New World (for example, Guadalajara, Toledo, or Santiago), the name Atacama was—like many Indigenously named places in the New World—somewhat exotic to the Spanish ear, its origins obscured in Amerindian folklore rather than rooted in Spanish history. Atacama was, in a word, *unique* to this part of the world. A hint of what was occurring here is reflected in the place name San Pedro de Atacama. Like the community of the same name, this is a hybrid term of both Spanish and Native origin. Some sources claim that the Pedro in San Pedro de Atacama honors not only the Catholic saint but also Pedro de Valdivia, who reportedly founded the town in 1540.[42] The word founded, though, is a euphemism for appropriated, in that an Indian settlement at the same place or very nearby might be reoriented and given a more familiar Spanish identity. Although that name reveals the fusion of culture that occurred here, it further emphasizes the cultural and political, rather than physical or climatic, nature of the term Atacama in colonial times.

The name Atacama itself is fascinating. It is relatively rare to find a four-syllable name/word in which each syllable has much the same sound, in this case "ah". That "ah-ah-ah-ah" combination helps set up a rhythm that makes the name easy to remember, or rather hard to forget. The region's fog or camanchaca is another such word, and it too is Indigenous. For the record, the Indigenous peoples did not have a written language, and so the Spaniards wrote it using their own alphabet. The name Atacama is pronounced phonetically in Spanish as "Ah-tah-KAHM-ah," the accenting of the second-to-last syllable being typical for words ending in a vowel. That is straightforward enough, but there are subtleties that I have discovered while traveling there. By listening very carefully to the way that some Chileans in the far north pronounce this name, one can sometimes detect a slight holding of that second-to-last syllable just a little longer than normal, so that it may effectively sound a bit more like "Ah–tah-KAHHMM-ah." However, I am not sure whether that represents an ancient tradition of pronunciation dating to Indian or early Spanish colonial times, or whether it is a more recent way of pronouncing the name.

Moreover, listening very carefully to the way some Chileans pronounce it suggests a more Castilian sensibility, for one can detect a slight lisp that sounds more like "Ath-ah-KAHM-ah." It is always worth listening carefully to how locals pronounce words, for they sometimes reveal the legacy of places far distant. The Atacama is no exception to the rule that even an Indigenous name can be hybridized to reflect the language of the colonizers who originally learned it from the Native peoples, but who often inflect its pronunciation in subtle ways.

While on the subject of Indigenous names in colonial context, I would be remiss if I did not mention a port in Ecuador named "Atacames," which is associated with the Indigenous Atacames peoples who lived in that tropical region at the time of Pizarro's explorations aimed at conquering and colonizing Peru. Although I have found no connection between the two widely separated locales and peoples despite both being located in western South America—and both being at the northern and southern edges of Inca rule—perhaps further research might find a connection beyond sheer coincidence.

Names can last a very long time, and so can aspects of material culture, especially in arid areas such as the Atacama. With this in mind, I want to remind readers that the landscape found here by the Spaniards was a palimpsest in that some things were written over, so to speak, and replaced by

Spanish features such as churches and plazas. On the other hand, many features from Indian times are durably imprinted. Along the many trails they traversed, the early Spaniards often had Native guides to whom the landscape spoke in a very different tongue. Still, it did not take long for that landscape itself to reflect the different cultures that interacted syncretically, adopting things from each other despite the vast differences in their histories and cultures. Driving through this desert today can reveal traces of the past that are etched into, or piled on top of, the surface of the land close to the roads and highways. The desert north of Chile and southern Peru is formidable but not trackless, for almost everywhere there are markers, usually made out of stone but less frequently of wood, that sometimes give directions and always tell stories to those who know how to read them. The wood in these markers comes from God knows where, evidently carried here because there is none to be had for miles in any direction. By contrast, stone is found everywhere, except far out on the salars or salt lake beds. In design, these markers may vary widely. Some are miniature structures patterned after chapels or churches and are clearly linked to the predominant religion—Catholicism— of colonial Spain.

Many of these markers are shrines that commemorate the dead, and their presence is sometimes a stark reminder of the accidents that have befallen travelers since time immemorial. Along heavily traveled highways in the Atacama, many shrines date from fairly recent times, but they are part of an age-old tradition in which Catholicism has been deeply embedded into the soul of this region (Figure 2.13). Along the lesser-traveled roads, some shrines are in such remote locations that travelers are left to wonder who maintains them. Whereas these miniature structures may be common in many parts of Latin America, nowhere in that vast realm do they have the impact that they do in the Atacama Desert. Fewer spatial contrasts are greater than that experienced by drivers here, for one's attention is drawn back from the unimaginably wide-open vistas of this desert, with nothing but the road and vastness of an almost Martian-like landscape, back to a shrine with its dimly lit recesses featuring heaps of small objects such as statuettes, candles, and photographs. In fact, nowhere else on earth do I know of a greater contrast between exterior and interior spaces.

These monuments in the middle of nowhere have long impressed travelers. In 1930 *Fortune* magazine featured an article about miners in the Atacama Desert, which hinted at the community-oriented nature of these monuments despite the almost overwhelming emptiness of their surroundings. Although the article focused on the mining economy, an observation about folklore

Figure 2.13. As in many parts of Latin America, the roadside shrine is common in the Atacama Desert, the one here situated along a highway in a mining area southwest of Calama. (Photograph by author)

was also made using a photograph of a shrine standing in complete isolation. Under the photo, the term *"HIC IACET"* was placed. For readers unfamiliar with Latin, this phrase means "here lies," or "here is buried." However, this may not always be the case; as the caption further noted, "When a worker is killed at his task, the place is marked with a wooden cross by his friends. Though he may be buried miles away, the cross is decked with paper flowers, and at its foot burn candles in lanterns improvised from Gringo oil cans."[43] This commemoration of the individual is noteworthy, but it also subliminally links a place to visitation and hence suggests pilgrimage. Although deeply embedded in Latin American Catholic religion, there is some evidence that this tradition, and these artifacts, may also be traceable to earlier Indigenous culture. In his insightful study of the spiritual geography of another arid region, the Pimería Alta in southern Arizona and northern Sonora, anthropologist James Griffith observed that although these objects of commemoration are associated with the Catholicism of the Spaniards, the concept of erecting death markers may predate the European presence. Moreover, he suggested that these markers meet an increasingly universal need today, as they "are apparently being erected...by non-Catholics and even non-Christians."[44]

Regarding this subject, we can again call upon the observations of geographer Isaiah Bowman, who was especially impressed with such vernacular sign making along the roads and trails he traversed in the Atacama in the 1910s. As Bowman noted, some of these signs might be religious in nature, but others might also serve other purposes, including helping people navigate the desert. Some signs were recent, but others had been erected in the distant past. Bowman began his discussion by noting that trails were long-lasting in the desert. To make his point, he compared them to trails in wooded areas, often marked by cutting branches but which would soon vanish as vegetation grew back to encroach upon them. As Bowman observed, "While in a few years a trail in the forest may be choked and even forgotten, a trail in the desert remains a trail even if there is passage over it only at intervals of several years."

There was another reason for the continued existence of trails. According to Bowman, "in desert country it is the *signos del camino*, or signs of the way (trail markers one might call them), that are kept in repair." These may be "rough piles of stone or may be recessed chambers and even mortared structures or may be nothing more than little wooden crosses such as are used to mark the graves." Especially impressed by the longevity of roads and roadside features, Bowman added that "the Inca road through the Desert of Atacama is said to be traceable over many leagues." In a superb sentence that I vividly remember from when I first read *Desert Trails of Atacama* in the 1960s, Bowman concluded: "The fixed climate of desert and mountains, the open character of the country, the thinness of settlement, and the limited population which the region can support tend to keep the trails in fixed locations, and we may read their history from the earliest colonial times, if not earlier, down to the present."[45]

Remembrance was on my mind as my son Damien and I traversed the highways and backroads of the Atacama in March 2015. Safety seems on people's minds here as they drive, as when one approaches a slower-moving semi-truck (we never saw one speeding) and wishes to pass it. These trucks often have their left-hand turn signals on, and in the United States that would be a warning that they were planning to turn and you shouldn't pass. In the Atacama, however, we quickly learned that this blinking signal means it is *safe* to pass—the signal is a courtesy of the truck driver. This was a bit hard to get used to, but it quickly became much appreciated and made sense, given that

Figure 2.14. As evident in this photograph taken along the Pan-American Highway at the edge of the Central Depression about 200 kilometers southeast of Antofagasta, travelers in the Atacama Desert find it irresistible to use stones to emblazon messages on the landscape. (Photograph by author)

there are so few side roads crossing the highway anyway. I suspect that this type of communication also relieves some boredom, as many travelers are lulled by mile after mile of desert scenery.

In the Atacama, sleep rather than speed can be the danger—as evident in the profusion of stark white crosses and miniature chapels along the roadsides. Most of these apparently mark the scene where someone had perished in a highway accident. Others, though, may mark places in the now-desolate countryside that were once the sites of mining camps; in other words, they commemorate a place that was once living but has passed into memory. Just as often along the roads and highways, we encountered stones arranged in the form of written messages, some of them simple and some quite elaborate. As we stopped to decipher these messages in stone, we noted that some commemorated love, some loss, and some both (Figure 2.14). These rock messages are found along all of the roads in the Atacama, from the heavily traveled Pan American highway to the backroads connecting small villages. To us, they confirmed that the impulse to leave one's mark in the Atacama is not confined to the past. The vastness of the place, and its subtle landscapes, seems to invite such inscription.

One afternoon, heading south along the main Pan Am, we decided to take a backroad toward the coast that twisted through a labyrinthine

Figure 2.15. In a dry wash at the bottom of a small valley traversed by the lightly travelled road between Las Bombas and Caleta Pan de Azúcar, one encounters this veritable city of stone cairns. (Photograph by author)

landscape of arroyos and steep-sided mesas. This road was a classic secondary byway, showing on our modern maps as traversable and even on Bowman's century-old map as a dotted line. How could we resist? On this shortcut north of Chañaral, it was amazing to see how quickly civilization disappeared, the road being our only lifeline through the maze of multicolored gashes and burnt-looking hills. My goal in suggesting this road was to locate various species of cacti that might be found closer to the coast near Pan de Azúcar, but as always in the Atacama, we quickly became sidetracked by something that was even more interesting than what we originally set out to find. In my experience, letting the landscape set the agenda never fails to yield something noteworthy. As we rolled westward along this road toward a lowering sun, we entered a place where the valley widened and encountered a scene that amazed us. Here, someone—but more likely a number of people— had assembled piles of stone that resembled an ancient city in miniature (Figure 2.15). Each of these stone cairns was interesting, but as a *tout ensemble* their effect was breathtaking. Naturally, we stopped to give it a closer look.

After a few minutes of scrutinizing the site, we asked ourselves many questions about this surreal rock city but could not tell whether it was created in one frenzy of activity or over a long period of time. To our knowledge, this was not part of the Inca road system, and yet it too was well marked—though how recently, or distantly in the past, we could only guess. More disconcerting, we could not tell if it was the work of one individual or a communal effort. In any event, this "cairn-scape," as we called it, was stunning.

After we got back in the car, Damien drove as I pondered what we had just experienced. As we continued to drive down that unpaved road, a poem—actually a paean—called "Desert Monuments" came to me. It asks three questions, all of which may never have an answer.

DESERT MONUMENTS

Who erected those piles
of carefully placed stones
alongside the remote roads
in the enigmatic Atacama?

Were they hoping we'd remember
their works for the ages,
and grant them immortality?

Or secretly pleased that they made
another traveler forever wonder
who would expend so much energy
investing in anonymity?

Later on down that lonely road, we stopped to stretch our legs and take some photos of the desert landscape. It was an especially desolate location, with smooth sand on the valley floor and huge piles of multicolored rock strewn about. As we often do when hiking in interesting territory, Damien started off in one direction and I in another, setting out to see what we could find. Upon hiking up a small hill, I looked back and noticed Damien gathering stones and placing them on the ground. Fascinated, I returned to where he was working and noted that he was positioning individually selected stones into letters that formed a message (Figure 2.16). In this spot about six thousand miles (10,000 km) from home, Damien was thinking about his daughter, Maya, for today was her birthday; he was also thinking about his

Figure 2.16. Down the road from the stone cairns pictured in Figure 2.15, the author's son Damien got into the spirit of marking the desert with stones, creating this message for his daughter Maya and son Leo. (Photograph by Damien Francaviglia)

son, Leo. Damien's message was simple and symbolic, the large letters "M" and "L" situated above a stone heart. With the stone message completed, Damien took out his iPhone and shot a picture of it; later that evening from our hotel in Caldera, Damien sent the photo to his wife, Sarah. By the time he skyped with Sarah and the kids after dinner that evening, they had already seen the photo and loved it. Raised on such technology, Damien took this instant communication in stride, but I kept thinking what it would have been like to travel those same roads centuries ago, not seeing one's family for weeks, even months, or perhaps even years at a time.

The stone monuments just described may have been made in the recent past, but much the same things can be pondered about the fabulous geoglyphs seen throughout the Atacama Desert of Chile and well northward into Peru. A geoglyph (literally "earth carving") uses nonverbal symbols to convey information or a message at a fairly large scale, usually at least several meters. This part of South America contains some of the most impressive geoglyphs on earth. Some are so large they can best be comprehended from a distance, or from the air. Geoglyphs may be made in a number of ways, as long as the technique chosen increases the contrast between the rocky background and the subject depicted—which may be geometric designs or more recognizable animals and human figures. Given the presence of desert varnish across broad areas, one common technique involves removing darker stones to reveal lighter ground underneath. The striking *Atacama Giant* at Cerro Unitas near Pozo Almonte is said to be the largest known human form in the world, measuring 119 meters (ca. 300 ft) in height. In other places, one can find designs that were created by chipping away the veneer-like desert varnish on rocks to expose lighter-colored stone underneath.

In his travels through the Atacama more than a century ago, Isaiah Bowman commented on petroglyphs featuring chinchillas and llamas. In most such rock art, the artistic medium is the lithosphere itself. However, pictographs can also be made by painting pigments onto rock surfaces. According to Chilean archaeologist Luis Briones, the rock art created in the Atacama Desert appears to date from about 1000 to 1450 CE and is a manifestation of Andean culture. Chilean archeologist Gonzalo Pimental noted that these geoglyphs may have served as route markers or signposts.[46] Geoglyphs are not freestanding sculptures but rather meant to be read as if one were looking down upon them. Given their size—some of them are thousands of meters long—imaginative observers speculate that they were meant to be viewed from outer space. Rather than such, they could be devices that indicate direction to water sources, or indicate best directions of travel. After all, humankind has commonly mapped places planimetrically (as if looking straight down on them) for thousands of years, and by so doing enabled other earthbound users to comprehend how these places were configured. With our road maps in hand, we traverse the Atacama Desert today much as we do other places, never thinking for even a moment that modern road mapmakers were motivated by, or their products made for, extraterrestrial travelers.

The interior of the Atacama Desert is serene and kaleidoscopic, and its roads and trails remind us that it has been traversed continually for thousands of years before the Spaniards arrived. These people knew and related to the interior desert between the coast and the Andes in ways that we will never be able to comprehend. Although the first Spaniards arrived here by land, they soon realized that accessing it by sea might be the more efficient way to penetrate the interior. As the proliferation of place names on Pedro de Cieza's 1560 map suggests, the coast was to become the major gateway through which people would experience it. Eclipsing the modest seagoing of the Indigenous people, the Spaniards' large sailing ships had now set the stage. Thus a pattern was well established in later colonial times; for the most part, access to and knowledge about the Atacama Desert was maritime. Lying between the coastal ports and the sources of riches and products that originated far inland in places such as Potosí (Bolivia) and Salta (Argentina), the extremely arid area gained a reputation as being difficult to traverse, not because of topography, but because of scarce water supplies. Early sources also note that those ports were places where small streams manage to reach the coast. Though meager, there was enough freshwater to be collected here to serve passing ships. The English, French, and others soon realized that anyone who could control the ports could control the country inland.

By the mid to late 1600s, a type of seaborne activity that would vex Spain was being developed into an art. This was called buccaneering, and its practitioners, buccaneers. Derived from the French term, which in turn originated in the Arawak language of the Caribbean, a buccaneer was someone who defied Spain's claims to areas in the New World. Although originating in the French Caribbean around 1630, buccaneering soon spread to the English and became a mode of operation along the Pacific shores of South America and Mexico. The buccaneers used force to either seize and capture vessels, and in the process make off with treasure, or even seize particular places, which they in turn used as bases of operations. About a century ago, maritime historian Sir Albert Gray astutely observed that buccaneering was a "polite West Indian synonym for piracy."[47] As such enterprises go, buccaneering was liminal, that is, in a vaguely defined zone between rugged individualism and established or sanctioned nation-building activity. Although technically unofficial, buccaneering served the interests of countries like England and France who preferred to sponsor effective raiding rather than conducting outright warfare, which was far more expensive. Buccaneering thus created a new breed of sailor, one who raided on behalf of both Crown and self.

The most tangible seventeenth-century British intrusion into what we today call the Atacama Desert was documented by one such sailor, William Dampier (1650–1715), who traversed this coastal desert region on several voyages into the Pacific Ocean. Whereas most buccaneering was well known by word of mouth, Dampier raised it into the public and literary consciousness by writing and publishing a book about it. Entitled *A New Voyage Round the World* (1697), it became a sensation and opened Europe's eyes to the riches of the entire Pacific Ocean south of the coast of Mexico, along the west coast of South America, and thence westward to Southeast Asia and Australia. For our purposes here, *A New Voyage* also contains numerous references to the desolate area along the coast of Peru and Chile, which Britain raided from time to time. Even at this early date, ports such as Arica and Coquimbo/La Serena were prominently mentioned, usually in relation to their respective position as harbors where goods from the Andes, or metals and minerals like sulfur from the adjacent desert, were loaded aboard ships for transport to Europe.

On the one hand, Dampier can be viewed as part of a cadre of swashbuckling men who terrorized the desert coast of southern Peru and northern Chile in the early 1680s. From Dampier's writings as well as those of others who also published recollections, we know that their ventures were simultaneously calculated and yet wildly opportunistic. Among those involved were John Watling and Bartholomew Sharp, whose names are well known in the annals of pirates and brigands. On the other hand, Dampier was different and rises head and shoulders above his shipmates. Rather more sensitive than most of the crew, Dampier wrote from a perspective that engaged him in the action but placed him safely enough at a distance to escape outright condemnation for its consequences. In other words, Dampier was both opportunistic and yet moral—a combination that may seem strange but is best understood in the context of his time. As Dampier and the buccaneers traversed the South American coast south of Guayaquil (Ecuador) to the island of John (Juan) Fernando, which lies far west of Valparaíso, they fixated on taking the venerable port city of Arica, much as Drake had done about a century earlier. As Dampier euphemistically put it, from Juan Fernando Island "we went back again to the Northward, having a design upon *Arica*, a strong Town advantageously located in the hollow of the Elbow, or bending of the *Peruvian Coast*."[48]

It is noteworthy that Dampier's passage is both narrative and graphic—which is to say cartographic—in that he helps the reader visualize Arica's geographic position on a map. To conduct buccaneering along the western coast of South America, maps were in fact crucial. As noted by Dampier's biographers Diana and Michael Preston, Captain Bartholomew Sharp would shortly thereafter capture the *Rosario*, a Spanish ship that was carrying an unexpected bonanza, identified as "a complete set of maps of the Pacific coast of South America," which Sharp brought back to England. As might be expected, the Spanish ambassador was indignant and demanded that Sharp be tried for piracy, but King Charles II pardoned him; moreover, Sharp was given a commission in the Royal Navy, though he soon returned to piracy.[49] Maps were clearly part of high intrigue, but now they also caught the imagination of an increasingly literate public. As might be expected, *A Map of the World. Shewing the Route of Wm. Dampier's Voyage Round It. From 1679 to 1691* accompanied editions of Dampier's book. The publication of *A New Voyage* actually began a revolution in such maps, and careers such as that of mapmaker Herman Moll were in part spawned by the public's increasing appetite for geographic information.

But to return to the tempting desert port of Arica, Dampier later described the trouble that taking this small but strategically important city presented. The Spanish not only captured some of the crew but apparently also gained some important information about their strategy. As Dampier, who stayed on board and thus out of the fray, recollected: "Had we entered the Port upon the false Signal, we must [that is, would] have been taken or sunk; for we must have passed close by the Fort, and could have had no Wind to bring us out, till the Land-Wind should rise in the Night."[50] This passage provides a clue to Dampier's powers of observation that transcended the typical buccaneer and placed him among scientists. Throughout his numerous voyages in the Pacific, Dampier compiled observations about both the winds and ocean currents, and he is widely recognized as the forerunner to today's oceanographers.

As Dampier repeatedly noted, the prevailing ocean currents and winds off the west coast of South America are from southwest to northeast. These winds, though strong, are rarely as powerful or changeable as the tempests that blow in other latitudes. Dampier makes frequent mention of the consequences of both currents and winds on navigation. He also masterfully described atmospheric phenomena, including clouds of various types. In April 1684, while sailing northward from Juan Fernando Island "along the Pacifick-Sea, properly so called" by mapmakers, Dampier commented that "there are no dark rainy clouds" such as those seen in other latitudes, though he quickly added that there was "often a thick Horizon, so as to hinder an Observation of the Sun with the Quadrant; and in the Morning hazy Weather frequently, and thick Mists, but scarce able to wet one."[51]

Several other passages in *A New Voyage* are noteworthy in reference to the natural history of the Atacama Desert. Of the topography seen from sea at from about 24 degrees south, sailing northward, Dampier noted "the Land…is of a most prodigious Heighth. It lies generally in Ridges parallel to the Shore, and 3 or 4 Ridges one with another, each surpassing [the] other in height; and those that are farthest within Land are much higher than others." Readers who recall my physiographic block diagram (see again Figure 1.14) can see that Dampier beautifully captured in words the stair-step character of a vast swath of territory that rises higher and higher the farther one travels inland from the sea. However, in this area the Andes lie even closer to the coast than shown in the block diagram. He then added a sentence that perfectly linked the topography and climate of these elevated ranges far inland from the camanchaca fogs: "They always appear blue when seen at Sea; sometimes they are obscured with Clouds, but not so often as the high Lands in other parts of the World, for here are seldom or never any Rains on these Hills, any more than in the Sea near it; neither are they subject to Fogs." Of these towering Andes Mountains, Dampier commented, "These are the highest Mountains that I ever saw." He rightly noted that "by all likelihood these Ridges and Mountains do run in a continued Chain from one end of Chili and Peru to the other, all along this South-Sea Coast, called usually the *Andes* or *Sierra Nuevada* [sic] *des Andes*."

Of the hydrology, here Dampier speculated: "The excessive Height of these mountains may possibly be the reason that there are no Rivers of note that fall into these Seas." As a mariner used to sailing into the channels of large rivers, Dampier noted no such possibilities here. Of the few river courses that reached the sea, all had channels that were "too little and shallow to be navigable." Most of these river channels were at least seasonally, some perennially, dry. Of the former, Dampier mentions the Ylo which is "dry at certain Seasons of the Year." In this passage, Dampier was likely referring to the Ilo River (called Osmore on current maps), which reaches the coast at Ilo, Peru. That small port city was often called or written as "Hilo" by the buccaneers, who sometimes used it to resupply as they attempted to take the prized Arica, which lies about sixty miles (100 km) to the south. Of special interest here is Dampier's desire to characterize the river's flow accurately. He noted that he had experienced this river's seasonality of flow twice, that is, on two separate voyages, and concluded from those observations that such was normal. Still, he added some corroboration, noting that "I . . . have been informed by the Spaniards that other Rivers on this Coast are of the like Nature, being rather Torrents or Land-floods caused by their rains at certain Seasons far within land, [rather] than perennial Streams."[52]

To Dampier's considerable skill as an observer of ocean currents, winds, and hydrology, we may also confidently count him among the earliest to record the general characteristics of the Atacama Desert's cultural ecology. In an insightful paragraph, he noted that "the Building in this Country of Peru is much alike on all the Sea-Coast." Of what we now call the regional vernacular architecture along this arid coast, he observed that "the Walls are built of Brick, made with Earth and Straw kneaded together." In a classic description of adobe construction, Dampier observed that the bricks "are about three Foot long, two Foot broad, and a Foot and a half thick: They never burn them, but lay them a long time in the Sun to dry before they are used in building." In keeping with the dry climate, Dampier noted, "in some places they have no Roofs, only Poles laid a-cross from the side Walls and covered with Matts." Dampier concluded that "The Houses in general, all over this Kingdom, are but meanly built, one chief reason, with the common People especially, is the want of Materials to build withal; for however it be more within Land, yet there is neither Stone nor Timber to build with, nor any Materials but such Brick as I have described; and even the Stone which they have in some places is so brittle, that you may rub it into sand with your fingers." This highly friable, powdery sandstone is commonly encountered along the coast, but in many places building stone can indeed be found. Likely, though, it was easier to make sun-dried bricks rather than quarry stone and haul it to the building sites.

Dampier added an important observation that linked architecture and environment. "Another reason why they build so meanly is, because it never Rains; therefore they only endeavor to fence themselves from the Sun." He attributed the antiquity of such structures to the fact that "their Walls, which are built of an ordinary sort of Brick, in comparison with what is made in other parts of the World, continue a long time as firm as when first made, having neither Winds nor Rains, to rot, moulder, or shake them." His expression "for a long time" is interesting, for these buildings made of adobe brick (a term Dampier did not use) derived from the Old World. The Indians used what has been called "puddled" adobe, in which mud was plopped on top of

Figure 2.17. A view of the small desert coastal settlement of Paposo, from Rudolph Philippi's *Reise durch die Wueste Atacama* (1860). (American Geographical Society Library, University of Wisconsin at Milwaukee)

mud that had been placed earlier, and then left to dry. With each successive addition of this mud, the walls slowly rose. By contrast, adobe bricks represented an introduced form of technology that had arrived on these shores with the first Spaniards. Hence, at that time, they were likely less than about 150 years old, though they may have seemed ancient indeed.

In what appears to be the first detailed and comprehensive description of this desert region that the general public encountered, Dampier noted that "This dry Country commences to the Northward, from about *Cape Blanco* to *Coquimbo*, in about 30 d[egrees] S. having no Rain that I could ever observe or hear of; nor any green thing growing in the Mountains: neither yet in the Valleys, except here and there water'd with a few small Rivers dispers'd up and down."[53] So destitute of timber was this desert region that Dampier observed that wood for construction had to be shipped in from far to the north or south. Dampier also noted that the local people were dependent on the sea for much of their sustenance. In his time, small sea-oriented villages huddled where nature provided a sheltered harbor. In places such as Paposo, human habitation was simple indeed, as it was about a century and a half later when other European explorers arrived to scrutinize the area (Figure 2.17). Between these small fishing villages for league after league, the stunning but virtually deserted coast of Chile's desert north impressed Dampier and

Figure 2.18. Today, portions of the Chilean desert coast appear much as they did hundreds of years ago—virtually uninhabited, serene, and seemingly timeless. (Photograph by author)

the English interlopers three centuries ago as it did later explorers. For that matter, much the same can be said of the region today: for mile after mile one can experience this splendid isolation and, in its small fishing villages, a continued dependence on the resources of the sea (Figure 2.18).

Although Dampier's account of this remote area was both informative and comprehensive, nowhere in it does he use the term Atacama Desert. In his book, and in his mind, the region remained nameless—a nominally bone-dry part of the wild coast of southern Peru and northern Chile. Nevertheless, Dampier helped open Europe's eyes to the rugged coastal zone of what would later be called the Atacama Desert. With Dampier's vivid descriptions in one hand and a map in the other, it was a short step for readers to connect the place name Atacama (which as noted was venerable indeed, as it referred to established places of Indian origin) with the evocative "desart" place that he described in such vivid detail. One of the keys here is authority. By the time Dampier published his book, in fact, he had made the transition from buccaneer to natural scientist, joining the Royal Society (founded in 1662) and dedicating his book to its director. By fusing authoritative geographic narrative descriptions with the rapidly developing maps showing such voyages, the world—and the world's deserts—would soon, but not quite yet, be branded with authoritative names.

In addition to William Dampier, many others wrote about these piratical exploring voyages into the South Sea, and some authors were far more explicit about the mayhem inflicted by the buccaneers. Most enduring is the writing of Alexandre Exquemelin (1645–1707), also known as John Esquemeling, whose riveting 1684 accounts of buccaneering in the Americas have been translated into many languages. These sometimes-lurid writings included detailed accounts of the December 1680 sacking of Coquimbo, in which the town was spitefully reduced to ashes. The English buccaneers justified burning down the entire town because someone had tried to set fire to their ship in the harbor. This was, as historians delicately put it, "an age when cruelty was commonplace ashore and at sea."[54]

Although Dampier's writings mention some of these events, he took the high ground, distancing himself from the meaner or more sordid of them. Through such distancing, Dampier's life represented a remarkable transition in which a buccaneer found a higher, which is to say more respected or respectable, calling—namely, as a naturalist. Nevertheless, at this time virtually all references to natural history related to the value that such resources might present to England. This, after all, was the early dawn of natural science as we know it, but Dampier undeniably helped usher in the transition from exploitation to observation. For the record, though, it is worth noting here that the Dampier who explored the Chilean coast in search of booty was not yet the respected naturalist that he would later become.

There is yet another reason, however, why Dampier said and wrote little about the nature of the desert here. At this time the European mindset still regarded deserts as forbidding, unredeemable places of virtually no value. The fairer lands were the greener lands, those where meadows, fields, and forests abounded. Not so coincidentally, Europe itself was the model of such a fair land. In the desert, there seemed to be little or nothing of value outside of possible mineral wealth. Thus, despite considerable reconnoitering along the coast of southern Peru and northern Chile, Dampier seemed to have overlooked much of what we today call natural history. It is noteworthy, though, that Dampier began to inquire more deeply about those aspects of the environment after he reached a cluster of brooding volcanic islands farther to the north and west off the coast of Ecuador. These islands were called the Galapagos, and they would transform the history of the natural sciences in the coming century. As Dampier thereafter experienced the abundant flora and fauna of the tropical world in the Pacific Islands and southeast Asia, his writings quickly filled with natural history references. There is irony here, for those are the lands—the real Indies, as it were—that Columbus originally sought, though never reached.

Still, it is important to give Dampier rightful credit for being one of the first non-Spanish observers to seriously and accurately describe what he had encountered along the arid coast of South America. Placed in context, Dampier's descriptions of ocean currents, winds, coastlines, and the like helped to initiate the long process of scientific articulation. His writings formed the link that would ultimately position this part of South America on the broader collective map, and in the growing imagination. As James Kelley correctly observed, "Dampier's likely influence on the history of imaginative literature awaits proper study. What is certain, however, is that the *New Voyage* precipitated the phenomenal popularity of voyages and travels in the eighteenth century."[55] Another important point to recall here is that the English were not alone in creating increasingly accurate travel narratives. The French were also exploring these same shores, and even working their way onshore, at this time. Moreover, both the Age of Exploration and the Age of the Conquistadors were ending, and a new era based on objective scientific observation was beginning. The scene was thus set for the instantiation of the Atacama Desert into the popular mind in Europe, an act of creation that depended on the amalgam of two elements: the authority of science, and the allure of romance.

Three

"THE GREAT
DESERT OF ATACAMA"

The Dawn of Science, ca. 1700–1865

By around 1700, climate and characteristics of the landscape began to be factored into depictions of western South America, much as they were in other places still being explored by Europeans. In regard to the Atacama at this time, historian of South American exploration Edward Goodman lamented that "maps of the desert were poor, and information on it was scattered and often contradictory."[1] Of course, much the same could be said of all the world's deserts, which were little explored, but in the eighteenth century that would begin to change. Maps during this period are especially revealing for what they convey about the Atacama region, for they began as fairly vague but became increasingly more specific as scientific knowledge developed.

Noteworthy among these maps is Guillaume Delisle's 1703 *Carte du Paraguay, du Chili*, which comprises the southern half of South America and represents a cartographic watershed in that it boldly delineates the "Desert d' Atacama" (Figure 3.1). This map also depicts a mountain named Atacama as well as uses the name in bolder letters to indicate either the port city below it or perhaps to demarcate the province of the same name. Delisle's map bears a date of 1703, but some sources including David Rumsey's website list its actual publication date as 1708. In either case, it appears to be the first popular map use of the name Atacama to signify a desert, that is, an area having an arid climate and devoid of vegetation, rather than a political jurisdiction. Its cartographer, Guillaume Delisle (1675–1726), was the son of the noted French intellectual Claude Delisle, who possessed a solid background in geography and history. As the expression goes, the apple did not fall far from the tree.

Guillaume Delisle became recognized as one of Europe's most respected mapmakers. Although he never traveled to most of the far-flung places that he depicted, Delisle sought the most accurate information possible, and his maps set a new standard.[2] Delisle's seminal map showing these features in the Atacama confirms that other European powers were aware of aspects of the geography of New Spain, and their interest was more than scientific.

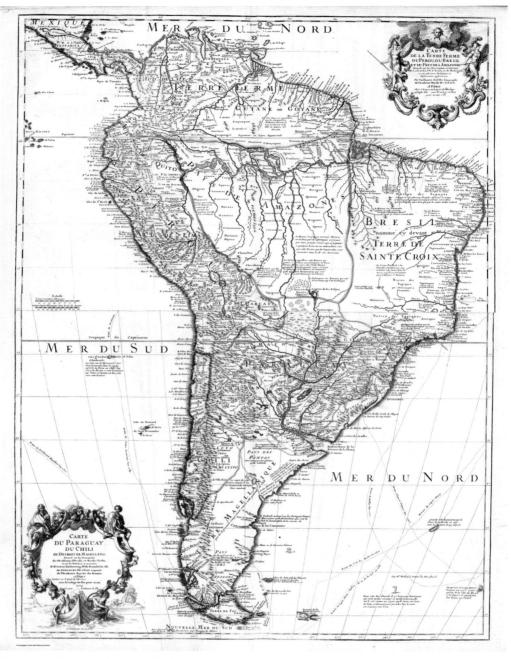

Figure 3.1. The southern half of this map of South America by French mapmaker Guillaume Delisle is entitled *Carte du Paraguay, du Chili* (1703/1708); it is among the earliest cartographic representations of the "Desert d' Atacama" (Atacama Desert) and also shows a port and mountain named Atacama. See also Figure 3.1 detail. (David Rumsey Map Collection)

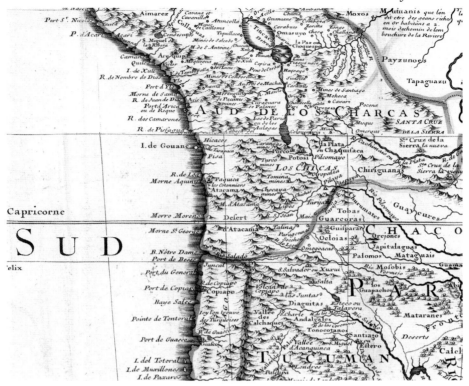

Figure 3.1 detail showing "Desert d'Atacama" straddling the Tropic of Capricorn.

In addition to trusted secondhand sources, some of this information was provided by expeditions aimed at plundering treasure. France took a strong interest in this region, as evident in a rendering of Arica in 1709 by Louis Feuillée on orders of the King of France.[3] The Pacific coast of South America was strategically important, primarily because it was the maritime key to inland treasure and wealth.

As knowledge was dispersing rapidly in Europe with the advent of printing and learned societies, the French public quickly discovered William Dampier's accounts of British exploits in the fabled South Sea. In fact, Dampier was prominently cited by French explorer-naturalist Amédée-François Frézier (1682–1773) in his book *A Voyage to the South-Sea, and along the Coasts of Chili and Peru in the Years 1712, 1713, and 1714*. Given Frézier's background as a naval engineer, it is no surprise that his book contains numerous maps of the ports that he explored at the behest of the French navy. Although Frézier is best known for his discovery of the Chilean strawberry, which he imported back to France, his book was soon translated into English

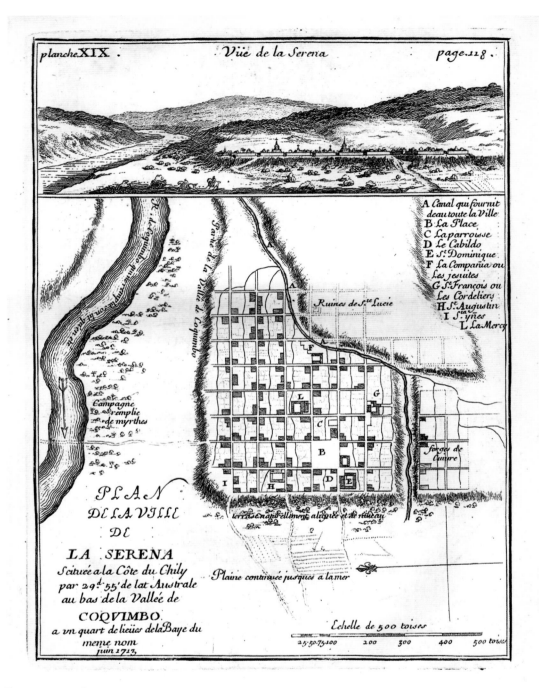

Figure 3.2. Reflecting France's interest in the desert regions of South America, Amédée Frézier's map of La Serena and environs was published in his popular book *A Voyage to the South-Sea and along the Coasts of Chili and Peru in the Years 1712, 1713, and 1714* (1717). (American Geographical Society Library, University of Wisconsin at Milwaukee)

and underscored how closely linked military and scientific exploration had become.[4] Like most Europeans, Frézier reached Chile by rounding Cape Horn at the southern tip of South America. After being buffeted by storms, most travelers took solace in putting in at Valparaíso and visiting the capital of Santiago. Frézier's goal, though, was to explore the entire coastline of western South America, so he headed northward toward the tropics.

Of special interest here are Frézier's experiences in the desert region of Chile and Peru, which he visited by stopping in numerous puertos intermedios (intermediate ports), as the Spaniards called them. Frézier's maps of the ports along the desert coast of Peru (e.g., Arica) and Chile (e.g., Coquimbo) represent the best of European mapmaking: accurate and beautiful, they are testimonials to Frézier's observational skills and his ability to glean information from other reports. His narratives are also insightful. Like Frézier, I shall begin with his descriptions in the vicinity of Coquimbo, and work northward. Upon reaching that rather wide bay, Frézier drafted a fine chart of it. While his ship was anchored, he visited the nearby town of La Serena, a place whose name means serenity but which had seen more than its share of mayhem in the form of Spanish brutality, Indian counterinsurgency, and subsequent piracy by the British. Frézier's map of La Serena shows that small city in planimetric view and identifies its important landmarks. As part of this map, Frézier also provides an eastward-oriented vista that shows not only La Serena and its scrub-dotted site, but also urges the viewer to gaze up the beautiful Elqui Valley, a verdant and lush corridor that reaches into the parched interior of the southern Atacama Desert (Figure 3.2).

Upon leaving Coquimbo, Frézier's ship attempted to sail northward but was unable to leave the harbor for almost a week because the wind had died down to nothing. After the wind finally rose again, and the ship got underway, he reached the Bay of Guasco (today's Huasco), which was little frequented "because there is no other Trade but that of a private Person, who takes Copper out of the mines." Frézier's frequent mention of copper is noteworthy, for France's appetite for useful metals was increasing, as was its gaze toward foreign sources. Continuing north, Frézier describes the rugged coastline and the place where the "River of *Copiapo*" reaches the sea. This was called Puerto del Ingles (Port of the English), named "because a Pyrate of that Nation was the first that anchor'd there," and is still known as Bahia Inglesa (English Bay) today. Frézier next coasted around the point and into Port de la Caldera. Finding easy anchorage, "There we unladed a little Corn for the Town of *Copiapo*, and laded Sulphur, which we found upon the Shore, where it had been laid against our coming."

Frézier here documented a trade pattern that continues to this day: food was shipped in, and minerals shipped out. This port at Caldera "is the nearest to *Copiapo*, but little frequented, because it affords no Conveniency: Wood is very scarce there, and they must go five or six Leagues up the Vale, thro' which the River runs, to get it." Clearly unimpressed with Caldera, Frézier recorded, "The Watering-Place is bad; it is taken in a Hollow 50 Paces from the Edge of the Road, where a little brackish Water meets." Caldera was, in a word, forlorn. As Frézier noted, "There is no Dwelling about it, but only a Fisherman's Cottage, at the Bottom of the N.E. Creek." In an interesting note about natural history, and a reference to earlier English observations, Frézier added, "All the Shore of *La Caldera* is covered with Shells, especially that sort they call *Locos*, so that Dampier is in the wrong to say, that there is no Shell-Fish along that Coast."[5]

Inland from La Caldera, Frézier next described the mining town of Copiapó in considerable detail. Although Spaniards had imparted some spatial order to this area as they colonized it, conditions had changed shortly before Frézier's visit. He noted, "*Copiapo* is an open Town, the Houses whereof do not stand in any Order, but scattering up and down." This, to my knowledge, is the earliest candid reference to the fact that rapidly growing mining areas in the Americas were subject to haphazard urban development that trumped careful planning. In booming mining areas such as Copiapó, time was of the essence, and development frenzied. Frézier observed that "the Gold Mines that have been discover'd there within six Years past, have drawn some people thither; so that at present there may be above 900 Souls." According to Frézier, the increase in the number of gold-seeking Spaniards resulted in the taking of both land and horses from the "poor *Indians*." At the time of his visit, the mines were located above the town as well as two or three leagues distant. In 1713, there were six trapiches (grinding mills) at which ore was crushed. In addition to gold mines, there were in the vicinity of Copiapó "Mines of Iron, Brass, Tin, and Lead, which they do not think fit to work: There is also much Load-stone, and Lapis Lazuli, which the People of the Country do not know to be of any Value."

Nearby, too, were many lead mines. In an early reference to nitrates and other mineral salts, Frézier noted that "In short, all the Country is there full of Mines of Sal Gemm, for which reason sweet Water is there very scarce: Salt-peter is no less plentiful, being found in the Vales an Inch thick on the Ground." Moreover, according to Frézier, there was additional mineral wealth of interest farther from town. In the chain of mountains southeast of Copiapó rather generically named "La Cordillera," Frézier described a mine containing "the finest Sulphur that can be seen: It is taken pure from a Vein two Foot wide, without needing to be cleans'd." Descriptions such as these piqued the curiosity of European entrepreneurs who envisioned benefitting from such rich, and seemingly underexploited, resources.[6]

Natural history was also on Frézier's mind, especially in relation to how useful it might prove. He described a "Sort of Rosin, coming from a Shrub, the Leaf whereof is like Rosemary; it proceeds from the Branches and from the Berries, which they cast into Cakes two Foot Long, and ten or twelve Inches thick." This resin was dry, though, and only good enough to use as a substitute for glazing on the earthen jars in which wine and brandy were stored. Nevertheless, although this resin had some monetary value, he concluded, "In other respects the Country is barren, scarce yielding enough to subsist [i.e., support] the Inhabitants, who have their Provisions from about *Coquimbo*." He was, though, impressed by the abundance of guanacos in the nearby mountains. These, he noted, were "a Sort of Creature between a Camel and a wild Goat." In a fascinating reference to the medicinal value of things encountered on journeys of discovery at this time, Frézier noted that it was in the bodies of these guanacos that "the *Bezoar* Stones are found; formerly of such value in Physick, that they were worth their Weight in Silver; but now it has been found out (i.e., learned) that Crab's Eyes and other *Alkalis* can serve instead of them; they have lost much of their Value in *France*; however the *Spaniards* give great rates for them still." These stones obtained from the gastrointestinal tracts of animals were thought to be antidotes for poison.

Frézier's description of the landscape is of special interest. He characterized the country between Copiapó and Coquimbo, which he estimated to be one hundred leagues, as having "no Town or Village, but only three or four Farms." Even today the landscape along the Pan-American Highway is mostly barren, though Frézier appears to have overestimated the distance. A league is normally considered to be about three miles (ca. 5 km), which is the distance a person could walk in an hour, but in reality these two cities are closer to two hundred miles (ca. 330 km) apart. Moreover, the town of Vallenar lies about halfway between them. On the other hand, in Frézier's time the road was likely far more circuitous. In the other direction from Copiapó, northward to Atacama, which he notes as being in Peru, the distance is perhaps double. Of this long route, Frézier memorably wrote that "the Country is so hideous and desart [*sic*], that the Mules starve for want of Grass and Water."

Here Frézier may have underestimated the distance, but his point is well taken. This is a very arid stretch, almost devoid of vegetation except in a few of the watercourses. Frézier noted that "in eighty Leagues Length there is but one River," and that it was a strange river indeed, in that it "runs from Sun-rising till it sets, perhaps because that Planet [i.e., the sun] melts the snow, which freezes again at Night." Several small rivers north of Copiapó do indeed flow intermittently from the mountain ranges that rise to the east, and this peculiar river could be any of these. Frézier recorded that "the Indians call it *Anchallulac*, that is, the Hypocrite." He next mentioned "the dreadful Mountains which divide *Chili* and *Peru*, where the Cold is sometimes so excessive, that Men are frozen up, their Faces looking as if they laugh'd; whence, according to some Historians, the Name of Chili is derived, signify-ing Cold; tho' beyond those Mountains the Country is very temperate." Of these formidable mountains, Frézier observed that according to "the History of the Conquest of *Chili*…some of the first Spaniards who pass'd it died there, sitting upright on their Mules." With these rugged, almost inhospita-ble interior deserts and towering mountains in mind, Frézier concluded: "A much better Way has been now found out along the Sea-Coast."

With the sulfur cargo loaded aboard his ship, Frézier continued sail-ing northward to Cobija, which he described as "the Port to the Town of Atacama." From the latitude given in his book (22 degrees, 25 minutes south), this appears to be the port north of present-day Antofagasta and close to Tocopilla in the vicinity of Punta Cobija on modern-day maps. About eighty miles (120 km) due east of here lies the inland city of Calama, and farther eastward beyond that the high Andes. Given the fact that the Rio Loa flows from the Andes, and thence to Calama, this may well be the area that Frézier called "Atacama." In a subsequent paragraph, in fact, he noted that "From *Calama* to *Chionchion*, or the Lower *Atacama*, six Leagues; being a Village of eight or ten *Indians*, 17 Leagues South from the *Upper Atacama*, where the Corregidor, or chief Magistrate, of *Cobija* resides."

I mention this because when the name "Atacama" was used in centu-ries past, it could have referred to any one (or more) of a number of places. Frézier's use of the name Peru for the location of this area was understand-able; even though the borders between provinces were indistinct, many maps also locate these desert communities in that province north of Chile. What distinguished Cobija to Frézier was a landmark called "*Morro Moreno*, or the Brown Head-land, which is ten leagues to the windward." Frézier's ship didn't even stop there, but he felt obliged to mention and even describe it. Why? The answer becomes clear. "Tho' we did not put in there, I will not omit inserting what I have been told by the *French* who have anchor'd there: They say it is only a little Creek, a third part of a League in Depth, where there is little shelter against the South and S. W. Winds, which are most usual on the Coast." Cobija was not easy to get a ship into. As Frézier warned, "They who will go ashore, must do it among Rocks which form a small Channel towards the South, being the only one where boats can come in without Danger."

Moreover, Cobija consisted of only "about fifty Houses of *Indians*, made of Sea Wolves [i.e., sea lions], or Seal's Skins," with relatively poor-quality water and only a few palm and fig trees that served as landmarks to those seeking a place to anchor. Frézier then added a startling sentence that may come as a surprise to those who think Spain was in complete control here. "This Port being destitute of all Things, it has never been frequented by any but the *French*; who to draw the Merchants to them, have sought the nearest Places to the Mines, and the most remote from the King's Officers, to facilitate the Trade, and the Transporting of Plate and Commodities." The word "Plate" here meant plata, or silver, in Spanish. Though otherwise a poor port, Cobija was "the nearest to *Lipes* and *Potosi*, which is nevertheless above a hundred Leagues distant, through a Desart Country." Frézier next outlined the routes through the Atacama villages well inland, through the desert and into the Andes to the highlands where silver abounded—routes along which wood and water was exceedingly scarce, and travel highly dangerous.[7] With Frézier's revealing sentence in mind, it is easy to see why the word "spy" was part of his confidential résumé, along with naval officer and naturalist. We can also see why the Atacama was crisscrossed by trails, many of them marked by graves, some no doubt for Frézier's compatriots who had ventured here.

Continuing northward along the coast, Frézier reached Iquique, where a large island loomed just offshore. He noted that "The Island of Iquique is also inhabited by *Indians* and Blacks, who are there employ'd to gather *Guana*, being a yellowish Earth thought to be the Dung of Birds." Frézier observed that some doubted this claim because "it is hard to conceive, how so great a Quantity of it could be gather'd there; for during the Space of a hundred Years past, they have laden ten or twelve Ships every year with it, to manure the Land.... and that besides what is carry'd away by the Sea, they load abundance of Mules with it for the Vines and plow'd Lands of Tarapaca, Pica, and other Neighbouring Places; which makes some believe, that it is a peculiar Sort of Earth." Being an astute observer, Frézier weighed in on the side of the birds. "For my part, I am not of that Opinion, for it may be said without romancing, that the Air is sometimes darken'd with them."[8]

Frézier's words were on my mind as my son Damien and I explored the rocky coastline and he was drawn, camera in hand, to the immense flocks of birds that cluttered the sky. Some plunged into the water to gorge on the seemingly endless supply of fish, only to light on rocks and add to the white coating encrusting everything in sight. As described by many earlier travelers, including Frézier, we found the awesome sight of these birds tempered by the strong odor of their droppings, which covered outcrops and jetties along the arid coastline. Here one stepped, and looked skyward, very carefully.

Frézier's arrival in Arica underscores the importance of that desert port city's landmark rock as well as more recent human-made landmarks. As he observed, "Entering the Road of *Arica*, Ships may coast the Island of *Guano*, which is at the Foot of the Head-land within a Cable's length, and go and anchor N. and by E. of that Island, and N.W. from the Steeple of *San Juan de Dios*, distinguishable by its Height, from all the Buildings in the Town." Ever aware of strategic location and fortification, and reflecting both his skills as spy and naval officer, Frézier commented on how the port was defended. "To obstruct the Landing of Enemies in that Place, the Spaniards had made Entrenchments of unburnt Bricks, and a Battery in the form of a little Fort, which flanks the three Creeks; but is built after a wretched Manner, and is now quite falling into Ruin, so that the said Village deserves nothing less than the Name of a strong Place given it by Dampier, because he was repuls'd there in 1680."

Frézier reminds his readers that the English buccaneers then landed at the Creek of Chacota and marched northward to plunder Arica in a surprise attack. According to Frézier, "Those Ravages, and the Earthquakes which are frequent there, have at last ruin'd that Town, which is at present no other than a Village of about 150 Families, most of them Blacks, Mulattoes and *Indians*, and but few Whites." One event, an earthquake on November 26, 1605, triggered a tsunami that bore down on a portion of the town, leaving large areas in ruin. Wisely, the residents chose not to rebuild there, but now lived on higher, safer ground. Still, Arica presented a sorry appearance. Its houses were flimsy, at least according to Frézier's standards. They were constructed of sedges or reeds held together by leather, though some were constructed of adobe. "The Use of unburnt Bricks is reserv'd for the stateliest Houses, and for Churches; No Rain ever falling there, they are cover'd with nothing but Mats, which make the Houses look as if they were Ruins, when beheld from without."[9] Frézier's book also included a map of Arica that could enable his readers to innocently enough envision the city's layout and site—but could also serve as a blueprint for those who might wish to invade it.

Frézier's book also contains a remarkable illustration of the landscape well inland from the coast at Arica, where mining enterprises existed in

planche. XXII.   page. 138.

N. Guérard le file fecit

A Llamas ou moutons du Perou   E Plan de la desazogadera
B Trapiche ou moulin a minerai   F Profil de la desazogadera
C Buiteron ou cour ou lon petri le minerai G La pigne
D Bassins a lauer   H Fourneau atirer le visargent

Figure 3.3. Amédée Frézier's depiction of a silver mine in the desert regions of southern Peru from his book, *A Voyage to the South-Sea and along the Coasts of Chili and Peru in the Years 1712, 1713, and 1714* (1717), underscores the importance of metals-mining in this isolated area. (American Geographical Society Library, University of Wisconsin at Milwaukee)

isolation and the Andes rose abruptly as a backdrop (Figure 3.3). At this time, Peru extended far into present-day Chile. The artist does not specify the exact locale, and it may even be a composite of several sites, meant to capture the essence of activity. In the text, Frézier noted that "There being neither Wood nor Coals throughout the greater part of *Peru*, but only the Plant they call *Ycho*...they heat the Masses by Means of an Oven placed near the *Desazogadera*, that is the Machine for drying the Silver and separating the Mercury, and the Heat is convey'd through a Pipe, which violently draws it, as may be seen in this Figure." Prominent in this illustration are llamas, which he called the sheep of Peru. These animals, which occupy the foreground, were beasts of burden. In the middle ground, the mining operation is depicted, along with miners who have diverted a streambed and created settling ponds (D). The source of this water was runoff from the Andes mountains. The trapiche or arrastra, which grinds the ore, is indicated by the letter B. The palisade wall marked by the letter C is a place where the ore is stored in preparation for being placed in the furnace, which is marked by the letter H. The insets marked E and F are details of the plan and profile, respectively, of the instrument used to draw off the quick-silver in the furnace. Although the letter G supposedly indicates the mass of silver, it is indecipherable on the drawing. The landscape is generally barren and slopes upward toward the rocky base of the towering Andes mountains, which rise nearly vertically. The mercury used in this process was likely mined elsewhere and occurred as cinnabar, a bright red mercury sulfide mineral found in portions of the Atacama Desert. The ore was likely any one of several silver-bearing minerals, including the silver arsenic sulfide proustite, although native silver was also found in this region. Needless to say, this was hazardous work, and it impacted the health of all involved, including the environment.[10]

For the record, I should note that Frézier was not the first to visually record the silver trade. In fact, more than a century earlier (1601), a scene by Theodore de Bry (1528–1598) was published showing a scene in which llamas (or at least an approximation of them) haul silver bars down from the Andes near Potosí (Bolivia) to the port of Arica.[11] These arrieros, or workers who transport cargoes using pack animals, enabled such commerce to thrive. De Bry and his sons specialized in illustrating New World expeditions, and many observers had plenty to say about what they experienced. Most mentioned imposing mountains, wide deserts, and grueling haulage of heavy silver bars down tortuous trails. The real challenge for illustrators lay in giving visible form to something that had only been described in words; naturally, the results varied. In Frézier's time, though, increased accuracy began to characterize such illustrations. The fact that Frézier either personally experienced the scenes that he depicted, or corroborated information from several reputable sources, helped increase their accuracy and credibility. His book was widely read and helped further refine the popular image of the region among European elites. Interestingly, Frézier did not call this region the Desert d' Atacama, as did DeLisle's map illustrated at the beginning of this chapter. I suspect that mapmakers were aware of the more venerable Spanish description, "Despoblado de Atacama," and freely translated that into wording that increasingly referred to climatic, rather than population, characteristics.

For its part, regardless of names used, Spain was understandably suspicious of such image making and intelligence-gathering activities by zealous outsiders, scientific or otherwise. At this time many travelers from European countries had little compunction about believing that the entire world was open to scrutiny and up for grabs. Nevertheless, although Spain was aware that its territory was coveted by England and France, conflicts eased somewhat as European powers realized that scientific exploration could be advantageous to all parties. In the period 1700 to 1740, maps began to become more accurate, which resulted from improvements in navigation and permitted scientists to better understand the true shapes and positions of landmasses. In Chapter Two, I noted that a major difficulty was determining longitude accurately, but this was about to change.

In 1730, Englishman John Harrison (1693–1776) developed the marine chronometer—a clocklike instrument that kept accurate time despite the motion of ships. Now one could determine not only the local time but the time at the place used as a prime meridian. This coordination of time at two distant points was key to determining longitude. Harrison's invention revolutionized mapmaking, for now the shape of the continent could be determined with great accuracy. This explains why maps from that period began to more correctly depict coastlines, including the Pacific littoral. That having been said, problems could still arise if errors were not guarded against. Of such mishaps, historian Glyndwr Williams soberly noted, "As late as 1741 [Commodore George] Anson's ships almost ran ashore on the rocks of the Chilean coast at a moment when their sailing-masters estimated that they were nearly three hundred miles out to sea."[12]

By the mid-1740s, when the search to determine the actual shape of the earth resulted in numerous expeditions, some dispatched to South America, Bourbon Spain found itself involved internationally. Although the French scientist and mathematician Charles Marie de La Condamine (1701–1774) spearheaded the French Geodesic Mission (also called the Geodesic Mission to Peru), it was a joint effort also involving Spain, as yet another name for it (the Geodesic Mission of France and Spain) confirms. And although the

Figure 3.4. Based on Guillaume Delisle's early-eighteenth-century two-part map of South America (see Figure 3.1), this similarly titled *Paraguay, Chili, [et] Detroit de Magellan* map by Covens and Mortier (1742) has much the same content, as this detail shows, and identifies the region as "Desert Atacama." (David Rumsey Map Collection)

main concern of the expedition was the equatorial regions, the explorers involved also investigated adjacent areas up to and including the Tropics of Cancer and Capricorn. Among the Spaniards who assisted were Jorge Juan y Santacilia and Antonio de Ulloa, whose sojourn in Lima enabled them to consult written sources about numerous places outside of the main area of interest, including the Atacama Desert.[13]

On many maps of the early to mid 1700s, the location of the Atacama Desert is confined to only a portion of what we consider the desert today. Noteworthy among these is the 1742 map published by the Amsterdam-based firm Covens and Mortier (Figure 3.4). Entitled *Paraguay, Chili, [et] Detroit de Magellan*, it depicts the "Desert Atacama," a mountain named "M.

d. Atacama," as well as a port named Atacama.[14] Observant readers will note that the contents of this map, and even its title, are familiar. In fact, it is almost an exact copy of the 1703–1708 map by Guillaume Delisle that was discussed earlier in this chapter (see Figure 3.1). Plagiarism was not an issue—Covens and Mortier openly credit Delisle on their map's cartouche—but their map is a reminder that mapmaking is often derivative and conservative. Its inspiration was a nearly forty-year-old map, albeit a highly respected one. The combination of cultural and physical features named Atacama on these maps represent a convention that would prove venerable indeed.

Dutch books as well as maps perpetuate the image of the Atacama as a marginal area with some resources but little else to redeem it. In the South American (second) volume of the *Hedendaagsche Historie of Tegenwoordige Staat van Amerika* (History and Present State of South America), which was published in 1767—the same year the Jesuits were expelled from the New World by order of Spain's King Carlos III—Arica is mentioned as part of a Catholic Diocese known for its fish and spices (viz., Capsicum or "Spaansche Peper"). The province named Atacama is said to lie south of Chuquifaca (Chuquicamata?) and is also administered as a religious district, but it is characterized as a sandy desert wasteland lying at the border of Peru and Chile. The foldout map of South America appended to this encyclopedic Dutch book depicts mountains parallel to the coast and uses the French name "Desert d'Atacama" for the area straddling the dotted line labeled "Keerkring van den Steenbok" (Tropic of Capricorn).[15] As in the other maps just mentioned, information from French and other cartographers appears to be the main source. A related point about these northern European maps is worth noting. In their application of names associated with the Atacama, they confirm that such branding may have been imposed by those residing far from the region. Just as the name Despoblado de Atacama was a Spanish term used by those in central authority, so too did the name Desierto de Atacama (or Desert d' Atacama) appear on European maps as a way of making sense of—which is to say simplifying—the complex geography of western South America.

By far the most detailed published map showing the Atacama Desert as such in the eighteenth century was the magnificent *Mapa Geográfico de America Meridional* by Juan de la Cruz Cano y Olmedilla. Published in Madrid in 1775 as a series of seven sheets that covered South America, this represented a cartographic tour de force.[16] For our purposes here, as seen in a version of the map published in London in 1799, sheet number five (Figure 3.5) shows not only the Desierto de Atacama but also delineates its topography in considerable detail. The Cruz Cano map, as it is widely called, confirms

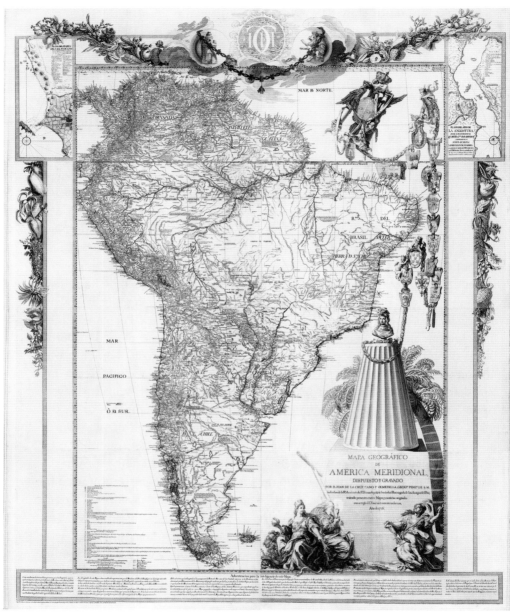

Figure 3.5. The Cruz Cano map (1775) depicts many features in the Desierto de Atacama and reflects Spain's considerable knowledge about its colonies in western South America, as seen in this later copy by William Faden (London, 1799). Ironically, identifying the desert as such may have encouraged Spain's Royal Botanical Expedition to Chile and Peru, 1777–1788, to avoid this area. (David Rumsey Map Collection)

Figure 3.5 detail of Atacama region on Cruz Cano map.

that Spain was developing a comprehensive understanding of its holdings at exactly the time that the British and French were also doing so. For example, Thomas Jefferys's popular *Map of South America* was published in 1776 in London. As it appeared in atlases, Jefferys's map was reproduced in a two-page spread that divided the continent into two sections (Figure 3.6). As might be expected, Chile's long shape necessitated portions of that country being represented on both the northern and southern sections. As a consequence, the Atacama Desert was cut in two; on the northern section, Jefferys depicts a city named Atacama. This is apparently San Pedro de Atacama, which is indicated by a stylized church symbol. On the southern section, the name "Desert of Atacama" runs north–south, straddling the Tropic of Capricorn in the process.[17] Here, too, we find an Atacama Desert, but it is much smaller in size, surrounded by mountains, and more bowl-shaped in outline than the linear desert we often see depicted today.

Figure 3.6. Thomas Jefferys's popular *A Map of South America* (London, 1776), though not quite as detailed as the Cruz Cano map, confirms Britain's interest in the continent. (David Rumsey Map Collection)

Placed side by side with the Cruz Cano map, Jefferys's is rather more rudimentary. Much of western South America was, after all, Spain's turf, and the Spanish Crown tried to maintain a tight grip on it. By 1783, though, the desert portion of South America remained fragmented politically. Though nominally all under Spain, two viceroyalties governed. The viceroyalty of Peru was governed in the north by the presidency (audiencia) of Cuzco, and in the south the capitanía general (audiencia) of Chile. In about the middle of these, the viceroyalty of La Plata stretched westward from eastern South America all the way to the Pacific coast in the vicinity of present-day

Figure 3.6 detail showing the Atacama Desert as well as a community named Atacama—likely the important cluster of villages collectively called San Pedro de Atacama today. (David Rumsey Map Collection)

Iquique. Given the emphasis on the more successfully settled areas in highland Peru and Santiago, the desert was a poorly governed, lightly populated nether zone of importance mainly for defense.

As the late historian David Weber noted in his monumental study *Bárbaros: Spaniards and their Savages in the Age of Enlightenment* (2005), the Indians in those more populated areas received the most attention and comment, but it was those in the fringes such as the Atacama (and even more so in the wilds of southern Chile) who really tested Spain and brought about reform. Weber's use of the term "savages" in the book's title generated considerable controversy, but in his defense, that was the term the Spaniards used for non-Christianized Indians, especially those who proved recalcitrant. Moreover, Weber knew that in the Age of Enlightenment, the term savage also began to represent something of an ideal—humanity that had not been sullied by civilization and its repressive or corrupt institutions. Understanding nuance and its delightfully subversive quality, Weber noted that the term savages came to have "heuristic value as symbols, both positive and negative." That may have gone over the heads of his critics, but few can deny that his book broke new ground in interpreting Bourbon New Spain geographically. As Weber observed, Indians on the margins fared differently: "Chile, governed from distant Lima by the viceroy of Peru, acquired greater autonomy as a capitanía in 1778."[18]

At just this time, Spain was beginning to fund high-profile expeditions. Nevertheless, in the literature of exploration, the Atacama Desert remains but a footnote during this period. Therefore, even though Spain unilaterally took a more active role in scientific exploration, the Atacama Desert was out of sight and out of mind. For example, even the much-heralded Malaspina Expedition of 1786–1788, which was nominally Spanish but headed by the ill-fated Italian explorer Alessandro Malaspina (1754–1810), missed the opportunity to explore the Atacama Desert, as its vessels left the fertile area in the vicinity of Santiago and Valparaíso, traveling northward from Chile into Peru and thence across the Pacific to the Philippines and Asia. On the way, despite pausing in the ports of Coquimbo and Arica in April and May of 1790,[19] the expedition did not explore inland, as the bleakness of the country there seemed to offer little or nothing for explorers interested in flora and fauna. This may seem surprising, but it should be recalled that early naturalists were drawn to places where plant and animal communities abounded. In other words, the desert was not yet in their sights as worthy of serious exploration. Emblematic of this desert-as-blank-space motif was Andrés Baleato's beautiful map entitled *Plano General de la Region del Perú* (1789). As a naval pilot and first director of the Academia Real de Náutica de Lima, Baleato carefully delineated the coast and important inland and Andean areas of Peru, but he left a large area at the right (southern) margin of his map blank except for the bold name "Desierto de Atacama."[20]

Worldwide, deserts were only beginning to be perused by science, often more or less accidentally as a result of the expansionist impulses of nations. Napoleon's 1798 foray into Egypt ultimately resulted in the world's first scientific report on a desert. Other deserts would soon follow, but this involved redirecting the gaze of scientists, who tended to be more impressed with lushly vegetated tropical regions. Maps and the literature of exploration in the late eighteenth and early nineteenth centuries confirm that many other parts of South America were of greater interest than this desert backwater that formed a vague boundary between the provinces of Chile and Peru.

This neglect is confirmed by the fact that the key figure in the process of South American exploration, the German explorer-scientist Alexander von Humboldt (1769–1859), entirely missed the Atacama Desert, getting no closer to it than the central Andes of Peru. Nevertheless, Humboldt helped set a new standard regarding graphic illustration of the environment, including topography and the distributions of plant communities.[21] As is well known, Humboldt was welcomed by Spanish authorities, though his published reports encouraged political independence movements that ultimately undermined the Crown's authority and resulted in the creation of republics that finally won independence from Spain, including Chile (1818), Peru (1821), and Bolivia (1825).

For these reasons outlined above, the Atacama Desert remained largely unexplored long after other places in South America had been studied in considerable detail. Through a combination of cursory travel and a consultation of written sources, however, the Atacama Desert was developing a reputation as a formidable place heard about by an increasing number of authorities but actually experienced firsthand by very few. Paradoxically, an observer on any ship plying the coastline of Chile and southern Peru could see the desert without even stopping to go inland; apparently, most were satisfied with doing exactly that. Like those of the Malaspina Expedition, the ships on which they travelled might put in at ports to resupply, yet few ventured inland to record the natural history—namely, the flora and fauna—sparse though it was. As Napoleon's majestic multivolume work on Egypt had proven, the desert was far more interesting than it appeared at first sight. Then, too, sight may be the operative word here, for another factor that may have helped keep the Atacama Desert in obscurity was meteorological, and related to visibility. At its western margin—the point from which most people encountered it at this time—this desert region was frequently shrouded in fog, especially close to the shoreline from which most observers glimpsed or attempted to glimpse it. As noted in the introduction, these overcast skies often hide the scrub-like plant communities called *lomas* that depend on such coastal fog as a source of moisture.[22]

Stepping back in time about two centuries, we can call on narratives and maps to help us understand how this remote coastal desert took shape in the popular mind. By about 1810, the Atacama Desert had become a relatively common feature on European and British maps of South America. Typical of these, perhaps, is Aaron Arrowsmith's atlas map *Chili*, which was published in 1812 and features the term "Desert of Atacama" in a blank space that lies just outside Chile's northernmost border. Based in part on these and

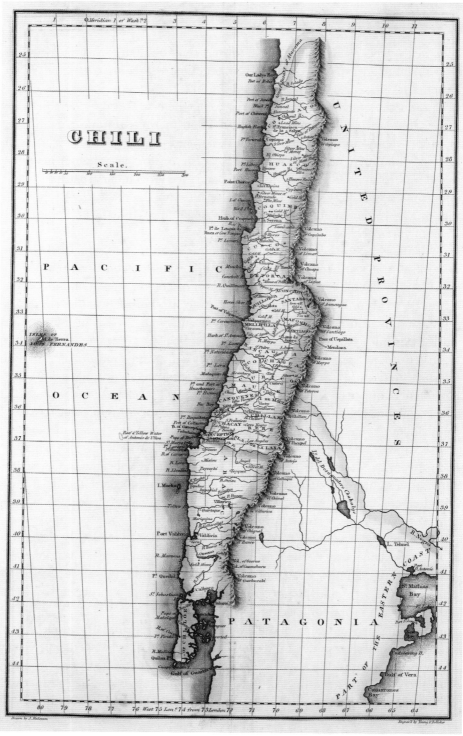

Figure 3.7. Beautifully engraved by J. Finlayson, the *Geographical, Statistical, and Historical Map of Chili* (Philadelphia, 1822) by H. C. Carey and I. Lea includes a superlative, identifying the region as "Great Desart [sic] of Atacama." (David Rumsey Map Collection)

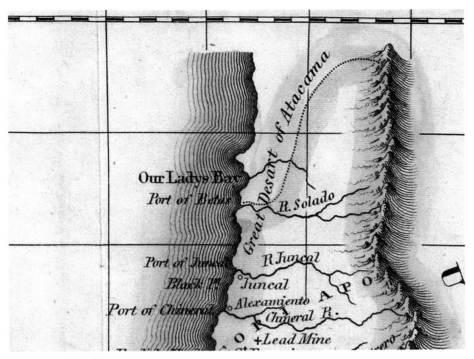

Figure 3.7 detail.

other maps, we also see the desert's name appear on American maps—that is, those that were made by mapmakers living in the United States, which was poised to become an important producer of maps in the 1810s based on efforts by John Melish (1771–1822) and others. These entrepreneurial cartographers recognized a growing market and began catering to it. Their patrons included the interested public, educators, and businessmen. Among these mapmakers were the artist/publisher Fielding Lucas, Jr. (1781–1854), and the writer/publisher Henry C. Carey (1793–1879), who began producing maps in Baltimore and Philadelphia, respectively. Like their European counterparts, they sold maps of individual countries but also published them in atlases.

Noteworthy among early American maps showing the Atacama Desert is Lucas's simply titled but beautifully produced atlas map *CHILI* (1817), which may have been inspired by Aaron Arrowsmith's but helped set an even higher standard for Americans competing to produce the most informative, and attractive, maps. Lucas's map shows the "Desert of Atacama" lying across that country's northern border in a way that emphasizes the desert's longitudinal position and latitudinal extent. Five years later in 1822, Carey teamed up with Isaac Lea (1792–1886) to produce a map that was similar to Lucas's, but if anything, even more beautifully engraved by J. Finlayson. Their map (Figure 3.7) would prove to be influential indeed. Ambitiously entitled *Geographical, Statistical, and Historical Map of Chile*, it was flanked by an elaborate text that provided an immense amount of information. Although Lucas's map had featured the name "Desert of Atacama," the map by Carey and Lea went a step further, enthusiastically branding it the "Great Desart of Atacama." At that time the word desert was still often spelled using an "a," but regardless of spelling, the name evoked considerable curiosity. The word great had at least two meanings—large and extreme—harkening to when the colonial Spaniards called it Gran Desierto, a place to be reckoned with. To Europeans and Americans alike, the world's deserts seemed not only forbidding, but also exotic and mysterious. The map shown here was tinted using three colors, and the Desert of Atacama occupies the northern part of Chile but reaches well into Peru. In explaining the climate, the text divided the country into three zones: rainy in the south, seasonally dry in the near north, while in the far north it was said to never rain at all. To emphasize the borders of Chile, this map outlined that country in pink, while those adjacent were rimmed in yellow and the Pacific littoral tinted blue. The statistical data was provided in the tabular information adjacent to the map. At that time Chile's mineral production was substantial, totaling more than $3,000,000 annually. Gold and silver were the major precious metals, but both the quantity and the quality of copper was noteworthy; in fact, this map claims that the copper from the mines near Coquimbo was the best in the world. Lea's input here was likely very valuable, for he was an amateur mineralogist with a strong interest in mining history.

This map's text referencing mineral production is significant, for at this time the United States did not extend very far beyond the Mississippi River, and it was lacking in the production of important metals such as gold, silver, and copper. Some metals mining had been conducted in colonial times and thereafter in the original states (e.g., Maryland, Pennsylvania, and Connecticut), but most of these mines were played out by around 1820. Additionally, few new deposits were being found despite considerable searching—a situation that encouraged prospecting farther west into lands originally claimed by Spain or France (e.g., Louisiana and Missouri Territories); gold was yet to be discovered in Georgia (1828), but even that find did not stop the search for mineral wealth elsewhere. By contrast, Chile seemed to be overflowing with mineral riches and was increasingly enticing American entrepreneurs who purchased such maps as a first step in anticipating investing capital in mines abroad. Much the same could be said of France, which was flanked by areas possessing silver and copper (viz. Saxony and Spain, respectively) and looked outside its borders for these mineral treasures—including, as noted earlier, locales as far distant as the Andes and adjacent Atacama region.

Unsurprisingly, a French version of the map by Carey and Lea was published three years later in 1825 by J. Carez of Paris. Its title—*Carte Géographique, Statistique et Historique du Chili*—is a direct translation of the Carey/Lea map that inspired it. Attributed to cartographer J. A. C. Buchon, this French version was also bordered by alluring descriptions and statistics. The name "Grand Désert d' Atacama" likely prompted an increasing number of imaginative people to wonder just what kind of opportunities might be found in that great and mysterious place situated in an otherwise nearly blank space on both the continent and the map.

Beginning about this time, those interested in such places might be motivated by something other than financial gain. In the early 1800s, the prospect of scientific discoveries in South America began to beckon a new type of traveler. A century after Frézier, the French explorer Alcide d'Orbigny (1802–1857) mounted an expedition to reconnoiter portions of Brazil and Paraguay. Orbigny considered himself a *naturaliste-voyageur*, and his specialty was mollusks, though he quickly diversified and became a well-known collector of natural history specimens generally. A multitalented observer of everything with which he came in contact, Orbigny was also interested in the varied people(s) he encountered. These he described in much the same detail— color, physiognomy, even the odors of their bodies—as he did other aspects of the natural history. Initially, much of Orbigny's effort focused on the vast lowlands east of the Andes, which he explored in 1826–1828. Although he also hoped to explore Chile, he was deterred in crossing the Andes for fear that the Araucaria Indians would attack. Undaunted, Orbigny finally made his way around the horn and landed at Valparaíso in February 1829. While sojourning in Santiago, Orbigny received the good news that president Santa Cruz of Bolivia, who was considered "a friend of the sciences," authorized him to help determine the unknown riches of that country.

On April 8, 1829, Orbigny sailed for Tacna, Peru, which lies in a river valley about twenty-five miles (ca. 35 km) north of Arica. From there, he traveled northeastward into the Andes, taking copious notes and suffering from *soroche*, or altitude sickness.[23] This provided Orbigny the opportunity to note that the Quechua peoples of the Andean highlands were very different from what he called the "Nation Atacama," who were similar if not identical to the

Changos.[24] Although Orbigny had only skirted the Atacama Desert, most of which lay to the south of Arica, he made a map of South America that depicted portions of it. In this regard, he was like many travelers who were aware of the Atacama Desert but had other goals in mind, and other places to explore first.

To French scientists, the young country of Chile seemed irresistible. A case in point is the transnational scientist and scholar Claude (Claudio) Gay (1800–1873). Born in France and initially trained as a medical doctor, Gay realized that his true passion was natural history. In 1828, he immigrated to Chile, where he taught natural sciences in Santiago. Gay also did some exploring in the desert north of Chile and Peru. Returning to France to develop a museum, Gay later returned to Chile to write the multivolume classic *La História Física y Política de Chile* (1844–1848). He then returned to France and even studied in the United States for a spell, but he returned to Chile for a third time before his death. Like many scientists of the period, Gay stood with one foot in Europe and the other in the Americas, including Chile. The former may have been the established center of learning, but the latter offered new opportunities and new discoveries. Like most scientists of the period, Gay was aware of the Atacama Desert region, but he did not spend much time there, as other places in Chile also captured his attention. Nevertheless, he is associated with the region and so beloved by his compatriots that a mountain range, the Cordillera de Gay, bears his name in Atacama province.

Although most early nineteenth-century travelers and observers might pause only briefly at the ports of the Atacama, that was about to change. By 1833, when Orbigny was in the highlands of Bolivia, a young British naturalist from England named Charles Darwin (1809–1882) had already arrived in Argentina after embarking on a trip around the world in order to record its natural history. In 1831, through good connections and a stroke of good fortune, the twenty-three-year-old Darwin was able to accompany the captain and crew of the *Beagle*. Darwin had grown up reading Dampier and even brought a copy of his book on this voyage. Darwin had also read about French naturalists who were exploring the New World. According to some accounts, including Wikipedia, Darwin was ambivalent about their presence; arriving in South America, Darwin learned that Orbigny—seven years his senior and gaining a reputation in the scientific community—had already arrived. Darwin was reportedly none too pleased that the French naturalist was already collecting, and that little might be left for him to accomplish. This concern, though, proved to be unfounded; ultimately, in fact, the two naturalists corresponded and came to hold each other in high regard. For his part, Darwin would find South America the perfect place for developing his unique insight into how natural history and geological history intersect.

As Patience Schell observes in her 2013 book *The Sociable Sciences: Darwin and His Contemporaries in Chile*, Darwin not only found a country that was scientifically interesting, but he also encountered scientific minds there that could further stimulate his interests in natural history. In her book, however, Schell emphasizes the competitive connection between Darwin and the French naturalist Claudio Gay, who was already well respected in Chile when Darwin arrived in January 1835. Schell confirms that Darwin was well received in central Chile, namely Santiago, which was then developing a scientific community and was already somewhat cosmopolitan in flavor despite its great distance from Europe. However, he and Gay "did not seek each other out," notwithstanding their similar age and shared passion for natural history. Among the reasons, Schell speculates, were national differences, and possibly because Darwin felt threatened that "Gay had already defined his career path in this new nation."[25] Speculation aside, it is safe to conclude that Darwin had many sources of information about the entire country, including its relatively less known far north, and that Gay's sources were among them. Before I describe Darwin in the context of those who influenced him, though, I would like to place his voyage in geographical context, as that can shed some light on his perceptions of the Atacama.

Darwin's first encounter with a desert-like landscape occurred in Patagonia, a bleak, wind-swept area in the southern part of Argentina and Chile where trees were rare and the land covered with stunted bushes and seas of grass. Here, toward the southern end of South America, Darwin was not favorably impressed with such arid land. Nor would his opinion change as he continued to encounter more desert areas in his travels. Of deserts in Chile's north, Darwin is famously known for harsh phrases such as, "I am tired of repeating the epithets barren and sterile." As a naturalist recently noted, "Darwin was ecstatic about South America's rain forest, but troubled by its desert." Troubled is a mild word. In describing the Atacama, Darwin warned his readers that "while traveling through these deserts one feels like a prisoner shut up in a gloomy court, who longs to see something green and smell the moist atmosphere."[26]

This is damning prose about deserts that might even border on xerophobia—the excessive or irrational aversion to dryness and dry places. That critical assessment of Darwin, however, would likely be made by a xerophile (someone who, like me, loves deserts). Looking deeper, though, xerophilia could be considered just as extreme as its opposite, xerophobia. Most people even today tend to be wary of deserts and view aridity somewhat negatively. As a fellow mining historian recently admonished me, "I hope you make this an interesting book, because the Atacama is a very dry subject." Whatever his reasons, Darwin certainly added an element of drama and judgment in his writing that did not favor the desert.

Here, some geographic contextualization may help. While in coastal southern Chile, Darwin was favorably impressed by its visually exciting landscape with robust pine forests, formidable fjords, and resplendent glaciers. Southern Chile was, after all, not unlike Switzerland or Norway, which were popular destinations for Brits seeking inspirational scenery. Moreover, after travelling northward along the Chilean coast, Darwin made important geological observations that helped him better speculate about the age of the Andes Mountains, and hence the age of the earth. He also visited the relatively bucolic areas in the vicinity of Santiago. While there, he enjoyed the company of many, including fellow naturalists, who were familiar with Europe. Even today, when Darwin is celebrated in Chile, it is for his accomplishments in the central and southern part of the country. Tours take visitors to places he experienced, including a hilly area of remarkable plant diversity near Valparaíso called La Campana, where forests of broadleaf evergreen quillay trees and scattered stately palms are found in what later (1967) became a National Park.

There are good reasons, perhaps, why far less is said about Darwin's experiences in the Atacama Desert. In his 1945 book about how South America was explored by naturalists, Victor Wolfgang von Hagen noted that Darwin found this coastal desert region to be off-putting at best. As von Hagen put it, Darwin's "trek along the coast was as dull as the barren countryside." According to von Hagen, the Atacama Desert to Darwin "was literally indescribable because there was nothing to describe—nothing that held the eye or exalted the spirit." No fan of the desert himself, von Hagen concluded, "Somber, dry, colorless, and flat, the Atacama Desert induced in Charles a strange depression or lassitude."

One wonders whether von Hagen had actually seen this desert or was merely dramatizing it for literary effect, but he was right about one thing: Darwin evidently met his match in the Atacama. It was, after all, this naturalist's first encounter with hyperaridity, and Darwin characterized what he saw as "very desolate." Indeed, this desert seemed so desolate that it prompted Darwin to contrast it with fairer places he had seen. In one noteworthy passage in his journal, Darwin opined that "it was almost a pity to see the sun shining so constantly over so useless a country; such splendid weather ought to have brightened fields and pretty gardens." There is a note of irony here, for the very thing that characterized the beautiful weather— day after day of plentiful sunshine, especially just inland from the coast—had also helped create the bleak desert scenery. Darwin's subsequent discoveries in the Galapagos Islands are usually highlighted. As von Hagen tersely concluded, "Darwin had drawn the first scientific portrait of Chile; now he was on his way to Peru" and then Ecuador to make his observations on natural selection.[27]

Although this is true, I think it shortchanges Darwin and demands a behind-the-scenes look at the *Voyage of the Beagle*, especially Chapter 16, "Northern Chile and Peru." First published as Darwin's *Journal of Researches* in 1839, and becoming so popular that it went into several editions, it was later rebranded as *The Voyage of the Beagle* in 1909. All versions have as their base the notes in his diary, which was first published in the 1930s and differs, in many places, from what he first published almost a century earlier. Combined, this material offers a revealing look at Darwin's trek into the Atacama—a part of his travels that has been oversimplified and needs to be amplified here.

The chapter on the desert begins in Valparaíso, with Darwin scouting the area thereabouts in preparation for his trip north into the (almost) unknown. It was an interesting area that had hints of aridity in that xerophytic plants, including what he called the "candlestick-like" cactus, thrived in drier locales. A bit romantically, he also called them "chandelier-like" in his book. In this, Darwin may have yielded to the realization that although there appears to be quite a difference between the singular versus arborescent or multi-stemmed forms of a cactus, sometimes the same species can indeed have *both* shapes; this provides a hint that classifying cacti was a challenging endeavor. It remains so today, though DNA is helping to provide some definitive answers.

After venturing northward, Darwin reached and explored the area around Coquimbo for several days, and here he described the lush Elqui Valley and adjacent hills. Even at this point, Darwin reveals a tendency to head for the hills rather than the plains. It seems that to Darwin, the more rugged the topography, the more opportunities for geological comparisons, as propounded by the brilliant geologist Charles Lyell (1797–1875). Darwin soon confirmed that the landscape here was the result of both geology and climatology. Upon starting out for the valley of Guasco, he became increasingly aware of the impact of drought and decided to follow "the coast-road, which was considered rather less desert than the other." His journal clarifies that this route was selected not for comfort but because it would provide more feed for his horses and mules. Forage was on Darwin's mind, but so too, it seemed, was the greener landscape of home that he felt he was abandoning with each mile traversed. As Darwin put it, "in the first part of our journey a most faint tinge of green…soon faded quite away." With a note of longing, Darwin added "even where brightest [green], it was scarcely sufficient to remind one of the fresh turf and budding flowers of the spring of other countries." That multi-country reference was for readers of the book, but in the diary he used only the name "England"—a specificity that suggests his homesickness. Darwin here sympathizes with an observer "who longs to see something green and to smell a moist atmosphere." His entry of a few days later hints at this longing for fairer, moister, more familiar, and perhaps more desirable scenes.

Of the camanchaca fogs he encountered along the increasingly desert-like coast, Darwin wrote that here "a uniform bank of clouds hangs, at no great height, over the Pacific." There is a touch of the sublime associated with this phenomenon. On a mountain near the coast, Darwin was presented with "a very striking view of this white and brilliant äriel-field, which sent arms up the valleys, leaving islands and promontories in the same manner, as the sea does in the Chonos archipelago, and in Tierra del Fuego." Those were locales in the south of Chile that especially impressed him, but Darwin was rapidly running out of such scenic places as he traveled northward. As it would turn out, seascapes were an apt analogy, for after he finally reached Copiapó about two weeks later, he observed that "the real desert of Atacama—a barrier far worse than the turbulent sea," began about here. Darwin was not the first to equate desert and ocean, nor would he be the last, but as metaphor for menacing indifference to humankind's feeble efforts, the driest and wettest surfaces on earth work hand in hand.

Along that same route in Chile, Darwin noted that the vegetation appeared to be withering with every mile travelled northward. To him, it seemed as if even the cacti were not up to the challenges. As he succinctly wrote, "During each day's ride, even the great chandelier-like cactus was here replaced by a different and much smaller species." Remarkably, he did not even attempt to describe the cacti that were found here, for his wording is colorful but vague. As we have seen, although written about three hundred years earlier by a far less scientifically inclined individual, even Gerónimo Vivar's description of a cactus was more explicit. Clearly, Darwin was dismissive of this region's botany in ways that seem surprising to us today. On the other hand, he was perceptive enough to note, albeit in separate passages, that portions of this region possess a range of plants, from "a filamentous green Lichen" in some locales, to "some Algarroba [*sic*] (a Mimosa) trees" in others. He also remarked on the large herds of guanacos. Perhaps Darwin's greatest contribution to the fauna, though, is his observation that "I saw only traces of one living animal in abundance; this was the Bulimus, the shells of which were collected together in extraordinary numbers in the driest parts." After noting that these mollusks thrive in the dampness brought by fogs, he concluded, "I have noticed in other places that the driest & very sterile districts are most favourable to an extraordinary increase in land-shells." He also specified that the soils in these locales are calcareous. Darwin here confirmed something that students of the desert learn sooner or later: looking close to the ground can be rewarding indeed.

Although von Hagen emphasized Darwin's displeasure with the aesthetic of the desert landscape, he neglected to note the importance of the Atacama's geology on Darwin's scientific thinking. Along with an occasional reference to spectacular cloud formations, Darwin also observed the pleasing, subtle coloration of the landscape, which, as I showed in Chapter One, is the result of the bedrock and how it weathers. Because the landscape here was unencumbered by vegetation, it was perfect for geologizing, as Darwin called it. Given his brilliant deductions about tectonic forces in central Chile, where his scrutiny of fresh earthquake scarps near Concepción led him to fairly accurately postulate the age of the lofty Andes mountains, it is unlikely that he was as totally numbed in the Atacama as von Hagen suggests.

The idea of Darwin completely dismissing any landscape, especially one in which unique geological specimens flourish and distinctive plants can indeed be found, is practically unthinkable—unless he was affected by physical illness or mental stress. This, after all, is a region where exposed geological strata abound and vegetation—especially certain succulent plant communities and numerous species of the barrel cactus *Copiapoa*—can be found.[28] And yet, Darwin seemed to have the bigger picture in mind. For example, he speculated that rainfall determines the character of vegetation in the long run, and perhaps even seasonally. Moreover, time was on his mind; in addition to the geological ages that were being revealed to him, he also hypothesized that the climate of this desert may have changed during human times, for example, from the Inca to the early Spaniards.[29]

Interestingly, Darwin apparently also drew a map of a portion of the Atacama Desert.[30] This map, which was "hand-drawn by Charles Darwin, date unknown," is worth some comment here. It represents a portion of the southern Atacama that Darwin traversed—from Coquimbo north to

Copiapó and then toward what was then the frontier of Chile—at a critical time in history. Contrary to what one might expect, it is not a detailed map of the area's physical features or natural history, but rather offers more of an outline and emphasizes varied towns he visited. On it, Darwin outlines the hydrology, shows various mountain peaks, and indicates a few towns such as Coquimbo and Copiapó by name. It is, in a word, a sketch, but it is well proportioned regarding the placement of those aforementioned features. However, it reveals little or nothing about the natural history, for example, where varied types of vegetation or rocks may be found. It is possible that Darwin acquired the information on this map from other maps—for that is the way maps commonly develop. However, it is also possible that the information could have been provided by his dependable guides, who traversed much of the area with him. Along the way Darwin conversed with locals, and they also may have provided input. Nevertheless, in his chapter on this desert region, Darwin was rather dismissive of some Chileans' abilities to recognize—much less make or even use—a map.

Above all, something other than natural history must have been on Darwin's mind. When he came to this area, portions of it were bustling, as silver had recently been discovered at Chañarcillo (1832), and other nearby prospects were yielding gold and copper. During the period 1800 to 1840, in fact, a thriving mining culture developed—or rather continued to develop—in the north of Chile. According to Chilean historian Luz María Méndez Beltrán, the desert north had a mix of ports, urban centers, and small-scale agriculture consisting of Indigenous people and newly arrived immigrants who conspired to encourage and sustain mining. It was during this period that Chile became a major exporter of copper and precious metals.[31]

It was into this vibrant mining frontier that Darwin strode, and it impressed him mightily. I suspect that Darwin's failure to discuss or map the natural history of the Atacama Desert in detail should be considered in light of the high visibility of such mining enterprises. By the time Darwin visited this desert, in fact, miners were not only well situated here but still flocking to it. Darwin seemed to respect, or at least sympathize with, these miners. His journal explicitly noted how hard miners toiled as they carried heavy loads of ore (ca. 175 pounds, or 75 kilograms) up nearly vertical ladders.[32] We should recall that Alexander von Humboldt, too, had been well aware of the harsh working conditions that he encountered in South America a third of a century earlier. Thus, in addition to Darwin's not yet having developed an eye for the desert's natural history, one can conclude that he was probably preoccupied with, if not distracted by, something captivating indeed: the first full-fledged, modern, market-driven mining boom in the Americas. In fact, while the *Beagle* rounded the southern tip of Chile about a year earlier, Darwin lamented that there was no mineralogist on board, as the prospects for mining in the mountains seemed so promising. Although we may imagine Darwin blazing new trails on behalf of natural history wherever he went, the portion of the Atacama Desert that he traversed (and possibly mapped) was best characterized as mining country, as reflected in the increasingly well-worn road from the mines near Copiapó to the port of Caldera, where Darwin had prearranged to meet the *Beagle* once his travels in northern Chile were concluded.

Before leaving Copiapó, though, Darwin had a lot of geologizing to do in the nearby foothills of the Andes, which brings me back to the map attributed to him. To a historian of the Atacama Desert, a map by Charles Darwin would appear to be the Holy Grail. After all, we associate Darwin with boundless intelligence, limitless talent, and especially a gift for using informative wording supplemented by occasional scientific illustrations. Moreover, his geological maps of southern Chile confirm that he spatially conceptualized landscapes very ably. And yet he seems to have done little of this in the Atacama. Some might be disappointed that his map hardly gives hint of his observational powers as a natural historian. After all, the map is rather prosaic and reveals only the general outline of the area, but that squares perfectly with his perceptions of the desert here. If, in a sense, the map is rather uninspired, that may be because Darwin's attitude toward this region's physical geography and natural history was much the same.

However, like all maps, this one deserves closer scrutiny to not only better authenticate its provenance, but to also place it in the context of other maps, one of which was serendipitously discovered by geographer Isaiah Bowman almost a century ago. As Bowman recounted in 1924, "At Copiapó I had the good fortune to discover a great mass of buried treasure in the form of records and correspondence extending over almost a hundred years, and pertaining to the affairs of the Copiapó valley and especially the business of the principal copper mining company here." This corporate treasure trove of the Copiapó Mining Company consisted of "forty or fifty large wooden boxes which contained bundles of letters and records of the originals which had been sent to the directors of the company in London." Bowman's discovery was, as he put it, "particularly fortunate because the successive general managers or superintendents appear to have been exceptionally intelligent men, and in addition to reporting on the mining properties they were of necessity obliged to report upon the state of the [Copiapó] river, the occurrence of rains and unusual snowstorms, damages done by flood and drought, the condition of the trails and the pastures along them, the state of the ports, and the conditions of land transportation and shipping."[33] Among these highly talented men was George Bingley, who served as company manager when Charles Darwin visited the area. In fact, Darwin himself noted that "I had a letter of introduction to Mr. Bingley, who received me very kindly at the Hacienda of Portrero Seco."[34]

The map that Bowman discovered (Figure 3.8)—which is entitled *Sketch of the Mines & Estates belonging to the Copiapo Mining Company, in the Province Of Copiapo in Chili*—was attached to a letter dated April 30, 1835, that Bingley had sent to the company's directors in London. This professionally drawn map is instructive, for it reveals the close partnership between exploration and economics. In addition to depicting the topography using hachures and indicating the various silver and copper mining areas as they existed in the mid-1830s, this map also delineates large estates clustered along the Rio Copiapó.[35] As Bowman described it, the original map's use of color was very effective, indicating the mining areas in red and the estates in green. As Darwin recorded in both the journal (Voyage) and in his diary, Bingley's estate was impressive and in especially good order. To Darwin it was a verdant strip in the otherwise useless and rocky desert. Not coincidentally, this company map covers an area similar to a portion of the map that Darwin sketched, and it leads me to wonder whether Charles Darwin might have seen it and perhaps garnered any information from it for his own map, or for his multiple excursions into the mountains east of the valley, where he found a silicified log, among many other geological specimens. The verified dates of Darwin's visit, as well as the production of this map dated April 1835, further bolster the conclusion that Darwin the scientist-explorer may have been influenced by the Copiapó Mining Company officers and records as he sojourned here in late June that same year. While there, his diary and book confirm that he spent much of his time exploring the mountains looming behind Copiapó, but he was tethered to that city for several weeks.

The characterization of Darwin as being interested in mining and the affairs of a mining company operative may surprise some readers, for it counters the popular belief that Darwin operated independently, venturing into unknown territory well in advance of economic development. My characterization of the map attributed to him as mundane is not meant to detract from Darwin's mapmaking skills—including his fine geological maps of southern Chile—but rather to note that he evidently did not envision his map, nor this area, as having much intrinsic value, at least above ground. One

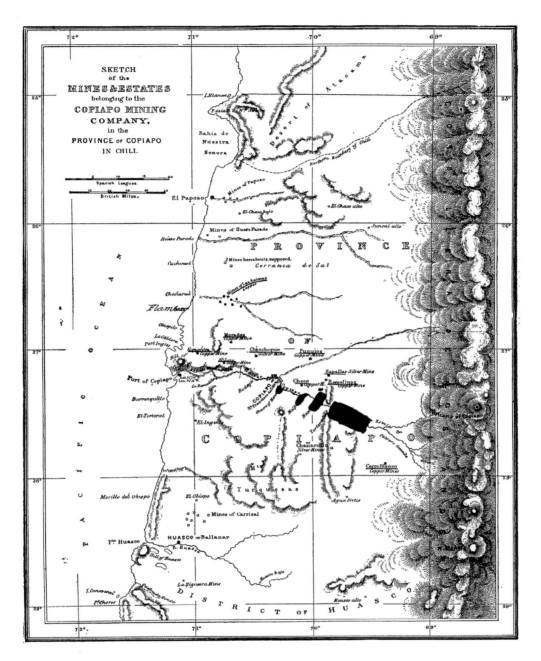

Figure 3.8. In *Desert Trails of Atacama*, Isaiah Bowman reproduced this map based on an original 1835 map of Chile's desert north that he discovered in the offices of Copiapó Mining Company—the map that Charles Darwin likely consulted during his visit to Copiapó that same year. (Author's digital collection)

can picture Darwin consulting this hand-drawn map on his way northward as he left the bleak area behind, perhaps breathing a sigh of relief when he reached the coast at Caldera to rejoin the *Beagle*. If so, the mining company map may have helped him better plan his travels than to make actual desert discoveries during them. For the record, I had hoped to obtain a digital color copy of the original Copiapó Mining Company map for reproduction in this book but the trail went cold. In contacting the mining historian Bill Culver, who is an authority on Chilean mining history, I learned that he too had hoped to locate the mining company records but now believes that they were destroyed in the Nazi bombing of Swansea, Wales, early in the Second World War.[36]

As Darwin was well aware, this desert region continues far to the north into Peru. Although he apparently did not draw any maps of the even drier areas farther north in the vicinity of Iquique, he left a vivid description of its austerity. At that time, Iquique was a small port located in Peru, which Darwin observed was in a state of disarray. Of Iquique, Darwin noted that "the town contains about a thousand inhabitants, and stands in a little plain of sand at the foot of a great wall of rock 2,000 feet high, here forming the coast." The climate was as dry as any place he had experienced, likely more so. As Darwin succinctly put it, "The whole is utterly desert." He was aware that the rarity of rain was evident from the look of the landscape, for he continued, "A light shower of rain falls only once in very many years; and the ravines consequently are covered with detritus, and the mountain-sides covered with piles of fine white sand, even to a height of a thousand feet."

Adding to the bleakness was the foggy weather. Darwin recorded that "During this season a heavy bank of clouds, stretched over the ocean; seldom rises above the wall of rocks on the coast." As I noted in Chapter One, that timing—Darwin visited in mid-July—coincided perfectly with the low sun (winter) period in which the camanchaca fogs are most common. To Darwin, the general aspect of this coast was "most gloomy." The barrenness of the land and the weather conditions did not favorably impress him. He penned that "the little port, with its few vessels, and small group of wretched houses, seemed overwhelmed and out of all proportion with the rest of the scene." Darwin wrote that the people of Iquique "live like persons on board a ship" because "every necessary comes from a distance; water is brought in boats from Pisagua, about forty miles northward." Darwin knew full well the confines of a ship, for he had spent months at a time on the *Beagle*. One thing that may have made his visit to this forlorn desert port tolerable, though, was the fact that he did not intend to stay very long. After commenting that "very few animals can be maintained in such a place," Darwin complained about the high price of "two mules and a guide to take me to the nitrate of soda works" which at the time supported the town.[37]

That trip inland was interesting indeed, for it took him over the mountains and into the wide-open pampa, where the nitrate miners toiled. Given Darwin's strong interest in geology, he found plenty to comment on in this part of the Atacama Desert. Four sentences from his notes reveal how comprehensively he thought. As he noted, "the appearance of the country was remarkable, from being covered with a thick crust of common salt, and of a stratified saliferous alluvium, which seems to have been deposited as the land slowly rose above the level of the sea." At this point, Darwin felt compelled to describe the mineralization: "The salt is white, very hard and compact: it occurs in water-worn nodules projecting from the agglutinated sand, and is associated with much gypsum." He next described the broader landscape comprised of these minerals. "The appearance of this superficial mass very closely resembles that of a country after snow, before the last dirty patches are thawed."

To Darwin, this scene and its mineral salts suggested something noteworthy about the climate. "The existence of this crust of a soluble substance over the whole face of the country, shows how extraordinarily dry the climate must have been for a long period."[38] These sentences reveal how a brilliant nineteenth-century scientist would begin with a comprehensive description of phenomena (in this case, a desert landscape), identify its individual components based on recognized classifications (salt, gypsum), and then speculate about the causes of what was discovered (sea level change, climate). At this time, scientists also made very effective use of similes that could help readers better identify with the new discovery—Darwin's reference to snow being a case in point.

This early nineteenth-century reference to nitrates is a reminder that mining, even more so than biology, was on Darwin's mind as he traveled through this forbidding coastal desert of South America. Scarce indeed were organic things that he commented on here, and for a description I turn to Darwin's diary. To him this was "a complete & utter desert" where "the road was strewed over with bones and skins of dead Mules & Jackasses." Here he noted, "I saw neither bird, quadruped, reptile, or insect." Of plants, all he saw was "an unattached greenish Lichen" strewn over the bare sand, and an eerie sight, "another minute species of Lichen on the old bones." Closer to the coast, he observed "in the clefts of rocks a few Cacti" that were "supported by the dense clouds which generally rest on the land at this height."

Whereas Darwin earlier thought he was experiencing the desert in Chile, he now came to realize that this part of Peru was "the first true desart [*sic*] I have ever seen." Darwin was now an old hand at observing this country, or so he thought, "owing to having been weaned to such a country whilst traveling from Coquimbo to Copiapó," which at that time seemed a "frightful desart" but was comparatively lush compared to the country lying behind Iquique.[39]

Above all, it was geology that Darwin's notebooks documented here in the interior basins of the Atacama. His entries include detailed comparisons of nitrates and considerable speculation about their origins ("it is almost certain that it has been deposited from the water, which was formerly contained in it").[40] I shall later return to the soluble minerals that Darwin examined here in the nitrate fields, for they would transform both the fortunes of Iquique and the map of South America in the next few decades. As Darwin and the *Beagle* sailed north out of Iquique with these desert wanderings behind him, he left a parting shot regarding this bleak landscape and the spirit of the country that now claimed it: "I cannot say I like what I have seen of Peru."

Within fifteen years of Darwin's visit, American and European mapmakers began competing to produce the most accurate map of this mineral-rich coastal desert country, a task made all the more challenging because parts of it occupied three separate countries—Chile, Bolivia, and Peru. Nevertheless, detailed mapping was inevitable, because the region's reported mineral wealth was now generating interest and luring capital. As hinted at earlier, this desert region was also luring scientific talent. In 1838, a Polish-born contemporary of Charles Darwin—geologist Ignacy Domeyko (1802–1889)—immigrated to Chile after escaping Russian persecution. Domeyko was a gifted earth scientist and held degrees from two European universities, Vilnius (Lithuania) and the School of Mines in Paris, where he obtained a degree in mining engineering. After arriving in Chile, Domeyko travelled extensively in Chile from 1840 to 1846. Given his interest in mining, he was drawn to Chile's northern frontier. Domeyko was also an educator who first taught chemistry and mineralogy at the mining school in La Serena. Located at the southern edge of the Atacama Desert, La Serena was the port for the mineral-rich area developing south of Copiapó. Putting his skills as an educator and mineralogist to work, Domeyko soon made an impact on two aspects of Chile's history. Almost simultaneously, it seems, he helped Chile develop its system of higher education, serving as rector of the University of Chile from 1867 to 1883, and he also became a major player in developing Chile's silver mines.

Figure 3.9. Forming the backbone of the Precordillera zone in the Atacama Desert of Chile, and snow-capped at times, the impressive Cordillera de Domeyko is a treasure trove of minerals named after the Russian-born scientist Ignacio Domeyko, who emigrated to Chile in the 1830s. (Photograph by author)

Copiapó and its adjacent silver mines especially impressed Domeyko, for they provided his first view of a booming mining frontier that was far different from anything he had ever experienced, either in his native Europe nor in the more well-watered areas of his newly adopted country. As Domeyko later recalled, "For the first time in my life, I saw what is meant by a...society without agriculture, without neighbors, without tradition or inherited ideas, whose principal and exclusive aim is to get rich." In Copiapó he concluded, "Wherever one pauses, whether in the street, a café or an inn, the only talk one hears is of money, of mines, of lawsuits."[41] The complex geology in the vicinity of Copiapó and Chañarcillo holds the key to the ores here, which are concentrated in veins running through ocher to orange sandstone and grayish-white limestone. These mines, it should be noted, were early examples of metallic mineral extraction on the Pacific Rim.

Domeyko's role in the development of both his adopted Chile and its Atacama Desert region is evident in his name gracing not only streets, parks, and plazas (for example, Domeyko Plaza in Coquimbo), but also an imposing north–south trending mountain range—the Cordillera de Domeyko—which bisects the desert northeast of Diego de Almagro and east of Antofagasta (Figure 3.9). Seeming to change color with the ever-changing desert light, this impressive serrated range stretches nearly one hundred and eighty miles (ca. 240 km) with a few peaks that rise to about fifteen thousand feet (4,750 km). In other parts of the world, such a range would be a major landmark not unlike the southern part of California's Sierra Nevada. However, the Cordillera de Domeyko lies in the shadow of the Andes, which frame it to the east and rise about another mile higher. Between these two ranges lies the impressive Salar de Atacama, a huge seasonally dry lakebed that sits at about a mile and a half above sea level. In the Atacama Desert, distance is measured both horizontally and vertically.

The Cordillera or Sierra de Domeyko is one of the many mountain ranges that proved to be storehouses of metallic ores. Its western slopes are dotted with abandoned towns and piles of multicolored rock that are the hallmark of mining's storied past in Chile. This past is venerable indeed, for the Indigenous peoples mined portions of this area well before the Spaniards had arrived. By about 1820, Chilean miners had developed their skills in the hills just east of Santiago, where mining camps had blossomed. Moreover, as Darwin had now confirmed, miners in Chile's far north were also helping to develop gold and silver mines. These norteños thus had a leg up on the competition to reach California ahead of other argonauts. Although gold had been discovered in southern California as early as 1842, the real rush began about six years later in 1848 and about five hundred miles (ca. 800 km) farther north at Sutter's sawmill. As Darwin was well aware, in gold mining areas, this soft but heavy metal is found in two types of deposits. When found in solid rock, often as crystalline masses in quartz, it has to be extracted through the crushing of that rock. However, because of its heavy weight (about twice that of lead), gold worn from the rocks by natural forces, including erosion, settles in the bottom of stream channels, where it is found as fine as dust and as large as nuggets. Observant miners often find this placer gold first, even in normally dry desert stream channels—such as Darwin had seen near Guasco—then carefully proceed upstream, testing along the way, to locate the original (hard rock, or lode) source of that gold. The miners then tunnel into these lodes and extract the ore. Being well versed in placer and hard-rock mining activities, and living close to ports such as Coquimbo through which they could rapidly reach San Francisco (Yerba Buena) well before those from the eastern United States arrived, some of the more adventurous among these Chilean miners were destined to board northbound ships and ultimately play a significant role in making California famous as the Golden State. Nevertheless, despite the appeal of California gold, opportunities remained in the Atacama Desert, where mines continued to produce silver, gold, and copper—and lure foreign capital.

The Sierra de Domeyko commemorates that brilliant naturalized Chilean scientist, but another topographic feature—Nevado Jotabeche, an impressive, often snow-covered 18,458-foot (5,626 m) peak in the Andes Mountains at the eastern edge of the Atacama Desert—honors a native son who knew the desert even more intimately. In Latin America, such native-born people of Spanish parents are called Criollos (or Criollas if female), and they became important players in the rise of national identities. Born in Copiapó in 1811, José Joaquín Vallejo (1811–1858) continued his education in Santiago, but he returned home in 1841 to practice law and develop mining properties. This contemporary of Domeyko adopted the nickname of Jotabeche and became one of Chile's most influential writers. It is widely claimed that the nickname Jotabeche is an abbreviation of Juan Bautista Chenau, an influential Argentinian of the period who also lived in Copiapó, for at that time the desert north of Chile was a haven for political exiles. In that same year, Vallejo began to write a series of articles about Copiapó and vicinity for the Santiago newspaper *El Mercurio* under his pseudonym Jotabeche.

By 1843 Copiapó had become the new capital of Atacama Province, and two years later, Vallejo founded the local newspaper *El Copiapino*. Chileans became enchanted by Vallejo's, which is to say Jotabeche's, evocative descriptions of Copiapó and its stark desert setting. Fascinated by history and authoritatively using folklore as his sources in a style that is called *costumbrista*, Vallejo opined that Copiapó had been a paradise at about the time the Spaniards arrived. According to an 1842 sketch (short article) that he wrote entitled "Copiapó," the town was once quite bucolic. In another 1842 sketch, "If They Could See You Now!," he noted of his hometown that "trees like the carob, *chañar* bushes, or *dadín* plants not only divided one property from the next, but they provided shade for buildings and spread to patios and sidewalks of the city." Vallejo added that "in the main plaza, we're told, these native plants grew in the same peace and freedom they enjoyed before Diego de Almagro came from Peru to bring turmoil to this previously quiet valley."

These wistful descriptions of Vallejo's hometown appear to place the plaza in pre-Columbian times, but that is understandable. Vallejo recognized that the village was turning into a city in his own time, and much was being lost as well as gained. Vallejo had good reason to sing Copiapó's praises. In that same article, he wrote that when he first experienced it as a child,

Copiapó was lying in ruins from a massive earthquake, its residents having "abandoned it almost completely and were wandering among the arid boulders scattered around town weeping over their lost homes and trying to placate the wrath of God by doing penance." According to Vallejo, Copiapó at that time "looked like the wasteland surrounding it," but he noted that it was now thriving and had developed its own distinctive character. Unlike many formerly colonial towns, with their monotonous white buildings, his city of Copiapó was different from place to place, a panoply of varied colors and diverse vegetation. More to the point, it was host to varied peoples.[42]

In addition to describing Copiapó, Vallejo also focused considerable attention on the "wasteland" that surrounded it. To Vallejo it was desolate country, but it had its assets that were also worth bragging about; in the vicinity of Copiapó lay mountains containing vast deposits of silver ore. In an 1842 sketch entitled "The Mines of Chañarcillo," Vallejo began by noting that "I have seen this settlement not of houses but of caves. I've seen a hill covered with round holes, like a piece of wood drilled by moths." He then situated Chañarcillo in time and place: "Twenty leagues south of Copiapó and at the end of a chain of mountains that extends over a long distance in several directions, its surface the color of maize or different metallic shades, a man hunting guanacos in May of 1832 discovered that deposit of silver, its value incalculable even today." This desolate locale became the site of some of Chile's most celebrated mines, including the Descubridora (Discoverer). In part through Vallejo's writing, the mine's founder, Juan Godoy, was to become one of Chile's legends. Ambivalent about mining, Vallejo predicted that although many had lost fortunes in Chañarcillo, "for many years it will go on being one of the most solid bases of wealth in the Republic."

In another 1842 sketch, "The Map of the Lode of the Three Mountain Passes," Vallejo describes in detail the plight of individuals who seek a lost mine using a tattered map. As related in his sarcastic and yet humorous narrative, hope leads the would-be miners to follow the map's tantalizing clues in the form of landmarks such as "a canyon that has two very thick carob trees at its entrance," or "a pass that has lots of thistles," and "an arroyo that has a lot of gamma grass." This vegetation provides shade and forage for man and beast, but the scene deteriorates into a more forbidding landscape such as "a plain that has a lot of thorn bushes" with "some very large rocks in the middle," and then "a steep cliff until you get to some sand hills." Alas, the route was supposed to take them to rich veins of silver, but instead they become hopelessly lost, nearly dying of thirst in the process.

In yet another 1842 sketch, "Vallenar and Copiapó," Vallejo wrote that between these "two sister towns" travelers "have to cross fifty leagues of sand plains, sand cliffs, and sand ravines" on ill-tempered Andean mules. Emphasizing the discomfort to readers in Santiago, Vallejo adds that an hour in the midday sun here is "hotter than the fires of purgatory, beneath a sun-blasted carob tree whose shade is scarcely wide enough to cover the hundred snakes and lizards who live among its roots." As if this were not off-putting enough, Vallejo confirms that the place names here are accurate, among them Devil's Point, Devil's Hill, Devil's Gap, Little Hell, and Demon's Water Hole. As he concludes of this stretch about one hundred miles (160 km) long, "all those places are consecrated to the same gentleman, because it seems all too true that they were all territorial sections of his dominion."[43] In sketch after sketch, Vallejo's colorful prose presents the most detailed, if somewhat overly dramatic, popular accounts of the Atacama Desert to date. Like Mark Twain's *Roughing It*, published thirty years later (1872), these sketches left an indelible imprint on readers eagerly learning about their nations' harsh and arid—but mineral rich—frontiers.[44]

By the mid-1800s travelers from the United States were about to experience the Atacama Desert and write home about it. During the years 1849–1852, Lieutenant James Melville Gilliss (1811–1865) directed the U.S. Naval Astronomical Expedition to the southern hemisphere, which was nominally oriented to the skies but whose report contained considerable information about the Atacama Desert. As an enterprising naval officer and astronomer, Gilliss also seized the opportunity to map part of the area, his premise being that no good maps already existed. As he dismissively noted, "it may be remarked that the map compiled by Arrowsmith, principally from information given him by a former British Consul General, can scarcely be considered a moderately good guide to the west coast" of South America, in particular Chile.[45] Gilliss was referring to John Arrowsmith's map *La Plata, the Banda Oriental, and Chile* (London, February 15, 1834), and Gilliss felt his own expedition's astronomical determinations, coupled with earlier surveys, could result in a better job. Accordingly, Gilliss's expedition prepared a remarkable three-section map of Chile, Sheet I of which depicts the "Desert of Atacama" in considerable detail (Figure 3.10).

This map was prepared from the surveys of Messrs. Pissis and Allan Campbell and confirms that the northern border of Chile was at that time considerably farther south than today. The name Allan Campbell is noteworthy; at just this time (1851) American entrepreneur and developer William Wheelwright, with design and engineering assistance from fellow compatriots Campbell and Walton Evans, were building Chile's first railroad, a steam-powered line that would connect the mines in the vicinity of Chañarcillo and Copiapó with the burgeoning new port of Copiapó

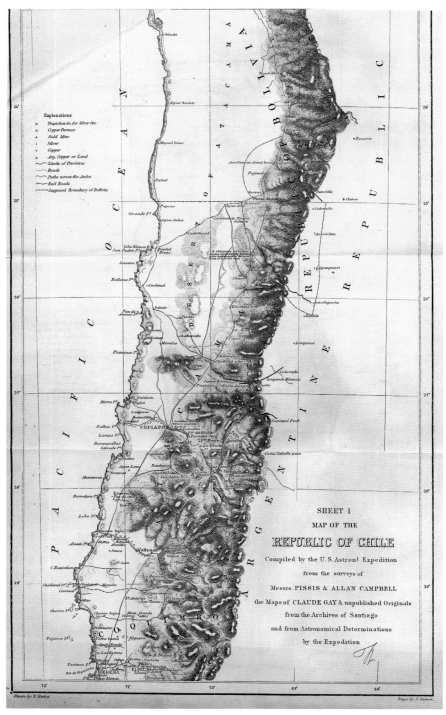

Figure 3.10. In the early 1850s, the American naval officer Lieutenant James Melville Gilliss oversaw the production of this map, which depicts the "DESERT OF ATACAMA" beginning about 27 degrees south latitude and stretching northward into Bolivia. (American Geographical Society, University of Wisconsin at Milwaukee)

Figure 3.11. The bustling but isolated desert port of Caldera, Chile, as illustrated in James Melville Gilliss's report, *The U.S. Naval Astronomical Expedition to the Southern Hemisphere* (1855). (American Geographical Society Library, University of Wisconsin at Milwaukee)

at Caldera. In Wheelwright's vision, this fifty-mile-long (81 km) railway would be the first link in an extensive rail system connecting Caldera with Argentina. Although that international rail link across the Andes did not materialize, the railroad connecting Copiapó and Caldera soon became a landmark on all maps of the region. From the 1850s onward, Caldera became the main port of entry into the desert region lying to the east, and most observers described and often illustrated it in their reports.

Gilliss was no exception. In fact, his report contains one of the earliest known images of this railroad, which it depicts at its western terminus at Caldera (Figure 3.11). At that time the railroad was only about half built, the track beginning at Caldera and reaching southeastward for about twenty-four miles, just about halfway to its proposed destination at Copiapó. Gilliss's description of both the railroad and Caldera are worth a closer look, for they reached an American audience. He arrived in Caldera on a vastly improved form of transportation, the steamship *Bolivia*, which reached Caldera at night. For those who today think the name Bolivia is odd for a ship sailing the waters of the Pacific, it should be recalled that Bolivia at that time was not landlocked; it controlled a portion of the northern Atacama Desert called the Distrito del Litoral (District of the Coast).

As Gilliss was aware, the new port of Caldera, Chile, amounted to little more than a colony of American workers. When Gilliss first saw Caldera in the light of day, it was not much to look at. As he put it, "when morning came, the aspect of the colony was by no means charming." His one-sentence description that followed confirmed that Caldera was a work in progress. "Two or three long rows of board buildings facing the bay and hastily put up, scores of rudely constructed ranchos, piles of lumber, coal, and iron rails, the skeleton of a large and better edifice intended as a custom-house, and a few tents—these were results of the first year, with untiring industry, of the party brought from the United States to build the road." In passages such as these, written at a time when race was considered a factor in intelligence, Gilliss contrasted what he saw as the laziness of the "descendants of Spain," as he called Chileans, with the "go ahead" attitude of Americans.[46]

Gilliss had set the scene a few pages earlier, noting of Caldera that "when Mr. Wheelwright first landed...there was neither house nor hut of any description; a fisherman or two, dwelling under sea-worn rocks, being the only inhabitants of its barren sand-hills." Being interested in history, though, Gilliss quickly noted that things had been different in the distant past. "Yet skeletons recently disinterred in cutting for the railroad, and not far from the beach, prove that there once existed a settlement here." He continued, "Nor...did the people belong to the present small race now in Chile; the crania being of unusual size and thickness, and the femur bones of extraordinary length." To all observers, these thirty skeletons being found in one place signified a community; mysteriously, however, no such community existed presently. Adding to the mystery, Gilliss observed, "Nearly all were found in an erect position, with implements of bone and copper buried beside them."

According to Gilliss, one should seriously question whether these were remains of the Changos, the known fishermen of the coast, who rarely congregated in such numbers. As to their antiquity, the mystery deepened and one could only guess. Whereas the preservation of their skulls suggested less than a century past, Gilliss noted that the niter found in this desert coastal region might suggest that they were much older. As he speculated, niter "is an admirable preservative." Drawing some geographical parallels, and implying some archaeological connections, he added that "this is proved by the mummies of Arica and Iquique, both towns of the coast of Peru; and the soil here also abounds with it." He concluded with a sobering and insightful observation, that the human remains "were in possession of the medical officer of the railroad company, who intended to take them to the United States. But what of the colony?—there remains not even a tradition respecting them."[47]

Gilliss left the matter there, for in the Atacama Desert of the 1850s mining was on everyone's mind, and he was no exception. Bound for the mining area in the vicinity of Copiapó, Gilliss nevertheless took time to explore the countryside along the way. In the company of Wheelwright, he used a railroad handcar to make observations about the geology and flora. At the temporary end of the rail line, where construction work was underway, he described the landscape as covered by immense quantities of shells, some polished to a shiny surface by persistent southwest winds, and in other places covered by a "superstratum" of silt. Gilliss thought that the shrubbery gave the Copiapó River valley a pleasant appearance in places, "but as we drew near, the bushes were found low, dwarfish, of few varieties, and covered with dust." Gilliss appears to be one of the earliest observers to record that the look of the desert landscape here depended on the amount of precipitation that had fallen during, as he called them, "fruitful years." As he observed, "If moderate showers fall on three days of the year, the land is considered to have been extraordinarily blessed, and prosperity ensues." Locals told Gilliss that "at such times even the sand-hills become like flower-gardens with an infinite variety of bulbous plants, which otherwise lay dormant until nature wills."

According to his informants, such good fortune came every eight or ten years. Gilliss was well aware that this part of northern Chile was wetter than the area farther north, and he took the opportunity to comment critically: "Soon one begins to perceive houses—but such houses! Thatched or mud walls, with almost flat and mud-plastered roofs!" Of such houses, he concluded, "Poor shelter these in rain-storms, one would think; and no doubt they would so prove, but it must be recollected, it never rains in northern Chile except in winter, and winter—at least such winter as is occasionally experienced even in Florida—is never known here."[48] This rather enigmatic sentence leaves the possibility of rain open, as Gilliss himself had noted earlier, but the double use of "never" seems to suggest that rainfall was unlikely to materialize. Then, too, reference to Florida may have been intended to help Americans better imagine a place with benign winter weather.

Of Copiapó, Gilliss wrote that in 1851 this mining center "was about three fourths of a mile long, half a mile wide, and numbered above 9,000 people, of whom nearly two thirds were males." The original street pattern in town was irregular, but subsequent development occurred with streets paralleling each other. "These last are crossed by others at right angles; and as is customary with Spanish founders, a public square has been left at the intended centre." His term public square instead of plaza is a reminder that he looked at everything through a North American lens. The church facing the plaza was worthy of comment as having a "Grecian front," which was "so unlike any ecclesiastical edifice in South America that it may readily be believed the architect studied his art in North America." On another side of the plaza, public offices and a barrack formed "a decent looking range of buildings of the usual style." What Gilliss called "insignificant tenements" faced the other two sides of the plaza, but there were signs of "adornment" underway as rows of "tall, straight willows"—Lombardy poplar-like trees—lined some of the streets. He also noted the presence of "*El Pueblo Indio*" at the eastern edge of town, "where there was quite an extensive village of the natives at [the time of] the invasion by Almagro."[49]

Even today, as its population exceeds two hundred thousand, one can recognize enduring aspects of Copiapó in Gilliss's conflicted description. The city remains one of contrasts: rows of ornate Spanish-style buildings stand cheek by jowl with modern boxlike buildings; the traffic pulses along avenues only to come to a dead standstill at the historic plaza; the common tongue of Spanish is supplemented by a dozen other languages; the aromas of fast food and ethnic cuisine waft through the congested streets; the clothes worn by residents span time and place from modern Santiago to the ancient Andes; and the ever-present desert is seemingly in control but constantly held at bay by dense greenery.

As writer Jotabeche (Vallejo) had observed a decade earlier, for a substantial part of its recent history, Copiapó's fortunes had been tied to mining. This was especially true in Gilliss's time. As Gilliss wrote in the romance-charged language of the mid-nineteenth century, "The products of mines in the Atacama almost make one believe the genii of Aladdin have still their favored mortals on the earth, one of them having yielded its proprietor within a single year more than half a million of dollars!" In discussing the burgeoning silver mines in the area north of Copiapó, Gilliss reported that almost all supplies and food had to be brought in from great distances. Like travelers of his time, Gilliss was prone to emphasize the desert as a hostile place of extremes in order to increase the drama of what he, the narrator, had personally experienced and what the reader could now fully imagine. "True," he observed, "some little pasturage is raised along the riband-like rivulet, and a few fruits and vegetables of native growth occasionally gladden the palates of millionaires; but in the district surrounding the newly discovered wealth, there is not a drop of running water nor a blade of grass—scarcely even a shadow in which to obtain shelter from the heat of a midsummer sun reflected from sand."

A closer look at his report, however, reveals that Gilliss traversed this area in June and July, which is winter in the southern hemisphere. Although days can get warm in this part of the Atacama Desert at that time of year, the daytime high temperatures normally average about 75 degrees (ca. 24°C). Gilliss may have experienced an unusually hot day here, one that reached the low nineties (ca. 33°C), but it should be recalled that he was competing with other naval officers who had experienced and reported on even harsher deserts which abounded in superlatives about hellish weather. Among these was Lieutenant William Francis Lynch's 1848 naval expedition to the Dead Sea in search of the cities of Sodom and Gomorrah. Conditions on that expedition were grueling, and Lynch described the oppressive heat in considerable detail.[50]

In his descriptions, Gilliss made frequent use of literary tropes. This, it should be noted, was not unusual in an age when scientific discovery was considered a dramatic enterprise. An example is found in a particularly colorful entry in which Gilliss recounted the rise and fall of Atacama Desert mining pioneer Juan Godoy (aka Godoi). This ill-fated miner had been amply discussed by Vallejo, but Gilliss ratcheted up the prose and the drama. In his official report, Gillis noted that Godoy as "sole master and owner of a donkey or two" in 1832 "was enticed from his legitimate trade of wood-hunting to the more exciting occupation of Nimrod's disciples, by the sight of a browsing guanaco, and, with dogs in advance and lasso in hand, gave chase to the nimble animal." Upon catching up with the furtive guanaco on the southern slope of a hill called El Bolaco, Godoy accidentally discovered what appeared to be rich silver ore. A less colorful but still dramatic version of the story is that Godoy was told about the mine by his dying mother.

Regardless of the original source, Godoy soon found himself on the brink of a fortune, as the ore he brought to Copiapó assayed as high in silver. However, having no knowledge of how to properly develop a mine, Godoy quickly spent his initial profits from the surface ore and signed over rights to the remainder for a quick profit that enabled him to entertain in high style and, even worse, recklessly gamble away much of what he had earned. Again using a famous story from the *Arabian Nights* that he had referenced earlier, Gilliss wrote that "the lamp of extravagance, for want of the precious oil, gave symptoms of expiring." And expire it did, as Godoy found himself as poor as he had been before he discovered the mine. As Gilliss soberly if romantically concluded in an Orientalism-inspired analogy, "Alas! poor

Godoi, thine was not the wonderful instrument of Aladdin!"[51] Though he died in poverty, Godoy is remembered not only in this colorful tale but also in the statue dedicated to him in Copiapó, which has long been a landmark. Godoy is also the name of a town that became important in the annals of Atacama Desert mining. But alas, that town has virtually vanished, although stark tailings piles—reworked from time to time like Godoy's story itself— stand as stark reminders of mining's lasting impact.

Like Darwin, Gilliss was struck by the profusion of shells in this desert. He used them effectively in propounding a theory about the region's geological history. As Gilliss began, "The making of this world of ours has been a wonderful process—or perhaps it would be more appropriate to say, *is* a wonderful process; for the changes still progressing in the upheaval of the broadside of this whole continent are here almost daily perceptible." By emphasizing the present tense, Gilliss not only reminded readers that these processes are in operation today, but he also cleverly, if subliminally, engaged them in the process of discovery. The fact that he emphasized this part of South America is especially noteworthy, for travelers since Darwin have long pointed out that Chile is a work in progress, geologically speaking.

Gilliss concluded his discussion of geological change in this part of the Atacama Desert with a dramatic example. "That all the land from the ocean to the city of Copiapó, and probably farther into the interior to an elevation of 2,000 feet, was buried under the waters of the Pacific within less than a thousand years, I have no more doubt than that my pen traces these lines." As evidence, Gilliss stated, "If the shells, which exposure to the action of a tropical sun and night dews has failed to decompose, are not sufficient proof, we may bring the testimony of their still thriving descendants from the waters of the coast, and the still existing indigenous trees found in a valley near the city where no water has flowed within the memory of man."[52]

Sweeping observations like this are only partly based on scientific observation of elements that are immediately visible. They also call upon accumulated knowledge of tectonic forces and hint at the deepening of time, from historic to geological. Alas, however, subsequent geological investigation sometimes derails such theories. Intrigued by the mystery of seashells in the desert, British geologist Oswald Hardey Evans, about half a century after Gilliss made his speculations, set out to resolve their origin. From the two years that he lived in the nitrate port town of Taltal, Evans seriously studied seashell positioning in the strata. This involved not only fieldwork on the immediate coast but also well inland near the nitrate towns (*oficinas*) served by the nitrate company railroad. After considerable fieldwork, Evans concluded that "shells of very ancient appearance are scattered at all elevations far inland, but these are almost certainly due to the habits of the former Chango inhabitants and to the shore-feeding birds."[53] Impressive indeed are these snow-white shells, which litter the land in many places, but perhaps nowhere more spectacularly than in the terraces lining the river valley between Caldera and Copiapó.

Understandably, landscapes this peculiar began to draw attention in Gilliss's time. In 1853 as Gilliss's expedition was concluding, German scientist and cartographer Rudolph Amandus Philippi (1808–1904) arrived in the Atacama Desert, and his work here would prove to be a watershed in the scientific understanding of the region. Originally, Philippi had left Germany for the sunnier climes of the Mediterranean, in part because of health concerns. While there on two separate sojourns, he engaged in research into his favorite subject—malacology (the scientific study of mollusks). According to a 2017 article by fellow malacologists Alan Kabat and Eugene Coan (see bibliography), the warmth and climate of Sicily not only reinvigorated Philippi but also resulted in his producing some cutting-edge research on mollusks of that locale. They also cite other factors that motivated Philippi to consider a permanent move to Chile, including worsening political conditions in Philippi's native Germany (1848) and the fact that his brother, Bernhard Eunom Philippi, lived in Chile, where he helped encourage German immigration.

By the early- to mid-1850s, Chile's mild climate, family connections, and professional opportunities proved irresistible. Fortuitously, Chile was seeking someone who could conduct a scientific study of its virtually unexplored desert north. They had in mind high-caliber European scientific talent, and Philippi was well qualified. For Chile and Philippi, it proved to be a match made in heaven, as it simultaneously enhanced Philippi's reputation and bolstered Chile's geopolitical position. To Philippi, the Wüste Atacama—as the Atacama Desert was called in German—beckoned as terra incognita, Darwin's cursory observations and Gilliss's recent speculation notwithstanding. Philippi succeeded admirably indeed, producing a comprehensive report at the behest of the Chilean government that compared well to the reports being done at this same time in the expanding United States by explorers such as John Charles Frémont.

Philippi was an astute field observer, avid reader, and quick study. Within two years he had not only read much of what had been written about the region, but he also consulted every map of it that he could lay his hands on. Wisely, he employed former student Guillermo Döll, who proved to be a competent mapmaker on the expedition. To assist with many chores,

Figure 3.12. The port of La Caldera under stormy skies, as illustrated in Rudolph Philippi's *Reise durch die Wueste Atacama* (1860). (American Geographical Society Library, University of Wisconsin at Milwaukee)

Philippi also hired two locals, Domingo Morales and Carlos Nuñez. His goal was to demystify the Atacama Desert by revealing its natural history in a concise but comprehensive report that could be read by scientist and lay person alike. To ensure that the report was read widely in both Chile and Europe, it was published in two languages. The Spanish version was entitled *Viage al Desierto de Atacama Hecho del Orden del Gobierno de Chile en el Verano 1853–54*. The German version was similarly entitled *Reise durch die Wueste Atacama auf Befehl der Chilenischen Regierung im Sommer 1853–54*, and both were published in 1860.[54] Regrettably, though, Philippi's report was never published in English, a factor in its relative obscurity to this day in England and North America.

In addition to page after page of information about aspects of the area's natural history and mines, Philippi's report featured dozens of line drawings of flora, fauna, and fossils, as well as panoramic landscape sketches. For many readers, the visual tour de force was its beautiful full-page lithographs of scenes depicting the places he encountered. As noted, one of these places was Caldera, at that time the newly founded port for Copiapó. Philippi candidly described the fledgling port as "safe and excellent, but its surroundings are summarily gloomy." This was no understatement. As Philippi noted, behind the bare sandy beach one could see only completely bald hills "without a single tree or a single bush." I should here note that Gilliss knew Philippi and respected his work, which was underway as Gilliss was completing his own report.

Philippi's illustration (Figure 3.12) captures the spirit of the place rather differently than Gilliss's, and invites comparison. In contrast to Gilliss's image (see again Figure 3.11), Philippi's is considerably more dramatic, showing not a placid harbor but one under stormy skies. In Philippi's illustration, seagoing ships are anchored in relatively deep water while a group makes for shore in a rowboat that serves as a lighter. Despite Caldera being a sheltered harbor, the waters are roiled by menacing waves. The ship closest to the artist appears to be the three-masted naval ship *Janaqueo* on which Philippi arrived, and that lighter bobbing in the surf reminds one that reaching shore had its perils. Oddly, Philippi does not show the railroad activities at the port, though in the far left, away from town, a train is seen emitting a trail of smoke as it steams inland. Behind the town, which consists of a few buildings huddled near the shore, the brooding sandy hills that Philippi described so well in words define the undulating horizon.

I have long been impressed by this illustration—I even have a framed version of it hanging in my home—but suspected that it was somewhat romanticized or exaggerated. However, with a photocopy of the original image in hand, I decided to explore the elevated section of Caldera that rises to the south. It did not take me long to find the spot where the artist—many say Philippi himself—paused to make the sketch. Upon reaching that location, which is along a twisting road that overlooks the port, I looked eastward and gazed upon a scene that looked much like what Philippi had encountered. Although the passage of time had changed a few things—for example, the port had modernized, the ships were huge, and the town had grown considerably—it was still unmistakably the same Caldera as depicted in Philippi's report, a low-slung town huddling on the harbor. Indeed, sweeping upwards from the town were the same sandy bare hills that form its stark background.

To this day, it is hard for me to imagine a more desert-like backdrop for a port, although some of the towns along the shores of the Red Sea in Egypt and Arabia come close. Elated that Philippi's image of Caldera met the test, I wondered if the owner of the nearby hostel that I was staying in had seen it. After all, this man's property was just two or three blocks away from where the artist had stood about 160 years earlier. I knew that many Chileans interested in history have an idea of Philippi's importance. And naturally, because

Figure 3.13. The plaza at Copiapó and the ornate church facing it, as illustrated in Rudolph Philippi's *Reise durch die Wueste Atacama* (1860). (American Geographical Society Library, University of Wisconsin at Milwaukee)

Figure 3.14. The dusty plaza at San Pedro de Atacama, with its adjacent church, as illustrated in Rudolph Philippi's *Reise durch die Wueste Atacama* (1860). (American Geographical Society Library, University of Wisconsin at Milwaukee)

Caldera was the subject of the image, I figured that nearly everyone in town had seen it. However, when I showed the picture to him, the hotel owner's reaction took me by surprise. Although he immediately recognized the scene as Caldera, he mentioned that he had never before seen the picture. Thanking me for showing it to him, he next asked where he might be able to get a copy. I was about to suggest that I send him a fine copy as a gift upon my return to the United States, but I then realized a copy was much closer, and could be had almost immediately. When I told him it could be found on the Internet, we went to his office, turned on the computer, and within a minute downloaded and printed one out. Overjoyed, he promised to show this prize to future guests, many of whom visit the Atacama Desert in search of its past secrets.

Of all the fine images in Philippi's report, several stand out. I have already mentioned one of Philippi's images of the Chilean desert coast in Chapter Two, as it featured the characteristic raft of the type used at Huasco (see again Figure 2.3). In that illustration of El Cobre, Philippi not only included the raft in the foreground but also a larger sailing ship (possibly the *Janaqueo*), lying offshore. The illustration thus nicely contrasts indigenous and more modern forms of transportation. By scrutinizing these images, it becomes clear that Philippi borrowed some things from other observers who had come before him (e.g., Claudio Gay), but this should not be too surprising. After all, those images were readily available and accurate. Generally, the images in Philippi's report are accurate if, in some cases, a bit derivative. They serve as a reminder that scientific observation is still a collective and cumulative process, as it was in the 1850s.

At that time, El Cobre was a port at the base of steep mountains that plunged down to the sea. In Spanish, the word cobre means copper, and this red metal has long been a major Chilean export. Of special interest in this chapter, though, is Philippi's characterization of the interior desert. In his report, he made a point of recording scenes at the dusty central plazas of towns such as Copiapó and San Pedro de Atacama. At that time, Copiapó (Figure 3.13) was a bustling mining town, but its plaza seems serene. The lack of trees in the plaza proper may seem surprising to those who know Copiapó today, but it accentuated the aridity in Philippi's time; mercifully, trees lined other streets and the nearby Copiapó River. Also at that time, as noted by Gilliss and other observers, the ornate church was the most elaborate building in town. About two hundred miles to the north and east, Philippi visited the central plaza at San Pedro de Atacama (Figure 3.14), which was also devoid of trees, though they lined streets and canals elsewhere in town; this,

Figure 3.15. This scene at Trespuntas, Chile, as recorded by Rudolph Philippi in *Reise durch die Wueste Atacama* (1860), was typical of locales where silver was mined in the mid-nineteenth century. (American Geographical Society Library, University of Wisconsin at Milwaukee)

Figure 3.16. Rudolph Philippi provided this three-in-one folding panorama of sweeping desert landscapes in his 1860 report titled, *Reise durch die Wueste Atacama*. (American Geographical Society Library, University of Wisconsin at Milwaukee)

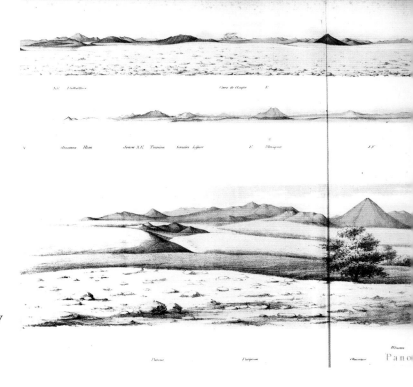

too, might surprise those who visit the plaza at San Pedro today and enjoy the shade of its pepper trees (see again Figure 2.5). The church at San Pedro was situated behind a wall, as it is today, though it apparently lacked a bell tower when Philippi visited in the mid-1850s.

Of the scenes Philippi captured in the smaller places, the one showing a typical mining operation in the vicinity of Trespuntas (three points, or peaks, though in this region the term can also refer to the bases of mountain spurs that jut out into the desert plains) is noteworthy (Figure 3.15). In it, Philippi depicted the town clustered at the base of the mountains, while in the foreground a mine head frame and wince shows the technology used to extract ore. In this scene two burros are at work, one either lifting ore up out of a mine at the wince or working its trapiche which ground the ore, the other hauling a load along the dusty road, led by an arriero wearing a tall cone-like hat or sombrero typical of this region. Other men nearby are also dressed in the characteristic colorfully fringed ponchos and also wear tall hats. Trespuntas is miles inland, situated north–northeast of Copiapó in some of the most forbidding country the Atacama has to offer. Described in detail as early as 1792, Trespuntas represented the typical Atacama Desert mining locale; when Philippi visited, it had long been known as a rich silver town. One of its mines, the nearby Inca de Oro, would become legendary. While in

Panorama de Rio Frio.

Panorama de Tilopozo.

Panorama de S. Pedro de Atacama.

this area, Philippi also visited the mining town of Chimba, where he met the fabulously wealthy amateur naturalist Doña Teresita Gallo, who had, among the many items in her collection, marvelous specimens of native silver.

While discussing Philippi's images of desert landscapes, I should note that his background in geology enabled him to describe and illustrate several portions of the desert in detail. Wisely, Philippi had hired an old prospector and miner named Diego Almeida, who no doubt shared insights about the mining activity that the expedition recorded here. Throughout the report's text are small sketches representing geological cross sections. In the plates section of his report, he includes one containing four geological cross sections along transects. These literally broke new ground, as they suggested what lay below the surface as well as above it. One of Philippi's most impressive plates contains three marvelous panoramic illustrations stacked one above the other. This remarkable plate (Figure 3.16) is, to my knowledge, the first illustration that visually depicts the Atacama Desert itself as an object worthy of study. Its three panoramas capture the subtle diversity of landscapes in the interior Atacama. They include 1) the Rio Frio panorama with its delightful guanacos for scale, this scene beautifully capturing the openness of the interior country east of the Precordillera and west of the Andean volcanoes at today's Parque Nacional Llullaillaco; 2) a sweeping view across the bleak Tilopozo area, which lies east of the Sierra de Domeyko and south of the great Salar de Atacama; 3) and lastly, the most settled and verdant place of the three in the vicinity of San Pedro de Atacama is shown with its trees indicating the namesake river that bisects it.

Conveying the sweep of the desert landscape, this three-in-one plate involved the reader visually and tactilely. Because it spanned three pages (that is, consisted of two pages joined by a third that folded out), the reader had to physically open it to behold its full scope. Visually speaking, this plate offered vistas that words alone could not capture. On the one hand, they revealed the almost agoraphobia-inducing openness of the landscape, and on the other depicted something of the variety or diversity of the landscape. Particularly effective is Philippi's use of objects in the uppermost panorama, including those three guanacos, each gazing in different directions much as the reader was invited to do. In these three vistas, Philippi demonstrated that the Atacama is lightly populated but not uninhabited. Nor is it completely devoid of vegetation, as evidenced by the lower vista depicting houses that huddle near some trees in the low-lying area along the San Pedro River. Collectively, these three scenes enabled the reader to experience the unobstructed vistas that have long awed travelers here.

It is tempting to think that the Atacama vistas that Philippi sketched were natural landscapes, but even in his time humans were having an impact that was sometimes dramatic. The agricultural areas in the vicinity of San Pedro de Atacama provide a case in point. In 2002, an interdisciplinary team of German scholars published an article entitled "The Development of the Oasis of San Pedro de Atacama, Chile." As they conducted their research, they were well aware of Philippi's report, which they cited. This was a case of a German team doing a follow-up study based, in part, on the observations that had been made by a fellow German about 150 years earlier. However, as they studied this much-smaller portion of the desert, they focused on one aspect of it, namely the geomorphology. As they did so, the modern team also had the benefit of sophisticated instrumentation that could measure sediment erosion. Note the use of quantification in their observation that "about 94% of the discharge from the two rivers at the northern rim of the Salar de Atacama was used during the past centuries to irrigate the oasis." According to their calculations, "from the mid-16th until the mid-20th century 37.5 billion tons of material were deposited to an average thickness of 180 cm [72 inches, or six feet]."[55] Thus, during the time of Philippi's visit, the process of soil accretion was well underway as humankind was still shaping the landscape into productive terraced fields.

In addition to sketching selected landscapes, Philippi also illustrated the region's natural history. The images in his report reveal the diversity of life in this desert as well as its geological history. Two plates are geological in nature, depicting a variety of fossils such as ammonites and other mollusks. These remind the reader that the desert has a long past. Compared to other mid-nineteenth-century natural history drawings, the flora is modestly depicted in a series of line drawings revealing the morphology of plants such as *Eulychnia breviflora Ph[ilippi]*, a cactus native to the area, and numerous shrubs. The "Ph" in that name indicates that Philippi was the first to discover and/or name them for science. Although these plants are drawn in a Spartan manner, and in a somewhat monotonous sepia tone, the illustrations of Atacama fauna are quite the opposite. Of these, three plates are especially noteworthy and are reproduced here. These include a trio of lizards that jump off the page as they are depicted in color washes (reddish, teal, and brown) among plates that are otherwise monochromatic (Figure 3.17). This plate, and one of Philippi's drawings of two small rodents, both species newly discovered by him, are rendered as skillfully as any natural history drawings of the period (Figure 3.18). Typically, artists drew these specimens long after they were dead, and based their work on taxidermy specimens as well as field notes, but these animals appear to be alive and well enough to scurry off the page.

Lastly, Philippi's color depiction of flamingoes in their natural habitat—the lakes of the puna at the base of the Andes—reveal the unexpected beauty of life here (Figure 3.19). As noted by Tony Rice in *Voyages of Discovery:*

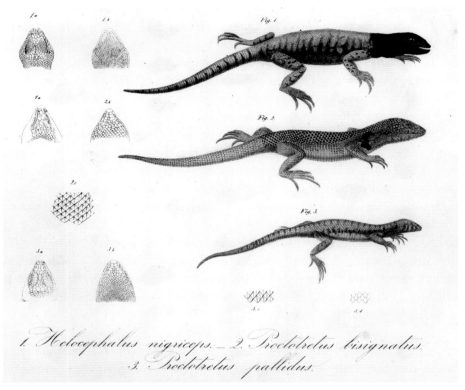

Figure 3.17. In the first scientific expedition to explicitly study the Atacama Desert, German-born Chilean scientist Rudolph Philippi identified and beautifully illustrated several species of lizards in *Reise durch die Wueste Atacama* (1860). (American Geographical Society Library, University of Wisconsin at Milwaukee)

Three Centuries of Natural History Exploration (1999), natural history illustration became more habitat oriented after the pioneering artist Maria Sibylla Merian published her color illustrations from Surinam in 1705.[56] In this one illustration, Philippi offers a composite view of animals and the type of landscape they inhabit. As we shall see in the following chapters, the flamingoes of the Atacama would become emblematic—colorful birds in a colorful landscape adjacent to the Andes to which tourists would someday flock.

In keeping with his desire to describe this desert scientifically, Philippi felt obliged to map it more accurately than his predecessors. In preparing the reader for this map, he first offered "critical observations about the [existing] maps that cover the Desierto de Atacama." Much like Gilliss, Philippi found great fault with Arrowsmith's map of 1843, noting that it was inaccurate in

Figure 3.18. In some locales, Rudolph Philippi encountered rodents, which he skillfully illustrated in *Reise durch die Wueste Atacama* (1860). (American Geographical Society Library, University of Wisconsin at Milwaukee)

Figure 3.19. This carefully designed plate in Rudolph Philippi's *Reise durch die Wueste Atacama* (1860) depicts flamingoes in their lake habitats in the eastern Atacama at the base of the Andes. (American Geographical Society Library, University of Wisconsin at Milwaukee)

mapping the coast, as it even showed a harbor near Paposo and a town near Tres Puntas that did not exist! Philippi's criticisms of Arrowsmith's map are too many to enumerate here, but they included misplaced river courses and incorrect distances. Philippi next dismissed the map that accompanied the travels of Alcide d'Orbigny as inaccurate in many regards, including the misplacement of mountain ranges and even the desert itself in reference to the west coast of South America.

The next map Philippi discussed was *Mapa del Desierto de Atacama por el Señor D. Constantín Navarrete*, which was in the personal collection of Ignacio Domeyko. This map showed rather accurately almost all of the watering places and roads that Philippi had experienced, but it inaccurately depicted portions of the topography and misplaced some rivers. The last

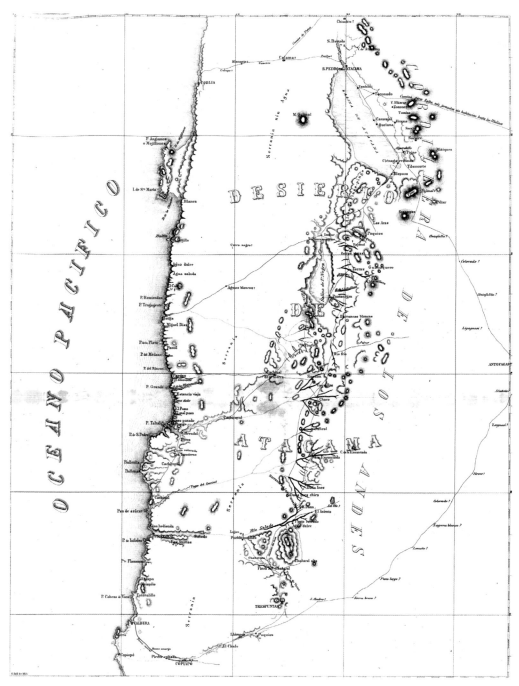

Figure 3.20. Given his penchant for accuracy and detail, it is no surprise that Rudolph Philippi's *Reise durch die Wueste Atacama* features a map of the Atacama Desert that set a new standard in mapping the region. (American Geographical Society Library, University of Wisconsin at Milwaukee)

map discussed by Philippi was John Arrowsmith's *The Provinces of La Plata, the Banda Oriental de Uruguay and Chile chiefly from MS. Documents, communicated by Sir Woodbine Parish, etc.*, which had been published in 1842, a year earlier than the one he had severely criticized. Stating that "this map is infinitely better than the maps of Bertes and of D'Orbigny," he nevertheless pointed out that it, too, included many errors, among them misplaced communities. Philippi offered faint praise by concluding, "A great advantage of this map is that the parts of the desert that are unknown are left blank and thus not filled with mountain ranges and hills invented by the sketcher."[57] By comparing those other maps to the measurements he was now taking, Philippi could keep what he deemed accurate and jettison those aspects he found incorrect. By such a process of elimination, maps generally become more accurate through time. In other words, although Philippi found much to criticize in the earlier maps, they did serve a purpose.

Those who anticipated Philippi's map were not disappointed. Based in part on his exploration from late 1853 into 1854 and subsequently fine-tuned, it represented a milestone in the cartographic history of the region (Figure 3.20). Unsurprisingly, given Philippi's penchant for detail and the competent cartographic support provided by Guillermo Döll, the report's map provided a more accurate rendition of both the topography and locations of communities than was heretofore available. Perhaps the first thing that most readers may have noticed about Philippi's map was that the largest and most ornate letters used emblazoned the name "DESIERTO DE ATACAMA" across the center of the map. When published, it represented a very comprehensive and accurate depiction of the Atacama. As cartographers had done to earlier maps, Philippi prominently depicted the railway from Copiapó to Caldera, which connected this rapidly developing region to the outside world. Especially noteworthy is the way Philippi's map delineated the topography, including prominent mountains and ranges of hills, as well as the distinctive *bancales* or terraces that loom above the entrenched river valleys. Compared to Gilliss's otherwise excellent map, Philippi's is even more detailed.

While on the subject of Philippi and cartography, I should note that I recently acquired another map of the Atacama Desert attributed in part to him and published in Germany (Figure 3.21). Its title is translated "Dr. R. A. Philippi's Exploration of the So-Called Atacama Desert [Wüste Atacama] November 1853–February 1854—after Döll's draft as redrawn by A. Petermann." The publication date of this map is 1856—four years *before* Philippi's report was published—and its place of publication is Gotha, Germany. The map's credited cartographer, August Heinrich Petermann (1822–1878), was far more famous than Philippi at this time. In fact,

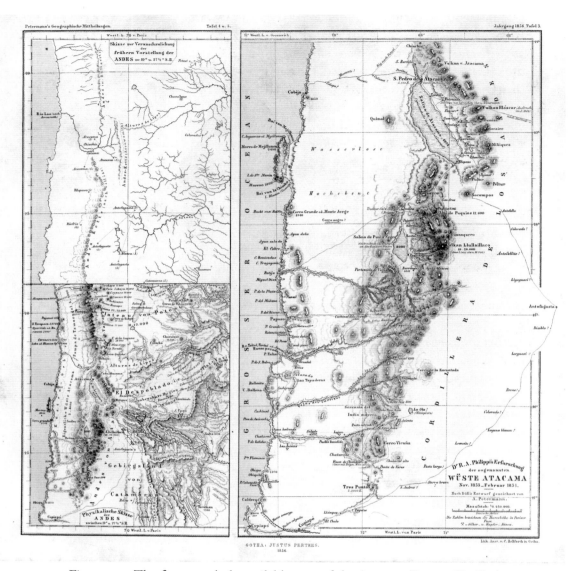

Figure 3.21. The first popularly available map of the Atacama Desert, Dr. R. A. Philippi's *Erforschung der sogenannten Wüste Atacama Nov. 1853–Februar* 1854, was published in 1856 by August Petermann based on notes provided from Philippi's expedition in progress. (Author's digital collection)

Petermann was recognized as one of the most remarkable armchair travelers, for although limited in his own travels, he had a penchant for representing the worldwide discoveries of scientific explorers. Even though he was well-known outside of his native Germany, Petermann centered his activity in Gotha, which was then a center of publishing. This map's publisher, Justus Perthes, was renowned for fine, meticulous products such as exemplified by this one. Although covering much the same geographic area as Philippi's, Petermann's lithographed map differs in many details, including its use of different lettering and its more ambitious depiction of the topography. As its legend confirms, this map also identifies silver and copper mines. Moreover, it shows certain routes of travel in more detail than Philippi's map, no doubt reflecting additional information available supplied by Döll.

Although this map's title mentions that it covers the Atacama Desert (Wüste Atacama), Petermann does not use that name on the map proper. Instead, he evocatively uses the words "Wasserlose Hochebene" (Waterless Tableland) to indicate it. Of special interest is this map's accompanying set of two smaller maps that flank it on the left; both depict the broader Andean region in which the Atacama Desert lies, the upper map showing the hydrography while the lower map emphasizes the topography. On the lower map, Petermann affixes the desert geographically using the term "Eigentliche Wüste Atacama." That word *eigentliche* is of special interest because it means "real" or "proper." On this map, Petermann thus distinguishes the genuine Atacama Desert from a) any other similar-appearing landscapes or, more to the point, b) locales on other maps that might be called Atacama Desert but with which Petermann (and by extension, Philippi and Döll) might disagree. This map thus attempted to provide a more carefully constructed positioning of the Atacama Desert than others that preceded it.

A case in point is seen in the northern part of the map, which Petermann identified as the Pampa del Tamarugal. On this map and many others to come, the Atacama Desert would be rather more restricted than it had been on some maps in the past and on some maps today. Also noteworthy is Petermann's subtle use of color on all three maps, the Pacific Ocean being rendered in light blue. The date of 1856 is especially important; to my knowledge, this set of maps is unique in being the first published cartographic representation of the Atacama Desert in and of itself. Typically, before this time the Atacama Desert was only coincidentally depicted as part of South America, or of Chile. With the publication of this authoritative map, the Atacama Desert is positioned carefully, and thus center stage, in the scientific imagination. While on the subject of the map's date, I should note that although it appears that Petermann essentially "scooped" Philippi, whose report was eagerly awaited in Europe, there is more to the story. As the map's title suggests, Philippi was the main source, and is prominently credited, in effect increasing interest in the report that would soon be published, the word soon being a relative term. Although Philippi's dedication page to the president of Chile is dated 1858 in both the Spanish and German editions, the reports would be published two full years after that. Seven years had elapsed since Philippi began his journey into the Atacama.

Despite any delays in publication, Philippi's report was well worth the wait. Reading the text of this comprehensive report, which is titled as a journey rather than an expedition, one can only marvel at his insightful observations. Philippi's discussion of the vegetation between Copiapó and Caldera, which he viewed from the train, is a case in point. Readers may recall that this was the same area dismissed as completely desolate by Darwin in 1835. In contrast to Darwin, though, Philippi observed in 1854 that "a great number of low plants grow in this extremely arid sand and they make agreeable the view with its gilded, blue, and red flowers." Enchanted by what he saw, Philippi added that "an infinite number of black Coleopteras [a species of butterfly] of the genera Gyriosomus fluttered alongside the [rail]road," and that these insects were nourished principally by flowers of the Malvas plants.

How might we explain the striking difference between what two independent naturalists encountered here? As also noted by Gilliss, this type of transformation depended on the area receiving two or three good rains. However, whereas Gilliss only wrote about such a transformation, Philippi actually experienced it firsthand, noting, "the vegetation is not always this rich [lush], I was assured that it was a consequence of the abundance of rains, which have been experienced this year in Copiapó; because there were three rain storms, one very great and two smaller." Philippi called these storms "chubascos," a term that is used to this day for storms moving off the Pacific. His statement is a reminder that the Atacama Desert landscape may differ from time to time, lying dormant for years but bursting forth into flower following rainy periods.[58]

In compiling his report, Philippi frequently conversed with Natives who had considerable knowledge of the flora, fauna, and geology. One of the enduring natural mysteries of the Atacama in Philippi's time was the enigmatic scattered deposit of dense iron fragments found near a small village, actually more of a watering spot, called Imilac. In fact, Imilac shows prominently on Philippi's map with an indication of its association with hierro meteorico (meteoric iron). The place was remote indeed and was situated in territory so vaguely defined that some were not sure whether it was in Chile or Peru. As noted in an earlier (1827) account by a Dr. Redhead, this iron was located "in the province of Atacama in Peru [sic] at a distance of about twenty leagues from the port of Cobija, in large masses embedded in a mountain, in the neighbourhood of the village of San Pedro, and scattered over the plains at the foot of the mountain in question for a distance of three or four leagues in fragments." That description was more or less repeated but somewhat clarified by Sir Woodbine Parish in 1839, who stated that the iron could be "found scattered in large quantities a little to the south-west of a small Indian village called Toconao, ten leagues from San Pedro, the capital of Atacama, and about eighty from Cobija, on the coast."

In the early 1850s, many searched for the iron-rich site but came up emptyhanded. In February 1854, it was finally located after Philippi tracked down the original discoverers of the site, José Maria Chaile and Matias Mariano Ramos. These two Indians had accidentally found it in 1822 while hunting guanacos. One reason that the site was hard to find was because the two had kept it a secret, thinking it was a deposit of silver. Despite their secrecy, locals knew about its existence, if not its exact whereabouts, because a few pairs of spurs had been made from this unique "silver." When Philippi and his companion Don Guillermo Döll reached the site in 1854, they solved part one of an enduring mystery, but much more research was needed. The largest piece of this iron weighed more than fifty pounds and was in the possession of Ignacio Domeyko in Santiago. As we shall see in a later chapter, it would take additional research to finally answer several crucial questions about this iron. In Philippi's time, though, some Native people in the region thought its source was either volcanic or the result of explosions from earlier mining activity, and hence called *reventazones*.[59]

It is tempting to think that because Philippi was a scientist, he was therefore a person with little interest in religion, as is the case with most scientists today. However, the mid-nineteenth century was a different time indeed. It was, after all, Philippi who wrote a startling one-page testimony regarding "the study of the Natural Sciences" in 1864. By this time, Philippi was well known as a well-respected Chilean scientist whose careful reporting had put the Atacama Desert on the map—which is to say, the most accurate map to date. Philippi began his missive by noting that "nothing is more sublime nor more religious than the study of nature." Lest some readers think this was a misstatement, and that he could not have meant religion, his next sentence clarified its meaning. "Through this work are known the master builder and in the marvels of the world are revealed its creator." For Philippi, studying nature brought him closer to the "Autor Supremo" (Supreme Author), as he called God.[60] Some people have claimed that deserts are perfect places for such soul-searching, if not for religion than for spirituality. As the Tuareg saying goes, "There are lands full of water for the well-being of the body, and lands full of sand for the well-being of the soul." As I shall show, the desert is increasingly regarded as a place of spiritual renewal.

However, during Philippi's time, there is little evidence that people deliberately went into the Atacama on such spiritual quests. Nevertheless, it was a time in which scientists often worked hand in hand with religion to better understand nature as one of the manifestations of the power of God. That said, it is worth noting that Philippi never stated his opinion on Darwin's theories about evolution—a topic that would have an increasingly polarizing effect on society.

While on the subject of religion in the context of mid-nineteenth-century Chile, I would be remiss if I did not mention a fascinating, albeit somewhat arcane, connection between the Mormon faith and the Atacama Desert. For readers not familiar with the Church of Jesus Christ of Latter-day Saints, it was born on the eastern American frontier in the 1820s, and most of its members traveled westward with it to Utah in the late 1840s. One of the key Mormon scriptures—the Book of Mormon—describes the journey of the Lehites, an ancient people from Jerusalem who supposedly traveled to the New World in 600 BC, after which they blended into the Native American tribes. Although the Book of Mormon is vague about where these people arrived, some devout Mormons in the 1840s and 1850s speculated that the Lehites' journey across the Pacific Ocean took them to a landing spot in western South America, more specifically at thirty degrees south latitude. That is the location of present-day Coquimbo and La Serena, at the southern margins of the Atacama Desert.

By this time, Lehi had died and his two sons Nephi and Laman (the former honorable and the latter dishonorable) would come to lead the peoples of the Americas. From this spot in Chile, some say, the Nephites travelled northward. Several maps by Mormons actually show a route traversing the Atacama Desert along what is widely regarded as the Inca road. Although this trip would begin in the verdant Elqui Valley, those who followed it would ultimately face the same inhospitable desert area that Frézier and others have warned travelers about. However, those Mormon maps do not name the desert as such, and perhaps those who drew them were unaware of the Atacama's existence. I have always been interested in how thoroughly this type of belief can captivate the imagination. When a Mormon explorer-missionary named Thomas D. Brown encountered Indians (whom he called Lamanites) in the Arizona–Utah borderlands in the early 1850s, he openly wondered if their origins were not in North America, but rather in "Chili [*sic*] and Peru."[61] When I asked a Mormon friend whether he believed that Nephi actually led his people through the driest desert in the world, he responded without hesitation, "Why not? They had already travelled through Arabia on their way to the New World!"

I illustrated some of these Mormon maps of South America in my book entitled *The Mapmakers of New Zion*.[62] Although other Mormons believe the landing place was elsewhere in the Americas, or refuse to take a stand on this controversial issue, some websites show Coquimbo as the point of embarkation. One of the more passionately worded websites is Nephi Code, which states, "In a quiet bay, such as Coquimbo (which means peaceful waters) Nephi could have beached his ship at low tide, or taken it into the lagoons where low tide would have made it easy to disembark for everyone—low tide falls to one foot or less just after noon in the bay."[63] As I also demonstrated in my aforementioned book, however, there are many other places that also might fit the geographically vague narrative contained in the scriptures. My point here is not to judge these Mormon narratives, which are based on deeply held beliefs, but simply to again restate an important premise of this book. What the Atacama Desert was thought to be is equally as important as what it actually was; hence I take very seriously the adverb "imagining" in the title of this book. Regardless of what did or did not happen, though, it is interesting to note that Chile became the exemplary twentieth-century success story of Mormon proselytizing in Latin America, for many Chileans were converted through the missionary program—a tough task indeed given the strong Catholic traditions and beliefs of the country.[64]

At the same time that Mormons were speculating about the arid Pacific coast of South America, the budding American writer Herman Melville (1819–1891) made numerous references to the same coast. Melville had first-hand experience on the west coast of South America from his whaling journeys in the 1840s and had picked up additional information from other mariners. Melville was inspired by the ill-fated encounter of the whaling ship *Essex* in 1820, which had set sail from Atacames, Ecuador, and came to grief after being rammed by a huge whale. Melville was also familiar with Charles Darwin's voyage on the *Beagle*. In fact, Melville had a copy of Darwin's book on board his whaling ship. In his classic novel *Moby-Dick*, Melville described the coastal desert city of Lima, Peru as slumbering under "the tearlessness of arid skies that never rain."[65] Biographer Herschel Parker notes that Melville was impressed with Lima as a mysterious white-veiled city positioned between the coast and the Andes, and that he had also put in at various ports in Chile. Although *Moby-Dick* flopped when first published in 1851, Melville reenergized his flagging career by penning a series of articles in *Putnam's Monthly Magazine* that drew further attention to the area.

Melville was especially intrigued by the "Encantadas," or enchanted isles, a romantic name for the Galapagos Islands, which Darwin had brought to the attention of the world in the 1840s. In his article on the Encantadas published in 1854, and two years later in book form, Melville described "The Isles at Large" as a fascinating volcanic wasteland. I had read Melville's writings on the Encantadas during my visit to Ecuador in 1992, but geographer Karen Koegler recently asked me if Melville knew about the Atacama based on the following quote: "On most of the isles where vegetation is found at all, it is more ungrateful than the blankness of Aracama [*sic*]." Melville's use of the name "Aracama" always puzzled me, for I could never find it on a map of the period. However, a clue may lie in Melville's next sentence, which further described that locale as "tangled thickets of wiry bushes, without fruit and without a name, springing up among deep fissures of calcined rock and treacherously masking them, or a parched growth of distorted cactus trees." Seeming to forget what he had written about arid Lima in *Moby-Dick*, Melville added, "the showers refresh the desert hills, but in these isles, rain never falls."[66] It is possible that Melville was referring to another place named "Aracama"; however, it should be recalled that Darwin's books not only described the vegetation of the Galapagos and the blankness of the region commonly called the Desert of Atacama, but he did so within just a few pages of each other.

Given Melville's familiarity with Darwin's writings, I now suspect that he was indeed referring to the Atacama Desert, which Melville or his publisher may have misspelled. Misspellings aside, Americans were becoming familiar with the Desert of Atacama in the mid-1850s, for it was appearing in reports and on maps such as those by Wheelwright and Gilliss. Of those maps that actually misspelled Atacama, most appeared earlier in the nineteenth century and seemed prone to abuse the first consonant— "t"—in the name: *Alacama* and *Attacama* are examples, the latter found on Carl Ferdinand Weiland's map *Sued-America* (Weimar, 1823). Moreover, as will soon become apparent, writers other than Melville would also take occasional liberties with the name Atacama, continuing into modern times.

With the mid-nineteenth century in mind, I shall now return to a locale about 120 miles (193 km) north of Coquimbo, namely, the Copiapó River Valley. With the narratives of Darwin and Philippi in hand, it is remarkable to ponder just how differently various observers found the same stretch of ground between Caldera and Copiapó to be. A scientist such as Philippi evidently came to appreciate the arid landscapes, intrigued by the discoveries they might yield. Others, though, seemed more like Darwin in their dislike for the desert. Of the many who despised the desert and yet wrote about the area along the Copiapó River Valley, none better portrayed the region as dramatically as an anonymous writer in 1862. Some attribute this article to the Irish novelist Anna Maria Hall (1800–1881), but this is doubtful. It is true that Hall published an article in the same issue of the *St. James Magazine*, but she did so under her own name, and that may account for the confusion. Moreover, I can find no record that Hall ever travelled to South America, much less to the Atacama. Regardless of the mystery surrounding the identity of this source, its content is noteworthy indeed, and so I hope that readers will pardon me for addressing the author by the now-acceptable singular "they" pronoun.

The unidentified writer began the article by noting that their travels were "at the commencement of the year 185_ [*sic*]"—the last digit was mysteriously left blank. In this four-page synopsis, entitled "A Railway Trip in Chili," the writer recorded that they traveled by the steamship *Pacific* from Valparaíso, and "disembarked on the 27th of April at Caldera, a town situate[d] in a waterless sandy flat, interspersed with bare rocks of quartz and granite, yet rising rapidly in importance as the shipping-port of the Chilian silver regions and the terminus of the Copiapó Railway." After spending the night in Caldera, the writer boarded the train to Copiapó and "travelled the first fifty miles of the route, from the coast to the city, across the dreary desert of Atocama [*sic*], formidable for its parching heat and choking dust." As I noted in chapter 1, daytime high temperatures are normally relatively mild here, and a 100° day would be very rare, so one can only wonder if the writer had indeed experienced the withering heat implied. Nevertheless, the relative humidity can be somewhat elevated along the Copiapó River Valley, so that may explain any discomfort.

Just as likely, this Victorian-era writer seemed bent on impressing upon readers that the Atacama Desert was a harsh, if not downright disagreeable, locale. Observing that "the ground is extensively coated with an efflorescence of white salt and soda, the glare of which half blinded me," they then waxed increasingly dramatic, adding that "scarcely a living creature, bird, beast, or creeping thing, was visible." In prose intended to simultaneously impress as well as frighten readers, the writer further recalled that "the skeletons of mules and other animals, which had perished of drought, whitened on the wilderness through which we passed, and vultures whirled occasionally over our heads, as if on the lookout for similar victims." Emphasizing the scarcity of fresh water, they added, "Not a blade of grass appeared; not a rill of water—every drop provided for the relief of passengers and for the use of the boilers being carried with us in tanks, having been distilled from sea-water at Caldera." The area, though, did have one redeeming feature. As the writer put it, "Yet, this desolate tract teems, beneath the surface, with boundless wealth—copper, tin, bismuth, mercury, silver—offering an allurement to human avarice stronger than the fear of danger or death."[67] The entire passage epitomized colorful writing in an age when the public craved adventure, albeit mostly vicariously.

Regarding scientists, though, I would now like to return to the comparison between Charles Darwin and Rudolph Philippi. Whereas the former missed an opportunity to interpret the desert, the latter seized it. To Philippi goes the credit of being the first naturalist to comprehensively study the region's flora and fauna. In fairness to Darwin, who traversed the same portion of this region twenty years earlier, it should be recalled that the world was only just awakening to the uniqueness of deserts generally. By the 1840s and 1850s, a decade after Darwin's time in the Atacama, the world's arid regions were coming into view as places where significant scientific discoveries might be made. The American West was one such place, and it was being explored and mapped by the likes of John Charles Frémont and Charles Preuss. Building on experiences such as field observations of British-born naturalist Thomas Nuttall (1786–1859) in the increasingly arid area east of the Rocky Mountains in the 1820s, Frémont had the foresight to involve naturalists such as the American-born botanist John Torrey (1796–1873), who aided in the identification of new species, including shrubs and cacti in the American deserts that Frémont explored.

Simultaneously, scientific interest had also awakened in Chile. Although Chilean scientists had taken an interest in the flora and fauna as early as the 1820s, with some of the cacti being classified, at least formatively, long before Philippi arrived in South America, most of their effort had been focused on the central and southern part of that long, narrow country. Of his many botanical accomplishments on his journey through the Atacama, Philippi is perhaps best known for his observations on cacti. Several were first described by him, although they have been subsequently reclassified as taxonomy has evolved. Philippi himself apologized to the "gentle reader" of his report for using a classification system that was in its infancy. Typical of this classification conundrum is what Philippi called *Echinocactus cinereus* in his 1860 report. We now know this cactus as *Copiapoa cinerea*, and it is one of more than a dozen cacti in this genus named after the community and province of Copiapó. As a genus, *Copiapoa* is very varied, its twenty-six known species being quite different in appearance; most, however, are fairly small (under a foot, or about 20 cm); of these some are squat, and others somewhat more erect in appearance. All but one species of *Copiapoa* have pale-to-bright-yellow flowers, red being a rare cactus flower color in the Atacama.

In Adriana Hoffmann's informative book *Cactaceas en la Flora Silvestre de Chile* (Cactus in the Wild Flora [i.e., natural vegetation] of Chile), first published in 1989, and its subsequent revision with coauthor Helmut E.

Walter (2004), Philippi is recognized among the great botanists of Chile. As the authors conclude, "With respect to Cactuses, Philippi described in diverse papers and books more than thirty-five new species new to science, whose type plant [specimens] are preserved carefully in the herbarium of the National Natural History Museum."[68] Founded in 1830, that museum in Santiago remains a treasure house of the country's natural history, including archaeology and biogeography.

Praise for Philippi continues. The stunning new online Guía de Campo (field guide) *Cactáceas Nativas de Chile* (Native Cacti of Chile) by Florencia Señoret Espinosa and Juan Pablo Acosta Ramos credits Philippi as the discoverer of *Copiapoa cinerea*, formerly called by its synonym "Echinocactus cinereus Philippi 1860." However, as hinted at in the mention of thirty-five cactus species earlier, Philippi deserves credit for far more than discovering this variety of *Copiapoa* cactus, the genus of which consists of plants that are fairly inconspicuous. To Philippi also goes the credit of discovering two other cacti that characterize two very different ecological zones in the Atacama. The first—*Echinopsis atacamensis*—is perhaps the most spectacular found in this desert. Philippi originally named these cacti *Cereus atacamensis*, but the scientific name was subsequently changed. Standing up to twenty-five feet tall (6 to 8 m) when mature and growing in favorable habitats, these columnar cacti stand like sentinels, sometimes as individuals but often with companions comprising a small forest (see again Figure 1.19). Some of these cacti have curved arms that protrude from the main stem, not unlike the famed saguaro cacti of the Sonoran Desert. *Echinopsis atacamensis* is indeed emblematic of the Atacama Desert and portions of the puna, for it grows throughout much of the region, although only in the elevations above about 2,000 meters (6,500 ft). It thrives in the Andes and is not restricted to the Atacama in Chile, also being found in Bolivia and Argentina, another reminder that desert conditions are on both sides of the Andes. Its bright yellow flowers enliven the subtle pastel colors of the desert.

In addition to discussing this distinctive cactus, Philippi also described another columnar cactus that is found closer to the coast and grows in multi-stemmed clumps not unlike an organ pipe cactus of the American Southwest and northwestern Mexico (see again Figure 1.16). Philippi called this *Eulychnia breviflora* and the name has stuck. This is a conspicuous plant and has a number of close relatives. Like the organ pipe cactus, its slightly curving erect stems are usually segmented and sometimes deeply furrowed or corrugated. Cacti in this family are common in the area near Pan de Azúcar, a beautiful and somewhat isolated area along the coast between Paposo and Taltal. Like all cacti, these are extremely sensitive to environmental stress.

Some are so dark in color that they appear brown and lifeless, though in wet years they may bounce back into the more typical dark green color.

For his part, Philippi sustained his interest in the plants of this arid region of South America, identifying other species of cactus as late as 1891.[69] He also took an interest in the desert's geology, including that mysterious "Meteoritic Iron of Atacama," for which he provided geographic coordinates. This supposed deposit of extraterrestrial iron had long captivated scientifically inclined Europeans, including the prodigious British scholar William Bollaert, who mentioned this phenomenon in his 1860 Andean ethnography along with several references to Philippi's work.[70] If the supposed Meteoric Iron of Atacama, as reported by Philippi, was of considerable interest to Bollaert, it was mainly as another of the superlatives associated with the adjacent Andes Mountains. Despite his crediting Philippi's work, Bollaert's tome ultimately relegates the Atacama Desert to a series of footnotes. To Bollaert, the Atacama was a vast, almost uninhabited region lying in the shadow of the great Inca Empire, which captured the lion's share of popular attention, as grand civilizations often do.

With Rudolph Philippi's accomplishments in mind, I would now like to again return to Charles Darwin, or rather Darwin's inability to see the desert and place it in context. Rather than being oblivious, perhaps Darwin could be better described as distracted, not in a negative way, but more deflected toward his main scientific concern at this time, which was the diversity of life. Of this diversity, Darwin saw precious little in the desert, and hence left no record of it. This is understandable in that hyperarid regions exhibit less biodiversity than better-watered locales. As naturalist David Rains Wallace surmised, "Darwin's *Beagle* journal had little to say about [the] desert's possible effects on life, or about desert life in general." There was, however, one exception. "He did note in northern Chile that only one animal, a snail, seemed abundant on a particularly barren plain." As Darwin had noted, "in the spring a humble little plant sends out a few leaves, and on these the snails feed." Hinting that this was not an isolated occurrence, Darwin added, "I have observed in other places that extremely dry and sterile districts, where the soil is calcareous, are extraordinarily favorable to land shells." For Darwin, though, the paucity of life here made the desert an "evolutionary backwater."[71] Darwin's contemporary Alfred Russell Wallace evidently felt much the same. In his recent book *The Desert*, Michael Welland concludes, "After all, the ideas about evolution developed by Darwin and, from the jungles of southeast Asia, Alfred Russell Wallace, were about competition, and the perceived lifelessness of the desert seemed hardly conducive to evolutionary excitement."[72] Then, too, it should be recalled that Darwin had avidly read Dampier's *A New Voyage* and accordingly was lured away from the desert by the abundance of life elsewhere, just as Dampier himself had been a full generation earlier.

However, Darwin must be credited, because he left some words of wisdom for future researchers regarding the desert. Although not interested enough in the flora here to give it serious study, he anticipated that someone, someday, would be. As proof, I cite a remarkable quote from Darwin's writings (it appeared in slightly varied form in *The Voyage* and his diary) where he observed that even in such "uninteresting country" one will discover that "there are not many spaces of two hundred yards square, where some little bush, cactus, or lichen may not be discovered by careful examination."[73] In that one prescient sentence—he even suggested a quadratic study area for measurement—may lie Darwin's greatest contribution to the future study of deserts.

In the Atacama Desert, though, it was geology that Darwin found especially noteworthy, for the lack of vegetation enables the strata to be all the more evident. Returning to the anonymous Victorian-era writer, it should be noted that educated people of the time had an intense interest in geologizing. Many even had personal collections of rocks and minerals and were familiar with their classification—as was hinted at in the writer's reference to quartz and granite rocks defining the harbor at Caldera. While on the subject of geology, I should note that Darwin and Philippi passed nearby but missed seeing one of the locales I had longed to see, namely, the striking outcroppings of orbicular granite, or "gránito orbiculár" as locals call it, which jut out into the Pacific Ocean about ten kilometers north of Caldera (Figure 3.22). This orbicular rock is spectacular indeed, its circular inclusions reminding one of polka dots. Most geologists think this rare phenomenon is the result of differential cooling of the diverse minerals in magma. In the varied concentric orbicules, one can make out the crystalline grains of distinct minerals such as hornblende and the lighter-colored feldspar. Orbicular granite is not only interesting to ponder scientifically but aesthetically stunning as well. Succumbing to the romance of Victorian writers, I might best describe them as leopard-spotted whales frozen in stone, seemingly petrified while unceremoniously stranded, or beached, in the shallow waters at the edge of this restless continent. More prosaically, geologists have corrected the very name of this feature, noting that this unique formation is not technically granite at all, but rather more properly called tonalite due to the minerals involved.[74]

Again returning to the anonymous Victorian writer, when they reached Copiapó at the end of that long and dusty train ride, geology presented itself forcefully. At the otherwise "large and handsome" home of a Spanish

Figure 3.22. Geological wonders abound in the Atacama, an example being this unique outcropping of granito orbicular (orbicular granite), which was missed by Darwin and Philippi but is now a geological landmark on the coast about eight miles (10 km) north of Caldera. (Photograph by author)

Figure 3.22 detail. A close-up of orbicular granite (actually tonalite) reveals the distinctive pattern of mineralization. (Photograph by author)

merchant in Copiapó, they noted that the structure had been "wrenched and distorted in all directions and its walls, disfigured by wide rents and fissures, required huge iron rods to keep them together." Although the city of Copiapó was reportedly as quiet as "a city of the dead," the writer noted that it came alive at night as earthquakes rumbled far beneath it. In a superb passage that will be recognized as accurate by anyone who has experienced such an earthquake, the writer was unnerved by the "low, rumbling moan, like suppressed thunder, in the ground beneath me, followed by the trembling of the earth, increasing to a heavy shock, as if the solid globe were rending asunder."

Upon leaving Copiapó, the writer travelled to "Tres Puntos" [*sic*] and made a visit to some newly discovered silver mines. There they noted that "the mountains inclosing the valley are all of the older formations." Attuned to geology and primed for adventure, the writer observed of the landscape that "the stratified rocks, having been greatly disturbed by some convulsions of nature, are generally thrown up in nearly vertical directions, presenting the most fantastic shapes and outlines." In this landscape nearly devoid of vegetation, the writer also noted a paradox. "Marvellous it is, here, in country where rain has not been known for centuries, to observe vestiges of violent water action." This explanation involving catastrophism, or perhaps diluvialism, might help explain this dilemma, as the writer concluded, "the whole vale, indeed, seems one vast river-bed; and enormous heaps of *débris* and gravel appear as if but recently thrown together, though, in reality, the product of currents and eddies which existed whilst the now bare and burning mountain-peaks around were yet submerged beneath the ocean."[75] Clearly, this writer's belief that it had not rained for centuries was a factor in reaching for other more dramatic explanations. Alas, geologists and geomorphologists now know that running water from the Andes and/or local storms were largely responsible for these deeply gouged landscapes. In that regard, the Atacama is a rather typical desert, where—paradoxically—the action of running water is a major factor in the look of the normally parched land.

Humankind also transforms the surface of deserts through varied activities, the most profound being mining. In Philippi's time, mining in the north was booming. The Chilean historian Luz María Méndez Beltrán introduced earlier in this chapter noted in her comprehensive study of early-nineteenth-century mineral exports from Chile that copper, silver, and gold from the desert north were shipped to destinations worldwide. Méndez concludes that this industry helped Chile make the transition from a colony to a nation. Drawings and paintings from the time helped Méndez tell part of this story, as she used images from Philippi's 1860 report to give her readers an idea of what mining locales looked like. She also used a somewhat later illustration that reveals how mining was changing around the mid-nineteenth century. This illustration of a silver mine near Chañarcillo (Figure 3.23) by John Marx confirms the adage that mining literally turns the earth inside out. We have already seen another image by Marx that shows a hillside mine above

Figure 3.23. John Marx's 1876 watercolor painting of the Mina Dolores reveals that larger-scale mining had come to the Atacama Desert, as evident in the large buildings and extensive terracing of the site. (Author's digital collection; original in the Colección Iconográfica, Archivo Central Andrés Bello, Universidad de Chile)

the dense camanchaca fog (see again Figure 1.12), but that mine was much smaller. By contrast, Marx's 1876 illustration of the Mina Dolores perfectly captures the specter of huge ore dumps and tailings piles that accumulate as a mine develops. As in the other illustration by Marx, watercolor (acuarela) was the perfect medium to portray the pastel tones of the desert as well as the accretionary features of mining. In discussing this illustration, Méndez notes that although the glory days of Chañarcillo had ended, mining was still important.[76]

In this painting, Marx shows the new mine buildings perched on a steep, barren hillside. Crowning this complex, which is painted white, a tall stack belches out a column of black smoke. To the left of the largest building, a gallows-like headframe and hoist house are the means by which miners enter and exit the mine, and ore is hauled to the surface. The complex sits above a landscape that has been terraced through the systematic accretion of crushed rock, along the base of which a mule-drawn wagon and three men are pictured. As more material is added, this process builds up a huge base of pulverized rock that appears light blue in Marx's painting. This striking color may be the material removed from the earth and crushed, or it may be a reflection of the blue sky (much like snow may appear blue in shadows)—or perhaps some of both—but in any case is a reminder that what lies underground may be very different in color than the oxidized and weathered surface of the desert. Although these mines began as small operations like the one in Figure 1.12, which was similar to the mine at Trespuntas illustrated by Philippi twenty years earlier (see again Figure 3.15), Marx's painting of this silver mine at Chañarcillo makes it abundantly clear that considerable capital had been invested in the mine. Above all, Marx's watercolor confirms that the industrial age had come to the Atacama Desert. Tellingly, this painting's almost picturesque quality reminds us that in the nineteenth century industry and nature were seen as compatible. In Marx's almost sublime scenes, they comfortably coexist, for nature seemed boundless and man's actions to transform it were regarded as heroic rather than rapacious.

By the 1860s, most well-informed readers in Europe knew that portions of South America were deserts and that mining was an important activity there. Nevertheless, although the name Atacama Desert appeared on many maps, it did not appear on a good many others. A case in point is the beautiful two-color German map *Südamerika* by F. A. Brockhaus. On this map, which was published in 1863, blue is used to depict water and a sepia brown indicates topography (Figure 3.24). Two sepia-colored stippled areas draw the attention of those concerned with the character of the landscape. The first of these stippled areas, which stretches along the southeastern (Atlantic) coast, is Patagonia, where Darwin had his first encounter with aridity. The second, which is positioned straddling the east–west line called "Wendekreis des Steinbock" (Tropic of Capricorn in German), is the Atacama Desert.

This map by Brockhaus is just as interesting for what it does not say as for what it does. Despite not being named on this map, the Atacama Desert is nevertheless represented as a region having a particularly parched quality. As a reminder that natural features other than deserts were still drawing attention, however, none of the striking landscape images or vignettes that surround the mapped area of the continent depict the Atacama or for that matter Patagonia; rather, it is the towering Andes (for example, Mt. Chimborazo in Ecuador) and dense jungles of the Amazon that are immortalized. The closest one comes to the Atacama Desert in these vignettes is a depiction of condors devouring the carcass of what appears to be a

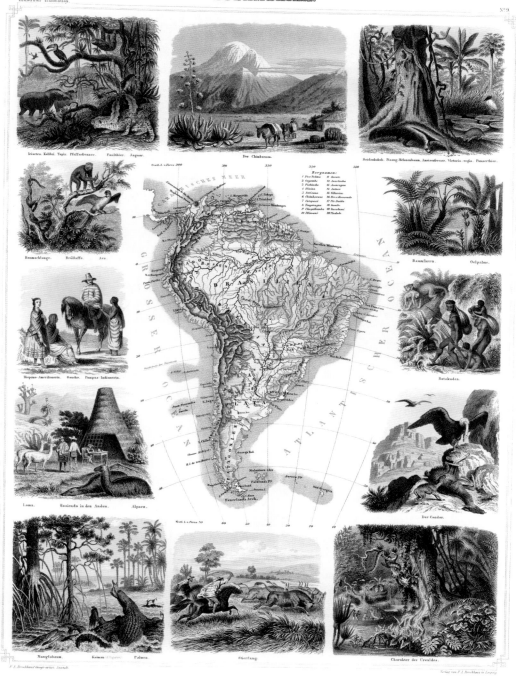

Figure 3.24. The beautifully illustrated map *Südamerika* by F. A. Brockhaus (1863) does not name the deserts of that continent, instead showing them in an evocative stippled sepia pattern; nor do the surrounding vignettes illustrate the desert, focusing instead on grander sights such as the jungles and the Andes Mountains. (David Rumsey Map Collection)

Figure 3.24 detail.

guanaco or deer in the Andes. At this time, depicting deserts was evidently not a high priority, in part because arid regions seemed bleak and did not possess the scenic grandeur associated with boundless forests and commanding mountains. Yet the absence of an illustration of the desert may have operated to increase the mystery about such sandy wastelands. As will become apparent in the next chapter, travelers would soon try to set the record straight about the economic and aesthetic value of deserts, including the Atacama.

Four

"A Centre of Enterprise...
and Civilisation"

*Transforming the Atacama Desert,
ca. 1865–1945*

This chapter begins with Rudolph A. Philippi basking in his recognition as one of Chile's great naturalists. When published in 1860, his report on the Atacama represented a landmark accomplishment, but it was the work of a generalist at a time when specialization began to reveal that one scientist could not describe, much less interpret, every aspect of an environment. Appropriately enough, the 1860s ended with another of Philippi's peers—the indefatigable Swiss-born natural historian Johann Jakob von Tschudi (1818–1889)—completing his own monumental five-volume work entitled *Reisen durch Südamerika* in 1869. This covered the entire continent, with Chile comprising the last volume. In his section on the Atacama, Tschudi described many of the subjects covered by Philippi yet also criticized some of his fellow naturalist's findings, including Philippi's classification of fauna and calculation of geographic distances. That type of criticism may seem harsh but that is the way science works, and other scientists soon critiqued portions of Tschudi's work. The point to recall here is that one subject Philippi knew and described best—geology—ensured that his Atacama report would be consulted for at least a generation.

As the increasingly frequent references to mining in the last chapter suggest, there is another Atacama Desert that lies beneath the surface, its rocks convulsed and shattered, enriched by trickling solutions of hot water pregnant with dissolved metals destined to become minerals such as atacamite, the resplendent green copper chloride that was first named in 1801. While traversing a secluded canyon in the coastal hills east of Paposo, which is situated about two hundred miles (ca. 320 km) northwest of Copiapó, I discovered the perfect place for geologizing. Nature here presented a complex display of geology, where right on the surface, rocks rich in copper and manganese oxides cropped out and lay there to be leisurely explored on a Sunday morning. Unintentionally assisting me, the owners of a mine had created

Figure 4.1. A handful of minerals found near a copper mine at Paposo, Chile. (Photograph by Damien Francaviglia)

a road here, and strewn about were ore samples that would have delighted Godoy and intrigued Philippi. Some of these rocks had evidently bounced off the trucks, making the task at hand much easier for me. A random handful of these (Figure 4.1) reveals their beauty and variety. Discovering some large, promising rocks, I took out my geologist's rock pick, aiming to do some serious investigation. After a few minutes of this, one of these chunky rocks miraculously yielded a small but beautiful section of emerald-green atacamite, the elusive copper chloride I had wanted to find since 1954, when I began my mineral collection. My 1955 edition of Frederick Pough's *A Field Guide to Rocks and Minerals* noted that atacamite is "a rare copper mineral" that nevertheless was "common under the extreme conditions of continuous dryness of the South American west coast in Chile (the Atacama Desert)."[1] I still own the copy of this popular book that my folks inscribed "To Richard— From Aunt Faye & Uncle Vic, 12/25/55." Although it is now well worn from its many trips into the field, it is my most prized book out of the many thousands that I've owned.

There is something magical about splitting open a rock and finding something never before glimpsed by another human being. In words that Darwin and Philippi knew well, the Bible's Book of Job noted more than three thousand years ago that miners in search of ore tunnel through rock and find things that are hidden, bringing them "forth to light" (28:9–11). Here in the hazy sunshine of the Atacama, I marveled that this was the first light to ever greet these treasures that were thousands, perhaps millions, of years in the making. Awed by this, I realized that this trip was rapidly helping me check off things on a very long bucket list. Wistfully, I realized I could do such field exploration *forever* and never tire of it. Like Darwin and Philippi, I too am especially impressed by mining, which in their time was helping to reveal the secrets of the earth and revolutionizing society. For his part, Philippi recommended the establishment of additional transportation and communications systems such as electric telegraphy, to help develop what he called the "fabulous mineral riches" of the Atacama Desert.[2] With this in mind, I now realize that Darwin and Philippi were working simultaneously in two dimensions—horizontal and vertical—as I have always tried to do. By this I mean that although they seemed for the most part concerned about what was happening on the earth's surface, they were also trying to decipher what was going on, or had gone on, *below* it. Theirs was, after all, the era in which the science of geology had just come of age, and was now increasingly coming to the aid of those who wished to develop mining properties into paying enterprises.

The time period now at hand, which began around 1865, witnessed major changes in the mining industry. These included a rapid internationalization of mining company activities, increased demand for minerals and metals needed by the manufacturing industries, and innovative technologies and processes (such as chloridization and cyanidation) that enabled miners to extract ores from previously uneconomical mining properties. The external forces that helped initiate new or increased mining included rapid population growth and innovations in other industries that in turn expanded the demand for metals. To cite just one example, the harnessing of electrical energy for lighting, household appliances, and transportation such as trolley cars and trams resulted in a higher demand for copper. As the world's population began to grow more rapidly, the amount of farmland increased; moreover, the productivity of such lands soon depended on inorganic (i.e., mineral) fertilizers, resulting in greater demand for nitrates. Lastly, as colonial powers expanded and came into conflict with each other, the machines of war demanded prodigious quantities of metals and minerals. Although quite distant from many of these developments, the Atacama Desert had ample quantities of strategic minerals.

At this point, I would like to address another force that made the development of mineral wealth possible, namely geology-related mapmaking. The modern exploitation of mineral resources was dependent on the coming together of the discipline of geology, which was relatively young, and mapmaking, which was venerable indeed. Only by around 1800 was the earth sufficiently understood for thinkers to grasp its incredible age—billions of years instead of thousands—and to understand that forces operating in the present were essentially the same forces that had always operated. By around 1800, it was understood that geological specimens from varied parts of the world could be compared to see how they fit into the bigger picture. In other words, by conceptualizing geological time geographically, one could then see broad patterns emerging—seas once covering entire portions of Europe at one time, for example, but leaving high and dry other lands at that same time. This helped put a crimp in the belief that a single flood had covered the entire surface of the earth, as a literal reading of Genesis might suggest. The fossil record helped immensely in this new premise, as specimens collected in varied locales were compared side by side. Thus, not only did the geological timescale develop, but maps could help show how that geological time played out spatially.

Pioneered by the early-nineteenth-century English and French geologists, geological mapping depicted the types of rocks that were found under the surface of the earth. Although the rock hammer was an indispensable tool in mining, efforts were helped immensely by miners toiling underground and construction crews building major transportation-related projects such as roadways and canals. The early-nineteenth-century contractor William Smith (1769–1839) produced the remarkable map that has been called "the map that changed the world."[3] Without such maps, modern systematic mining developments could not have occurred. The alternative would have been the hit-and-miss mining such as that which had pockmarked the Atacama since time immemorial. This worked well enough in finding the biggest or least concealed lodes, but now—with increasingly accurate geological maps in hand—miners and mining enterprises could effectively predict where the next major activities would occur.

In 1867, a geological map entitled *Carte des Terrains Metalliferes du Chili* (Figure 4.2) was published in Louis Simonin's classic *La Vie Souterraine, ou Les Mines et les Mineurs* (*Life Underground, or Mines and Miners*). By that time, Simonin (1830–1886) had completed his training as a mining engineer and was recognized as an authority on mineral deposits and their successful exploitation. He was well travelled and well connected, having explored many of the deposits he wrote about. The inclusion of Chile in Simonin's book is a reminder of how widely recognized this country's mineral riches were at this time. As Simonin's map reveals, Chile's metals-producing areas were restricted to the northern part of the country—from Santiago northward into what he labeled as "Desert d' Atacama." Silver, copper, and gold were the

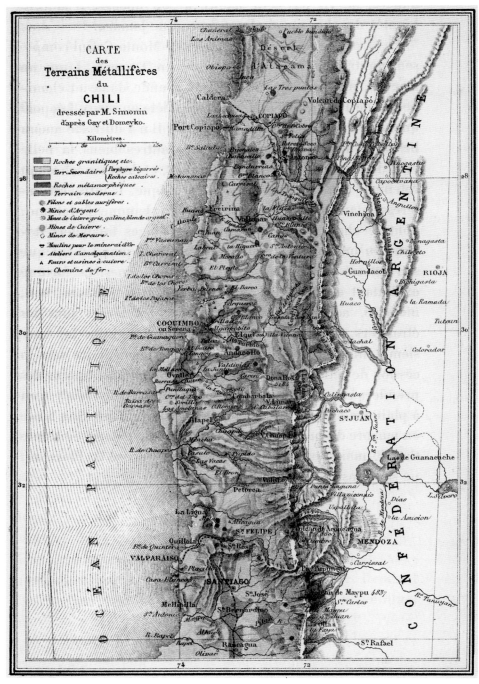

Figure 4.2. *Carte des Terrains Metalliferes du Chili* (1867) by Louis Simonin identifies varied mineral deposits in the country's desert north. (Author's digital collection)

major metals mined, and this map also depicts the mills and furnaces (smelters) where ores were processed.

As indicated in the map's legend, Simonin relied in part on data from Ignacio Domeyko to inform his readers about Chile, which was discussed along with other famous areas such as rich ore deposits of southern Germany and California. This map revealed the close relationship between mines and certain types of rocks, including porphyry and calcareous rocks such as limestone. For our purposes here, the area from Coquimbo northward is arid and dotted with mines. First named in 1841, the rare iron sulfate mineral coquimbite often occurs in showy purple crystals and helped immortalize this city in the scientific literature. In 1869, the Harvard University geological museum received a specimen of Chilean lazurite (lapis lazuli) along with the L. Liebner collection. This striking blue mineral, the only gem mineral known to occur in Chile, was found in the desert region north of Coquimbo. In 1865, Domeyko—for whom the rare copper arsenate mineral Domeykite was named—provided Harvard with about twenty mineral and ore specimens, mostly of copper and silver, from this desert region.[4] Through such transactions, the Atacama Desert came to the attention of American mining engineers.

North of Coquimbo, as indicated on Simonin's map, communities such as Vallenar and Copiapó were important mining centers. In an isolated area north of Copiapó lies Tres Puntas, a significant metals producing area that, as we have seen, was described and illustrated in Philippi's report. Note that although the railroad connecting Copiapó with the port of Caldera is shown, this map also indicates Port Copiapó south of it.[5] In time, that port would further dwindle in importance. From today's vantage point, the fact that the northern part of Simonin's map ends well south of where we know Atacama extends may seem odd. At this time, however, the Atacama Desert was nearly synonymous with metals mining in the minds of European capitalists. That perception would soon be amended to include the immense nitrate deposits that were found in portions of the northern Atacama Desert well into what was then Peru and Bolivia. These, as will be recalled, had been visited by Darwin, but their full extent had yet to be determined. Here, in parched country just east of the coastal cordilleras, lay vast deposits of *caliche*, the term widely used for nitrates. These deposits would soon transform not only maps of mineral deposits, but maps of the countries bordering Chile on the north and east.

Before discussing this transformation, though, I would like to say a bit more about geological mapping, which dovetails beautifully with other types of maps but has its own idiosyncrasies. On many historical maps of the Atacama Desert, mining properties are simply depicted by a named dot, or sometimes even more explicitly by a symbol of crossed shovels or rock picks. However, other maps may show them in far more detail, for example, as distinct parcels of land bounded by lines that separate a particular mine from its surroundings. One mine may thus be mapped, but others nearby can be mapped simultaneously, as in the case of mining districts where several mines exist side by side. The typical mining district appears as a crazy-patch quilt with different sizes and shapes. Each of these patches is a separate property, but collectively they comprise the district bearing a particular name—for example, Trespuntas (or Tres Puntas)—to distinguish it from other districts. In other words, maps showing mines lay at the juncture between scientific mapmaking and cadastral mapping, which is to say the actual delineation of property parcels. Like other nonproprietary maps, these show the surface extent of the mining property and are available to the public, who usually consult them in the records of a particular political jurisdiction, for example, the Provincia or Departamento de Atacama.

At a larger scale, the next type of mining map goes one step further, showing the actual workings of a mine such as tunnels and shafts below the surface. Although these are also drawn based on surveys, they effectively strip away the surface to show the spaces excavated by the miners. In the process of actual mining or exploration, miners may dig or blast out these tunnels, sometimes through non-ore-bearing rock in order to reach the ore body. Once they reach that ore, miners excavate it as economically, and hopefully as carefully, as possible. Given the cost and effort involved, plus the strategic value of this information to competitors, such proprietary maps were rarely seen by the public. This mapping might begin simply enough, but given the fact that mines often work at several levels under the same place on the surface, they become complicated indeed. It is virtually impossible to depict more than a few of these vertically separate tunnels on the typical map, even using varied colors, and so levels may be drawn on separate sheets. Moreover, to better show the vertical positions of these complex operations, the surveyor may draw them in cross-sections. Much like a geological cross section, and sometimes containing similar information about the type of rock that has been excavated, these maps are absolutely essential in keeping modern mining profitable and safe. Should trouble be experienced in one part of the mine—for example, a fire or cave-in—this complex mapping permits rescuers to find alternate ways to deliver supplies or evacuate stranded miners.

By now, it should be apparent that an underground mine is a complex, multistoried operation that works horizontally and vertically, and often at angles in between. To understand what was happening deep in the mine, map users such as mine operators had to compare the maps with the cross sections, in other words, think in two dimensions simultaneously. Some mining companies went beyond this by actually constructing three-dimensional models of the mine. So that each level could be seen in relation to the others, the individual levels were often made out of sheets of glass, each sheet having the information painted on it. Today, with the advent of computers, much the same kind of model is available, only virtually. With the click of a mouse and the touch of a key on the keyboard, the three-dimensional image of the mine appears on the computer monitor. With the right program, one can effectively travel through the mine virtually, or as a hologram. The experience of using these devices is much like moving through terrain in a video game. The maps in this category are remarkable indeed, intimate depictions of ore bodies in relation to the geology. They are the counterparts of the computer-generated images used by doctors to navigate their way to trouble spots in the human body.

Maps showing mine locations became common in the Atacama Desert as mining companies entered the area with an eye toward developing profitable operations. To make these maps, modern companies hired mining engineers, most of whom were educated in Europe or the United States. The creation of the school of mining and metallurgy at the University of Chile in Santiago dates to 1853. By the 1860s, it was apparent that Chile's future would become increasingly dependent not only on mining but on the mining engineer, whose talents by definition included surveying as well as mapping. The maps that they created served many purposes, not the least of which was impressing would-be investors as well as being used during litigation when one company was accused of impinging on the property of another. Their main purpose, though, was to ensure that mines developed and operated as efficiently and effectively as possible. Additionally, as noted, maps could mean the difference between life and death if a mishap occurred.

As I hope is clear by now, the Atacama is a place with a remarkable surface sculpted in sand and stone, but what lies below that surface has long held the attention of desert dweller and visitor alike. In the late-nineteenth and early-twentieth centuries, this desert was scoured by mining engineers. It is difficult to overstate the importance of mining here. Mining helped build the Atacama Desert's cities and transportation system as we know them today. It also helped shape the character of the Chilean people. With

so many foreigners coming and going, the desert north came to see the outsiders as more or less inevitable, even desirable. Though somewhat ethnocen-
tric and perhaps even a bit chauvinistic, Isaiah Bowman felt an affinity with
Chileans. As he put it, "the Chileans are well aware of their own progressive
qualities and are so proud of them that they call themselves the Yankees of
South America." Even though the term Yankee had negative connotations in
other parts of Latin America, the patriotic Bowman claimed that Chileans
felt differently. As he concluded: "To them, the word 'Yankee' means a person of energy and ingenuity." I will leave to social critics the task of scrutinizing Bowman's comparisons, but speaking personally, I am pleasantly surprised
that his words seemed to capture something positive about the enterprising
spirit that is still evident today.[6]

Much of the mineral wealth of the Atacama Desert now lies in Chile and
is related to metals mining. However, the portion of this desert controlled by
Bolivia in the 1870s had some promising copper deposits and was even better
known for its vast reserves of sodium nitrate, or niter. Miners in the Atacama
had long known about niter, including its explosive properties and its potential as a fertilizer. Regarding the latter, the Inca reportedly used nitrate-rich
guano—the excrement of birds that was found in immense quantities in the
desert areas along the coast. When strewn on fields, guano greatly increases
the productivity of crops. Throughout much of the early nineteenth century, guano was shipped to Europe after being mined from deposits on the
Chincha Islands that were up to 50 meters (ca. 150 ft) thick. So profitable
was this trade that thousands of Chinese workers were recruited to collect
it in 1849; in 1864–1866, Chile and Peru formed an alliance to keep Spain
from further exploiting it. Early miners were also said to have known about
nitrate's explosive properties, but regardless of any local use, it was international demand that made mining it profitable after Europeans discovered
nitrate's twofold uses. From minimal production in 1820, mineral nitrate
grew to challenge, and then eclipse, guano by about 1860. By the 1870s,
these nitrate deposits were drawing European capital to Bolivia and northern Chile.

Most of the nitrate mineral production centered on the coastal area
where the borders of Chile, Peru, and Bolivia intersected. In the dry climate,
huge beds of nitrates awaited exploitation. Some of this ore, called caliche,
was mined and shipped as is, but in other places it was mined, put into solutions, and boiled to increase the concentration of nitrate. In either case, it
took considerable labor to produce and get to market. Because the nitrate
was not found on the coast, but rather inland from the coastal mountain

ranges, it required considerable transportation from mine to port. Typically, mining enterprises not only developed the mining property but also helped improve roads, some of which amounted to little more than trails in the desert. In September 1868, two Bolivian businessmen—José Santos Ossa and Francisco Puelma—formed the evocatively named Sociedad Exploradora del Desierto de Atacama, one of its goals to build a road to reach nitrate deposits.[7]

Although the term exploradora implies exploration, which was indeed one of the society's goals, the ultimate objective was exploitation, that is, the commercial development of mineral resources. The term sociedad (society) signified what we might call a corporation, a group of people unified by a belief in developing something. As with all corporations, this one was granted the right to exist and carry out its goals by a government, in this case Bolivia. That decision made in Bolivia's capital, Sucre, had international repercussions. First, additional funding and expertise began to flood into this part of the Atacama Desert from England and elsewhere. The impressive four-story Casa Gibbs in the city of Antofagasta, which was at this time in Bolivia, dates from the time that William Gibbs was involved in this enterprise. Second, while Bolivia nominally owned this part of the Atacama Desert, its neighbor to the south—Chile—had a stake in it; since 1866, Chile and Bolivia had agreed to share the tax revenues generated in the zone between 23 and 25 degrees south latitude. In an 1874 treaty, Bolivia collected full tax revenues but agreed to a fixed tax rate on Chilean companies. Most maps showed this area as belonging to Bolivia, but Chile also had a stake in the outcomes of economic ventures here.

Among the many Europeans drawn to Chile at this time was the French scientist Pierre Joseph Aimé Pissis (1812–1889). Arriving in Chile in 1848, he adopted the Spanish pronunciation of his name—Pedro José Amado Pissis—married a Chilean woman, and lived in his adopted home for twenty years before returning to France in 1868. In Chile, Pissis is recognized for his survey work that helped Gilliss's expedition map the country, including its desert north (see again Figure 3.10). However, Pissis is best known for his comprehensive book *Geografía Física de Chile*, which was published in Paris in 1875 and used by students and scientists alike. Containing detailed descriptions of the topography, climate, vegetation, and mineral resources, it made numerous references to the Atacama Desert where, as he noted, "rain almost never moistens the soil," and the vegetation changes in aspect: trees disappear, the only ones remaining being the algarrobo and chañar. Like many observers, Pissis speculated that the Atacama Desert may have been wetter in the past, and he emphasized Chile's mineral resources, particularly silver and copper minerals found in the Atacama Desert. Among the latter was atacamita (atacamite).[8] In 1877, Pissis published *Mapa Mineralógico del Desierto de Atacama*,[9] which was no doubt influenced by Simonin's 1867 map. Pissis is also well known for his beautiful shaded physiographic maps of portions of Chile, including the Atacama, though his most impressive work involved depicting the landscapes of a physiographic province that had a special appeal to him—the spectacular Andes.

At the time Pissis's book was published, Bolivia also had grand aspirations, in part fueled by the mineral wealth that was shipped out of its coastal desert region lying to the west of its Andes-cradled highland city, La Paz. One of the more fascinating nineteenth-century maps of Bolivia was published in Germany in 1876 by August Petermann's Gotha-based enterprise, which was now becoming widely known as *Petermanns Geographische Mitteilungen* (Petermann's Geographic Communications).[10] Entitled *Skizze des Litorals von Bolivia* (Sketch of the coastal region of Bolivia) this map (Figure 4.3) accompanied Hermann Wagner's scholarly article on that area's geography and resources. Although considered Wagner's map, it was drafted by Otto Koffmahn, one of the many talented mapmakers employed by Petermann.[11] This map makes explicit the assumption that Bolivia's greatness was dependent not only on its natural resources but also its access to the coast.

The first thing that may strike the reader of this map is how effectively it conveys both a sense of the topography and the cultural features such as towns, ports, roads, and railroads. The latter are shown as either complete or projected. Minerals, though, are also one of its strong suits; it shows copper mines using crossed picks, and light-orange-colored areas represent the salitreras or nitrate beds. This map was published at the height of enthusiasm about how scientific knowledge and cultural development could interact to facilitate progress. Industry was regarded as the force that could bring about betterments, and such improvement was predicated on transforming the natural world. The resources of the natural world abounded in this remote portion of South America, which was almost halfway around the globe, yet at its doorstep now that steamships shrank that distance. The German steamship line Kosmos was among these. At this time, Bolivia was not a landlocked Andean nation but also swept westward to meet the coast, where three ports—Antofagasta, Mejillones, and Tocopilla—connected the hinterlands to the rest of the world.

On this map, the intervening country is beautifully articulated as a series of mountain ranges and basins, one of which bears the name Desierto de

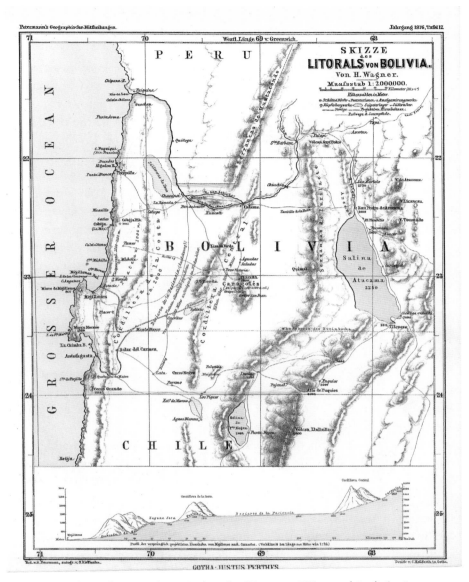

Figure 4.3. *Skizze des Litorals von Bolivia* by Hermann Wagner (1876) depicts a part of the desert region that now belongs to Chile. (Author's digital collection)

la Paciencia, with a parenthetical phrase in German questioning its actual extent. The term paciencia, meaning patience or forbearance in Spanish, is noteworthy. It could have many derivations but subliminally suggests that those in this region should take it on its own terms. In that sense, paciencia conveys a sense of respect, if not submission. The name Desierto de Paciencia is yet further evidence that what many people today broadly call the Atacama Desert was not so clearly conceptualized as a single desert; these interior deserts could be designated by any number of local names. Even though many others might have used the term Desierto de Atacama in the 1870s, most of them restricted it somewhat more than we do today. Wagner's map clearly suggests that a desert region is found within the coastal portion of Bolivia, but by not calling it the Atacama Desert (or Desierto de Atacama) he may have distanced it from Chile, where that name had long been used.

When Wagner's map was published in 1876, Chile and Bolivia were very visible contenders in a simmering, decades-long dispute that was attracting attention in Europe. England, especially, had a longstanding interest in Chile, which was now gaining a reputation as a place to make investments that would be protected by a pro-business government. In addition to wariness about Bolivia, Chile was concerned about Peru, which had long claimed a large part of the northern Atacama Desert in the vicinity of Iquique and Arica. In 1877, a map by Josiah Harding showing the nitrate-rich northern part of the Atacama was published in the *Journal of the Royal Geographical Society of London*. According to geographer and cartographic historian Karl Offen, this document, *Map of Part of the Desert of Atacama*, was both a record of what was thought to exist and also catalytic in shaping perceptions. As Offen perceptively characterized it, "this map helps construct the geographic reality it seeks to describe."

Josiah Harding, who drafted this map, was a British engineer in the employ of the British-Chilean Antofagasta Saltpetre and Railway Company (ANRC). The words "and Railway" in the title were both symbolic and symptomatic. As noted, transportation is an integral part of mining, for without delivery systems there is no way to reach a market. At just this time, railroads were poised to enter the scene in coastal Bolivia. This, after all, was the time when small industrial steam locomotives were being developed in Europe and the United States, many of them using narrow-gauge (often just two or three feet between the rails rather than the standard gauge of 4 ft 8 1/2 in). Such lines were cheaper to construct and operate than standard gauge, and their open-top gondola cars could easily carry as much as several horse- or mule-drawn wagons. After being hand shoveled into sacks made of jute from India, the nitrates were transported from mine to port via these narrow-gauge lines. Harding's map clearly showed the twisting line of the boldly named ANTOFAGASTA RAILWAY connecting that port city with the nitrate fields inland. Although this map shows the nitrate deposits and notes that they are in Bolivia, many Chileans claimed otherwise. According to Offen, Harding's map was available in Chile and might have helped "generate political momentum" for military action.[12]

That momentum was further galvanized by Chile's outrage as Bolivia imposed new taxes—after promising that taxes would not rise—on the Chilean mining company officially known as Compañia de Salitres y Ferrocarril de Antofagasta y Bolivia. Again, looking more closely at a corporate name is profitable, for the "and" in the company's name implies that Antofagasta is separate from, and hence not part of, Bolivia. That long but revealing company name is yet another reminder that mineral development and transportation often go hand in hand; in fact, many mining companies at the time were chartered to both develop mineral resources and transport them to markets. To do otherwise would be to rely on others to do the hauling, which could increase costs or cut profits. Given the rugged terrain and difficult operating conditions in the Atacama Desert, as well as its great distance from markets, transportation is a key factor in success or failure. With claims that a major Chilean mining and transportation enterprise was being hamstrung by higher taxes imposed by Bolivia, the international political situation in 1879 was nearing a boiling point. As an interesting aside, readers of newspapers worldwide were now familiar with the name "Atacama," for a steamship bearing that name under the ownership of the Pacific Steam Navigation Company (PSNC) "ran aground and wrecked at the mouth of the Copeapo [*sic*] River, Chile on November 30, 1877. 100 lives lost"; the *New York Times* claimed that seventy-two perished.[13]

With increasing frequency, this desert region was also appearing on maps. A fascinating German map by Augustus Petermann, entitled *Karte der Salzwüste Atacama und des Grenzgebiets zwischen Chile, Bolivia & Peru*, is one of the prizes in my personal map collection, and to my knowledge has not been featured in any scholarly book or article since its original publication in 1879. It depicts the geopolitical situation at exactly the time that disagreements between Chile and its northern neighbors escalated (Figure 4.4). Readers will recall that Petermann had published a map of the Atacama based on Philippi's expedition in 1856, and he also orchestrated the publication of Wagner's map of Bolivia two decades later (1876); now, with all eyes on the contested area between Chile and its neighbors, he enlisted the talents of his cartographer Bruno Domann to set a new standard in the region's cartographic history. This map accompanied an article entitled "Die Salzwüste Atacama,"[14] but the map's title is longer and more revealing, translated as

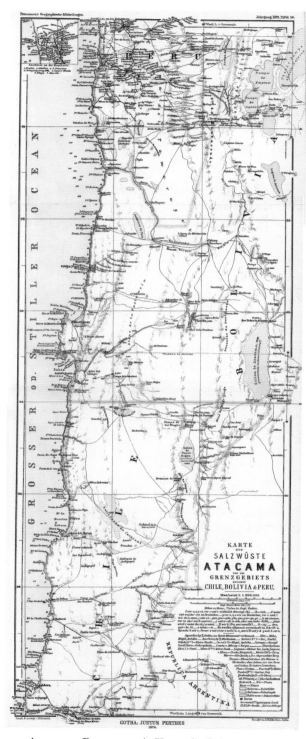

4.4. Augustus Petermann's *Karte der Salzwüste
Atacama und des Grenzgebiets zwischen Chile, Bolivia &
Peru* (1879).

*Map of the Salt Desert of Atacama
and the boundaries between Chile,
Bolivia, and Peru.* On it, the
mapmakers took great pains
to show the boundaries of the
three countries referenced in
the map's title. In fact, Chile
is outlined in pink, Bolivia
in green, and Peru in orange.
For the record, Chile's northern boundary had always been
vaguely defined, even in colonial times. Now, however, that
vagueness could be turned to
Chile's advantage.

At this time, Petermann's
color-coded differentiation took
on considerable importance, for
these were the very nations that
were now contesting claims to
territory, and increasingly suggesting that a war might be the
only way to resolve differences.
On one level, the map's title
simply refers to "the Atacama
Salt Desert" as a physically
defined territory, that is, an arid
region that contains surface
deposits of salt. However, more
than common table salt comprises these deposits, though
that can also be found here
as the mineral halite (sodium
chloride). Rather, the term salt
here refers to many water-sol-
uble minerals, including borax,
nitrates, and even salts of
iodine (yodo in Spanish). In
the Atacama, the term salitre

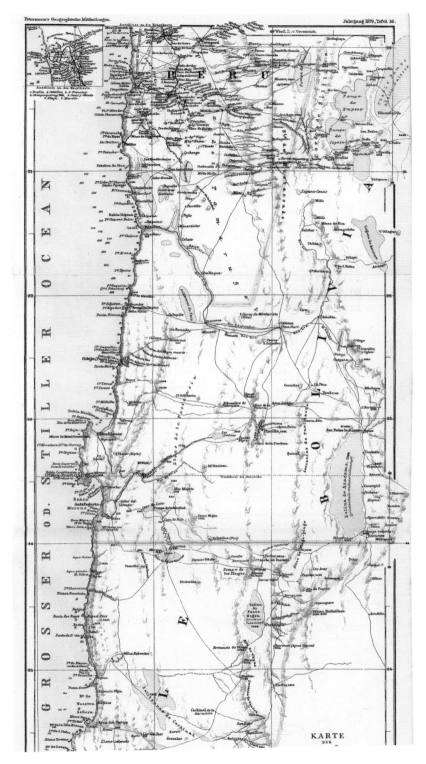

Figure 4.4 detail a. Northern portion of *Karte der Salzwüste Atacama und des Grenzgebiets zwischen Chile, Bolivia & Peru* (1879) reveals German mapmaker Augustus Petermann's attention to detail and his fascination with this desert. (Author's digital collection)

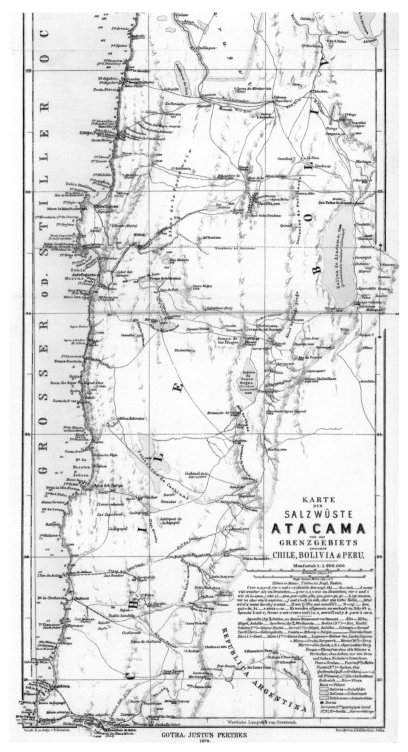

Figure 4.4 detail b. Southern portion of *Karte der Salzwüste Atacama und des Grenzgebiets zwischen Chile, Bolivia & Peru* (1879).

was used in much the same way, but especially for niter. With characteristic German thoroughness, Petermann distinguishes between and represents three different types of "salt" deposits using a combination of Spanish and German wording. The first two types—Salares-Salzfelder (salt fields) and Salinas-Salzsumpfe (salt marshes)—are shown in grey and designated by type in the text. In pink, Petermann depicts the Salitreras-Salpeterfelder (saltpeter fields), which were the primary areas containing nitrates. Saltpeter is the common name for potassium nitrate. Additionally, Petermann shows deposits of borax as purplish gray spots. He also makes several references to the hydrology, depicting streambeds and indicating numerous aguadas, that is, wells or springs where freshwater is said to exist.

The shape and format of this map—tall and lanky, not unlike Chile itself—is evocative, and its content is worth careful scrutiny. It shows the topography in considerable detail, identifying ranges of hills and mountains, including the evocatively named Morros de la Soledad (hills of solitude) that rise out of Peru's sweeping Pampa del Tamarugal. In Bolivia, the Llano de la Paciencia occupies much of the interior, while south of latitude 24 degrees lies Chile and its numerous nitrate deposits, such as the Salitreras de Cachinal, that were just beginning to be developed. As time would later prove, these deposits northeast of the port town of Taltal were rich indeed. Of special interest is Petermann's depiction of the transportation network. Reaching the inland pampa areas was a developing web of roads and railroads. Of the latter, several are noteworthy, including the rail line from Antofagasta that runs diagonally northeastward. Its ultimate destination would be highland Bolivia, but at this time the line ended inauspiciously at a dry lake east of the Llano de la Paciencia. To the north, the rail lines running inland from Iquique reached the interior nitrate fields in a series of twisting curves. Petermann indicates and names the many nitrate communities that were springing up here. Far to the south, in the vicinity of Copiapó, his map also shows the rail lines in the vicinity of that city as well as the line to the port of Caldera. For a German cartographer who never experienced the Atacama firsthand, Petermann was evidently obsessed by this faraway desert and came to know it intimately, at least on paper. In a sense, this map is one of the most impressive ever done by Petermann's long-lasting geographic enterprise, though alas, it was published a year after his death.

In that fateful year of 1879, many entrepreneurs in Europe and South America were also interested in the Atacama, as they knew that considerable economic reward awaited those who could exploit the desert's mineral resources. As tensions mounted, Chile confronted Bolivia over the treatment of the Compañia de Salitres y Ferrocarril de Antofagasta y Bolivia. Undaunted, Bolivia stood its ground. The situation grew even more dire when Bolivia refused to submit to arbitration or mediation that might prevent hostilities. Further complicating matters, and suggesting that war was a distinct possibility, Peru and Bolivia several years earlier (1873) had secretly agreed to defend each other against Chile should hostilities break out. The secret, of course, was difficult to keep, and all three countries increasingly hardened their positions despite diplomatic efforts. With no one backing down, the hostilities that Karl Offen alluded to now materialized.

The Guerra del Pacífico (War of the Pacific) lasted from 1879 to 1884. At the time, it was also nicknamed the "Ten Cents War" in reference to Bolivia's imposition of a small but troubling export duty on its products. What began as blustering about who could properly claim seaports such as Antofagasta and Iquique soon turned into a series of battles on land. In fact, the earliest mass casualties occurred at the inland oasis city of Calama, where Bolivian troops marching down from La Paz to Antofagasta were defeated by Chilean forces in a grisly encounter at the site of a bridge over the Río Loa. The nitrate fields literally became blood-soaked battlefields on occasion. Maneuvers took place in the vicinity of salitreras Dolores and Agua Santa. As historian William F. Sater grimly notes, Chileans overwhelmed the Bolivian and Peruvian forces near Oficina Germania, "killing more than fifty to sixty cavalrymen, whose unburied bodies the Chileans left to rot on the nitrate pampa."[15]

Closer to the coast, the pitched battles at Tacna and Arica brought back memories of the invasions by England in the days of Sir Francis Drake about three centuries earlier. A painting of the Battle at Arica features a foreground filled with furious hand-to-hand fighting, the backdrop to the mayhem being the fabled Morro de Arica, the tall, rocky prominence that looms over the port. That distinctive hill, which plunged to the sea and had three nearly vertical sides, seemed defensible. However, Chilean forces were able to storm up the fourth more-gradual slope and inflict heavy losses on the Peruvian troops. Ultimately, Chile's superior navy assured victory, not only in direct warfare but also logistically as they conveyed supplies where needed. Ground troops were resupplied by these naval forces and sometimes marched great distances overland in desert conditions. In vying to keep or gain the prize—the mineral riches of the Atacama Desert—these Latin American nations had fought each other in full-scale engagements. These pitched battles hinted at the brutal warfare that would characterize the carnage elsewhere in the early- to mid-twentieth century. Then, too, as historian and diplomat Bruce

W. Farcau observes, Chile's naval advantage in this war seemed a harbinger of later "gunboat diplomacy" between nations.[16]

Despite the military pact between Bolivia and Peru, Chile emerged triumphant from this five-year-long war. Chile's victory effectively pushed the borders of those countries far to the north and east. Thus, Chile now claimed all of the nitrate fields and hence the entire Atacama Desert. Another consequence was that Bolivia became—and still remains—a landlocked country, and is resentful to this day. In Bolivia, one can still hear the lament that the country is a "beggar on a throne of gold," or words to that effect. A student of history recently posited that Bolivia's ongoing political instability is one of the "eternal ramifications" of the War of the Pacific,[17] despite the Treaty of Peace and Friendship (1904) that permits Bolivia free access to the port at Antofagasta. Peru is also wary of Chile, though it still has ample access to the Pacific. A takeoff on an old expression is sometimes heard among Peruvians as advice to both their own government and anyone concerned about Chile's intentions: "When Chile builds a warship, build two."

Although the borders forged in the War of the Pacific have held despite minor adjustments in the vicinity of Tacna, tensions surface from time to time, as the bitterness persists. For example, in October 2015 Peru and Chile made news as they publicly disputed each other's claims to borders offshore in the Pacific Ocean, Chile solemnly reminding its northern neighbor that it will fight to maintain its borders.[18] Adding fuel to this lingering fire are recent studies that consult earlier maps in order to discredit Chile's victory in a war that took place nearly a century and a half ago. A case in point is found online in César Vásquez Bazan's May 15, 2013, website and blog, which use a series of historical maps to make the point that Chile "robbed" Peru and Bolivia of their rightful Atacama Desert possessions. Vásquez Bazan's solution—to reset the northern border of Chile at 25 degrees south[19]—would essentially restore the status quo of the mid-1870s. However, given this region's strategic and economic importance to Chile, that scenario is highly unlikely.

The maps immediately discussed and illustrated confirm that high stakes were involved in the Atacama Desert, and the mining industry was a key player in shaping geopolitics. In the region acquired by Chile after the War of the Pacific, huge nitrate and copper deposits awaited further development. Regarding nitrates, some go so far as to suggest that England was a major catalyst in the War of the Pacific, and that the "Nitrate King" John Thomas North (1842–1896) played a role. North had purchased large nitrate holdings in what was then Peru, and he profited mightily when Chile honored their conversion after the war.[20] Some have speculated that the companies precipitated this war, but that is simplistic and does not account for the Chilean government's growing economic instability before the war.

One thing is certain: with the close of the war, an immense amount of British capital rushed into Chile. If Chile had proven itself a defender of corporate rights, and capitalists around the world took heed, few acted on it with more enthusiasm than the British. Ironically, after the close of the war, Chile raised export taxes to forty *centavos*, four times more than the ten *centavos* Bolivia hoped for! One more thing is noteworthy about the War of the Pacific on the Chilean mindset. Although the origins of the terms Norte Chico and Norte Grande are debated, it is likely that they are a result of Chile's victory. After all, the name Norte Grande essentially covers much the same the territory that was ceded by Peru and Bolivia; in a manner of speaking, that was Chile's great victory, and not only helped enlarge the country but contributed to its greatness as an economic power.

The period just after the War of the Pacific is a good place to introduce one of the classic works describing life and landscape in the booming nitrate fields. Written by William Howard Russell (1820–1907), *A Visit to Chile and the Nitrate Fields of Tarapcá* (1890) represents one of the most engaging nineteenth-century Latin American travel narratives ever written. This is all the more surprising because it is not well known, perhaps due to its being overshadowed by Russell's other accomplishments. As one of the first great war correspondents, the Irish-born Russell covered conflicts from the Crimean War (1853–1856) to the American Civil War (1861–1865). He was well read, well travelled, and well connected in British Victorian society. However, before discussing Russell's book about this important part of the Atacama Desert, I should offer a caveat: although Russell attempted to be objective in his accounts of the War of the Pacific, his partisan stance is apparent in many places. For the most part, Russell characterized Chile's victory as a manifestation of the corruption and pride of Chile's neighbors and the indomitable spirit of Chileans, who had resisted on other earlier occasions. As Russell put it, "The Chilians, claiming the land as their own, flocked into the desert of Atacama, just as the Americans flocked into Texas before the war with Mexico, which resulted from their immigration." According to Russell, Chileans were "active and comparatively industrious, as well as adventurous" and "actuated in their dealings with Bolivia by a natural desire to obtain possession of the land."

Russell also noted that the consequences for Bolivia were harsh indeed, and he blamed their political system for "the insecure isolation in which it is at present situated, distrustful of its neighbors, without a port for its commerce, its voice almost unheard on the American Continent and altogether silent in Europe." Observing of Chileans that "they have an innate restless spirit which is developed and maintained by the discovery of new fields of labour," Russell concluded, "Wherever mines are to be worked, cities to be built, railways to be constructed, and industries developed, the Chilian is to be found, strong in arm, independent in opinion, violent in his cups, but on the whole a valuable agent in the work of civilization."[21] Russell considered the desert north of Chile a frontier, and the Chileans consummate frontiersmen. Nevertheless, he did not shy away from stating that the war was ultimately about mineral wealth. In a passage that revealed his British penchant for understatement, Russell speculated, "If the desert of Atacama, which had a traditional reputation for mineral wealth, had not blossomed into cuprates and nitrates, Chile, perhaps, would not have found any particular charm in the region, nor would she have sought, under any compulsion, to remove her neighbour's landmarks." With an air of inevitability, he concluded: "Why should it have been otherwise?"[22]

Before reaching those conclusions, though, Russell had learned a great deal about Chile through firsthand experience. To reach this recently war-torn part of Chile, Russell and his wife sailed around the horn on a steamship, spending time at the coal mines of southern Chile and seeing the sights in Valparaíso and Santiago. As his writings reveal, Russell had an interest in engineering, including civil and mining engineering, mainly as they pertained to transforming society and advancing what he called "civilisation." When this sojourn amongst the fineries of central Chilean society ended, they headed north from Valparaíso on the Pacific Navigation Steamship Company's liner named *Serena*. Built in 1881 by Robert Napier & Sons of Glasgow, *Serena* was one of two virtually identical propeller-driven steamships purchased by PNSC for the coastal trade between Valparaíso and Callao (Peru).[23] After passing numerous small ports, they put in at Caldera. Not especially impressed as he gazed at the city from the *Serena*, Russell wrote that this port "is purely commercial, and its prosperity depends on the mines near at hand."

Caldera had grown since Gilliss and Philippi described it, but some things hadn't changed much. Russell observed, "There are the usual factory chimneys, towering above the low houses of wood disposed in streets ankle-deep in sand, at right angles, with the universal background of sandhill and brick-coloured mountain." Northward from Caldera, the *Serena* steamed past numerous small ports, including Taltal, where activity was underway to develop a narrow-gauge nitrate hauling railway. Shortly thereafter, the *Serena* stopped at Antofagasta, where Russell stayed on board but noted the city's numerous buildings such as clubs, hotels, and churches. For good measure, he added that Antofagasta contained "squares of gaily painted wooden houses, built upon sand, a garden, in which a few flowers and plants are sustained at the cost of the municipality in the Plaza." At this time Antofagasta was booming, as the nitrate railways hauled their sacked mineral treasure into the port city from about fifty miles (75 km) inland. North of Antofagasta, the *Serena* steamed past ports such as Mejillones, Tocopilla, and others that were also handling nitrates, but Russell wrote little about these, his ultimate goal being Iquique and the nitrate country inland.

As the *Serena* dropped anchor in Iquique, the port city was shrouded in fog. Well aware of how Darwin had described this city, Russell was not surprised by the thick mist that enveloped it. He was, however, completely taken off guard by what came into view as the blanket of fog began to lift. As Russell recounted, "the veil of cloud or vapour drifting upwards revealed what I was not prepared to see—a town with pretensions, an imposing sea frontage, public buildings, the brightly-hued cupola of a Cathedral, a Custom House." Forty-five years after Darwin's visit, the city's population had grown more than twentyfold to twenty thousand. Moreover, a busy railway connecting it to the nitrate fields of the pampa, about sixty miles (100 km) to the northeast, was a prominent part of the city's waterfront. To Russell, this railway was a key feature of the landscape. Its switching and car storage yards were bustling with activity, and its large unloading cranes loomed near factories "giving out [an] abundance of smoke." Russell next described the "well-defined line of railway striking from a mass of great magazines and workshops, right up the side of the mountain, the summit of which was hidden by the clouds." He wrote with pride about this railway, which featured the most powerful English-built steam locomotives available and would be his mode of travel into the nitrate fields.

Before making that trip, the Russells spent several weeks in Iquique with hosts. In a sense, Iquique seemed part of the sprawling British Empire, for enough British capital was evident here to make the Russells feel at home. Nevertheless, the place faced some logistical challenges. At this time, as in Darwin's, Iquique still relied on the shipment of freshwater from Arica. As Russell observed, "Out at anchor beyond the Railway Pier you see one of the three water-ships belonging to Colonel North; vessels of 800 or 900 tons burden, some of which make fourteen trips a month between this and Arica and the coast, where they take in water to supply Pisagua and Iquique." According to Russell, Colonel North at one time controlled this

The Nitrate Railway. Iquique to Mollé.

Figure 4.5. Woodcut in William Howard Russell's *A Visit to Chile and the Nitrate Fields of Tarapcá* (1890) shows the Nitrate Railway track ascending the cliffs adjacent to the Pacific Ocean south of Iquique. (American Geographical Society Library, University of Wisconsin at Milwaukee)

water business but had sold it a few years before Russell arrived. In a footnote, Russell noted that a water main was being built to the city from Pica, which was located on the west side of the Tamarugal pampa.[24]

Russell's goal was to write about the works of British capitalists such as Colonel John Thomas North (the Nitrate King) and James Thomas Humberstone (1850–1939), both of whom he visited on this trip. Their home port of Iquique was impressive enough, but Russell needed to visit firsthand the nitrate fields that they were developing. Here, in the pampa, Colonel North's skills at transforming capital into more capital, and Humberstone's considerable mining engineering expertise, were improving production. Both of these nitrate moguls were anxious to show Russell their handiwork. For his part, Russell knew relatively little about this industry but was learning quickly. As he boarded a train to experience the action, much as he had ventured into battlefields, he was primed to inform readers almost half a world away about why this place mattered. For weeks he had seen these trains moving up and down the mountainside above the city, and now he and his wife were finally on one heading toward the goal of the trip.

As their train slowly chugged its way up the mountains behind Iquique, Russell described the changing views. Along one stretch of track where the railroad line momentarily headed southward, Russell related what glimpses to the left and right would be revealed to the traveler. "On the left hand as you mount there is the oolite rock of the mountain range seamed with ravines filled with drifted sand; to the right there is the steep slope of the hillside, descending sheerly to the sandy strip between its base and the beach." As Russell continued this description, he made clear its significance. "As we mount the view becomes more picturesque, in the true sense that it is very like a picture." A century later, one might have called this a "Kodak moment," a vivid memory of something noteworthy captured in both the mind and on paper for posterity. In pages adjacent to his description, Russell provided two wood-cut illustrations of that experience to leave little doubt that it was worthy of preservation in both word and image (Figure 4.5). One of them reproduced here shows the line clinging to the barren mountainside above the coast.

As the train surmounted the grade and rolled into the wide-open pampa, which Russell called "the Nitrate Kingdom," his prose became even more poetic. He not only called upon the earlier description by Charles Darwin, but he added some accounts of his own travels to the borderland of Egypt and Arabia. "Assuredly there is nothing to charm the eye in the scenery around us. I remember nothing like it save the bed of the Bitter Lakes at Ismailia before M. de Lesseps opened the flood gates and let in the sea water on the basin of salt and sand." Readers were familiar with Ferdinand de Lesseps (1805–1894), the French diplomat and developer of the Suez Canal. This major civil engineering feat had been excavated through a similarly flat and parched country between 1858 to 1869. In words clearly inspired by Darwin's earlier description, Russell added that "the plain was exactly like a field covered with the 'dirty patches' of snow which are left by a thaw in level country at home."[25]

Of the towns that clustered on the pampa, Russell had much to say. In page after page, he informed readers about life and landscape in these remote communities that were connected to the outside world by the umbilical-like rail line. At a place called Central Station, Russell noted that "29 miles from Iquique and 3,220 feet above the sea level, is a veritable surprise." Central Station consisted of the more or less predictable trappings of commerce, such as railway buildings, telegraph service, post office, a hotel, and restaurant, but "there are around the station—marvellous to see in such a place!—Norwegian frame houses—wood and zinc with high sloping roofs—and the

Figure 4.6. Fruit and cake sellers outside a pulpería on the nitrate pampa, as seen in this woodcut from William Howard Russell's *A Visit to Chile and the Nitrate Fields of Tarapcá* (1890). (American Geographical Society Library, University of Wisconsin at Milwaukee)

walls are covered with familiar names and advertisements, which recall the old world beyond the sea." Not far from this scene, though, he observed that "we see the Pampas spreading out on either side of us like an inland sea, or the bed of one, with ochre-coloured shores, above which, on the right, rise cloud-like Andes with snow patches on the summits, roseate in the rays of the declining sun."

Here, in a forlorn locale with evening coming on, Russell wrote that "the ruins of adobe walls and chimneys around us marked the sites of abandoned native manufactories of nitrate, but without help I could not have guessed what they were." Someone on board the train apparently informed Russell that the ruins were from earlier days "when the raw material was boiled, under the old system, in large shallow cauldrons, one or two of which were still to be seen lying about." The desert is a wonderful place to ponder the remains of earlier activity, for people often leave without cleaning up, and nature does little to remove what remains except to give it a patina of antiquity. It is indeed tempting to construe such remains as ancient ruins. Robert Lewis Stevenson did much the same in the American West when he noted that a California ghost town reminded him of ancient Palmyra. Russell, too, was prone to such flights of the imagination. Of his trip into the Chilean desert, he penned, "I have been continually reminded of the African desert and of the western shore of the Red Sea since I visited the pampas."[26] As I noted in *Go East, Young Man* (2011), these references to Syria and the Sahara are a reminder that Orientalism was a potent force in the discovery and exploration of the American frontier—in both the northern and southern hemispheres.[27]

As Russell and his wife encountered them, nitrate oficinas were well-organized company towns—planned settlements consisting of housing for workers and all the services they might need. Russell described the town of Primitiva as a typical oficina and devoted several pages to it. As in many company towns, there was a striking difference between the houses of the workers and the homes of the managers and owners. The common workers lived in dwellings that Russell described as "a patchwork." Elaborating on that term, he explained it consisted of varied materials: "Pieces of zinc, corrugated iron, matting, and shreds of sacking do duty as walls to some of the shanties." What Russell modestly called "Mr. Humberstone's house" was different. "It is a quadrangle with Spanish patio or court in the centre, on a solid base raised six feet from the ground, covered with zinc, a verandah all around, in the style of a good Indian bungalow." This house of "many rooms" was positioned to make a statement, and guests who stayed here, as did the Russells, were bound to be impressed. Gazing out from the verandah enabled Humberstone's guests to better understand the nitrate operations. As Russell explained, "The works are not a hundred yards away, and as the house is on an elevation, you command a view from it of the main railway, the oficina line, the boilers, tanks, nitrate shoots and canchas, and the large village built of one-storeyed houses numbered in order in regular streets, a few hundred yards further on, where the workmen of Primitiva and their families live."

Humberstone was master of all he surveyed, as the workforce consisting of Chileans, Peruvians, and Bolivians was well aware. At the center of many nitrate oficinas was the *pulpería*, the equivalent of the company store but which sold many items made by Native peoples as well as imported from Europe and even the United States. In some *pulperías*, alcoholic drinks were also served, which added to the frontier quality of these nitrate towns. Russell noted that he inspected a number of *pulperías* on his trip through the

Figure 4.7. A woodcut showing a calichera at Primitiva, where nitrate is mined from the pampa, as illustrated in William Howard Russell's *A Visit to Chile and the Nitrate Fields of Tarapcá* (1890). (American Geographical Society Library, University of Wisconsin at Milwaukee)

nitrate pampas. The illustration by his artist Melton Prior (Figure 4.6) shows the bustling activity as people traded both inside and outside the store-like building. Although the original illustration is in black and white, by most accounts the varied merchandise marketed and the dress of the miners and their families made *pulperías* colorful places on the otherwise drab nitrate pampas.

Describing this full-fledged mining landscape was especially challenging given the technical complexities of the nitrate operations, which Russell sought to clarify in numerous detailed woodcut illustrations. He observed that Primitiva occupied twelve square miles, only three of which were fully developed. Two illustrations epitomize this contrast between open country and densely settled space. In the first image (Figure 4.7), which shows a *calichera*—the area outside of town where the nitrate ore (caliche) is collected— the artist and viewer are positioned out on the pampa. The bare hills in the middle ground and distance frame this panorama, and the lighter-colored flat area in the foreground is being prepared for exploitation. A hint of industry is indicated as the railroad can be seen in the distance, but the overall mood is serene, a bit reminiscent of Philippi's sketches of metals-mining areas such as Trespuntas, far to the south.

In contrast, the second illustration in Russell's book depicts the bustling and densely settled portion of Primitiva (Figure 4.8). Although also a

Figure 4.8. A woodcut panorama showing the nitrate operations at Primitiva, with their tall smokestacks (center), and ornate home of James Thomas Humberstone (left). (American Geographical Society Library, University of Wisconsin at Milwaukee)

panoramic view, we now see how thoroughly Victorian culture and industry dominated the scene. In the background atop a barren ridge are clusters of workers' houses and other company buildings. In the middle ground, the three "chimneys" that Russell described in the text belch clouds of smoke that stream like banners on the wind. Closer to the artist/observer, rows of bins anchor the evenly spaced trestles that reach to the loading area adjacent to the main railroad tracks in the foreground. On those tracks, a distinctive double-ended Fairley steam locomotive pauses, ready to couple onto another train of gondola cars loaded with nitrates.[28] Standing prominently at the far left margin of this illustration is Mr. Humberstone's impressive home with its sprawling verandah. Russell recognized the pivotal role this splendid home played in the affairs of Primitiva; in several other passages, he dramatized the views of the town that visitors experienced from its windows and porches.

In addition to describing the town and its surrounding countryside, Russell's book also contains many illustrations of mining machinery such as the crushers which hammered away at the stubborn caliche. Collectively, as Russell concluded in an appendix to his book, "The illustrations give a better idea of the 'plant' at and of Primitiva than words can convey, and attempts to describe machinery without the help of wood-cuts are not generally interesting."[29] Russell was being modest here, for many of his narrative descriptions introduce vivid color and use evocative similes from his war correspondence. For example, he noted that the three tall chimneys in Primitiva were "bright red," and that the bateas or vats were "drawn up like a battalion of soldiers in company columns" and "painted scarlet." Russell also described the sound of the blasting in the mines as if it were "some cannonade, and when it became quite dark it needed no great stretch of fancy to the imagine that one was looking at a bombardment of some town from the sea."[30] He was aware of the fact that the war had been fought here too, but for most readers the war-inspired prose simply enhanced the drama of Russell's mining descriptions. Although his similes might appear coincidental, they actually served as the perfect metaphors for what was happening in the Atacama: nitrate mining left a pock-marked and overturned landscape that appeared as a battlefield—the aftermath of a regimented, almost military approach to conquering a resource on behalf of progress.

Before leaving Russell in these nitrate fields at the height of their boom cycle, I would like to examine one of his passages about Primitiva that is simply stunning, and prescient, in the manner that it weaves image and sound. The time of day—sundown—adds an element of drama. As Russell put it, "by the time we got back to the house, the sun had set in a golden haze over the low ridge of hill [*sic*] which bounded the Primitiva estate on the west." The following poignant description conveys the best of Russell's wartime correspondence with his penchant for long, information-packed, and emotionally charged sentences. From his vantage point, Russell observed that "the rays of the electric lights of the oficinas were piercing the gathering darkness, which seemed to roll down from the Andes on the plain—fiery flashes, followed by dull reports, sounded through the gathering night from the exploding *tiros*— the fires of the *maquinas* glowed on the Pampas around—in the deep vault of heaven above the stars shone and glowed in the sky, with the brilliancy which the air of the desert seems to give to their glory—the notes of a piano came through the open window of the drawing room, which was lighted up as bright as day—from the workmen's quarters the sound of a guitar and a chorus were borne on the night wind."

This description is not only vivid in fixed imagery, but it captures the passage of time and the meeting of cultures. With good reason, I used the word "prescient" to describe it; with almost no revision, in fact, it could serve as a page from a movie script, in which the director would command cameras to move from element to element, and sound technicians to dub in music that flowed seamlessly from piano to guitar, which is to say refined (or high class) culture and folk (or lower class) culture, respectively. Russell left little doubt about the deeper meaning of this passage as both a record of what he witnessed and a cue for what the audience should remember: it was a profound lesson about progress. As Russell himself noted in the next sentence, which could either be narrated in voiceover or read by the audience in words that lingered on a motion-picture screen: "What was a lifeless waste when the century opened was now a centre of enterprise, capital, science, and civilisation."[31]

Although Russell's description of Primitiva came just before the dawning of silent film, and about thirty years before talkies, I would still be tempted to use it today if I were making a documentary film about the nitrate towns of the Atacama. But the message would be bittersweet indeed for two reasons. The first involved economics and politics, which are often inseparable. At just the time Russell returned to London to publish this book, and British corporations controlled 68 percent of Chilean nitrate production, the nitrate industry was showing signs of weakening as prices began to drop and economies headed toward depression in the early 1890s. In fact, just a year after Russell's book was published, Chile was thrown into its 1891 Revolution, in part caused by its president alienating nitrate interests. Once teamed to win the War of the Pacific, the Chilean army and navy now fought each other for control in battles, including two (Caldera and Pozo Almonte) in the desert north. The outcomes included a parliamentary government and the resumed—but now warier—interest in nitrate investments.

The second reason related to technology, which could also change fortunes. By the early twentieth century, experiments began to determine that nitrogen could be extracted from the air (of which it comprises about 78 percent) instead of from the earth. This development could hardly be imagined by Russell, but it would affect the entire industry and deal a blow to Chile's far north. As I shall show later in this chapter, nitrate was awaiting a seeming rebirth in the early twentieth century, but that rebirth would not last. Equally humbling was what happened when those nitrates reached their final destinations. As fertilizer, they could greatly increase the crop yields of tired fields, but as explosives they could either be controlled to create engineering marvels such as canals but also be turned loose into deadly weapons taking countless lives.

A visit to these nitrate fields even today gives one much to ponder. After a daylong trip through those southwest of Calama, I reflected on the enduring legacy of the nitrate booms. That night I slept fitfully, realizing that the entire industry had resulted in considerable good but just about as much evil. Tossing and turning, a poem called "Caliche" came to me in the middle of the night.

CALICHE (NITRATE)

It took nature ages upon ages
to make this white blanket
that lured so many souls here,
digging, loading, shipping
the skin of the desert
to a hungry, uncertain world,
making distant fields blossom,
but claiming so many lives
in blinding flashes of light
on equally distant battlefields.
Like treasure, or life itself,
nitrate was all too easily squandered.
Too good to last, they used it up
and then made it out of thin air.

Travelers have long recognized that mining has transformed the land-scape of the Atacama. In 1889, Lazarus Fletcher, who served as Keeper of Minerals in the British Museum, noted that "since Philippi's journey vast changes have been brought about within the northern portion of the Desert through the discoveries of silver ores and nitrate of sodium within its bor-ders." Fletcher, though, had other minerals on his mind. His main objec-tive was to place in perspective the many meteorites that had been found here. As hinted at in the last chapter, meteorites were of intense interest throughout the nineteenth century. In fact, it had long been rumored that the arid conditions in the Atacama had preserved the remains of gigantic meteor showers that pummeled the earth from time to time. Of meteorites found worldwide, most (about 95 percent) are "stony," that is, composed of glassy minerals such as olivine or darker and more basalt-like rock material, the latter sometimes containing iron silicates such as hornblende. The other type of meteorites is relatively rare (less than about 5 percent) and metallic in composition, often consisting of solid masses of iron and nickel. Using maps by Philippi and another map by Professor Ludwig Brackebusch of Cordova, Argentina, who had conducted expeditions into the Andes, Fletcher plotted the distribution of meteorites of various kinds on a map in his October 1889 article "On the meteorites which have been found in the Desert of Atacama and its neighborhood."

Fletcher was skeptical of some claims, which had been circulating as early as the 1820s, that vast metallic deposits there were the result of meteors. Imilac, a small Indian hamlet in the central Atacama, was one of these places associated with iron and thought to be of meteoric origin. In one colorful aside, Fletcher noted that considerable speculation had resulted when a miner found iron scattered about and thought he had made a rich silver strike—apparently a version of the scenario related forty years earlier by Philippi. Imilac was found on Philippi's 1860 map, which indicates its position at the north end of the Salina de Punta Negra (see again Figure 3.20). Analyzing Atacama Desert meteorites in the museum's extensive collection and consulting reports from many sources, including those by Gilliss, Philippi, and others, led Fletcher to three sobering conclusions, namely: 1) that "the peculiarities of the Desert are entirely terrestrial, not celestial"; 2) that "there is nothing extra-ordinary in the number of meteorites which have fallen in the region"; and 3) that "the discoveries of meteorites in the Desert fail to prove the occurrence of widely spread meteoric showers." Fletcher had also finally put an end to speculation about the meteoritic iron of Imilac. In a rather grammatically fragmented sentence, he described it in terms of its mineralogy and physical characteristics, thus hoping to end the speculation about its origin. "The sponge-shaped nickel-iron, with its numerous olivine-filled cavities, being at once so definite and so rare a meteoritic type, there can be no difficulty in inferring that all which has been brought from a region of small area is the product of a single meteor."[32]

Today we use the term pallasite for such meteorites that consist of a combination of native metals and glassy minerals such as the iron-magnesium silicate olivine, which is known as the gemstone peridot. These are not only extremely rare but stunning visually as specimens; when cut and polished, they appear to be blobs of greenish or yellowish colored glass set into a mirror. Some collectors call these "Atacama meteorites," though they are very rare here, as elsewhere, and fetch high prices on the market. On the map included in his article, Fletcher pinpointed Imilac. Although this might have laid speculation to rest for a while, Atacama meteorites would be rediscovered more than a century later by scientists reaching new conclusions.

If the passages by Russell and Fletcher give the impression that the British were singlehandedly discovering, developing, and interpreting Chile's mineral wealth and mineralogical curiosities, I should correct this by showing how Chileans also played an important role. As noted, Chile's geopolitically charged mineral exploitation was coupled with increasingly sophisticated mining engineering. The latter depended on detailed mapping conducted by private and governmental expeditions aimed at encouraging mining ventures. Throughout the 1880s, Francisco J. San Román (1838–1902) was involved in

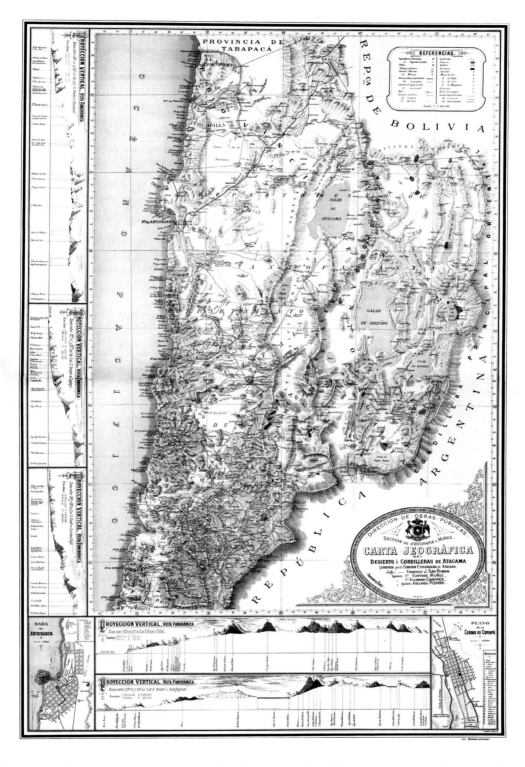

Figure 4.9. The impressive *Carta Jeográfica del Desierto i Cordilleras de Atacama* by Francisco J. San Román set a new standard when published in 1892. (Author's digital collection)

the exploration of the Atacama Desert. Born in Copiapó just three years after Charles Darwin's visit, San Román became an authority on mining technology as well as the desert itself. San Román is perhaps best known for the role he played in bringing the Atacama Desert's mineral resources to the world's attention. He conducted extensive fieldwork in mining areas, which was key to his producing a landmark map, *Carta Jeográfica del Desierto i Cordilleras de Atacama* (Figure 4.9). Drafted in 1890, this map combined state-of-the-art cartographic techniques with a vast amount of information on the distribution of various kinds of mines that he classified by their main resource (for example, gold, silver, and copper). San Román's map is not only a useful document but one that is aesthetically pleasing. It no doubt became a centerpiece of the presentation on the Atacama Desert that he made to the International Congress of Geology in Washington, D.C., in 1890.[33]

Published copies of *Carta Jeográfica del Desierto i Cordilleras de Atacama* appeared in 1892 and represented the efforts of the Comisión Por La Exploradora de Atacama under San Román's leadership. As readers may see if they compare the maps side by side, it is very likely that San Román's map was in part inspired by Wagner's beautiful 1876 map of Bolivia discussed earlier in this chapter (see again Figure 4.3). Both maps use color and topographic profiles to great advantage. The similarities between these two maps is yet another reminder that Chile in general, and the Atacama Desert in particular, was strongly influenced by northern Europe at this time.

After its arresting imagery, the next thing that may strike a person consulting San Román's map is the area covered, or rather not covered; instead of representing the entire Atacama Desert, San Román's map depicts only about the southern half. It begins in the south at the recognized southern edge of the Atacama Desert, but by using the serpentine Rio Loa as a northern border, a number of developing nitrate deposits northward were excluded. Nevertheless, the promising mine at Chuquicamata is shown north of Calama. Overall, San Román's map covers what is generally considered the heart of the Atacama, the two provincias (provinces) of Antofagasta and Atacama. These provinces were at that time, and still are, extremely important in terms of mining. The second noteworthy aspect of this map is how it depicts the physiography. Based in part on data from the commission's explorations as well as data from Chilean surveyors and private mining company engineers, this map depicts the topography, including the varied and complex mountain ranges and adjacent salars, in stunning detail. Although physiography is the map's strong suit, so, too, is its comprehensive depiction of mining areas. This should come as no surprise given San Román's knowledge of mining, but it also reflects the agenda of its publisher, Chile's Seccion de Jeografia y Minas (Geography and Mines).

Above all, perhaps, San Román's map is most striking for its lavish use of color and its meticulously articulated hachuring techniques. These techniques work in tandem to give the topography a three-dimensional feel. Employing state-of-the-art late-nineteenth-century lithography, this map also uses the tradition of the tableaus or vignettes to show other details and views. These include two city plans (Antofagasta and Copiapó) as well as five beautifully drafted topographic profiles, the latter reminiscent of those tableaus pioneered by Alexander von Humboldt about eighty years earlier. San Román's masterpiece confirms the maxim that mapping is both an art and a science. Although this map did not cover as far north as Iquique and Arica, it was a *fait accompli* that confirmed Chile's possession of this mineral-rich region following the War of the Pacific. The inclusion of Copiapó is understandable, as it was an important mining community and San Román's birthplace. The inset map of Antofagasta was prescient: although only founded in the mid-1860s, the city had become a major shipping port in San Román's time, and today has a population nearing half a million. Much of the copper mined in the vicinity of Calama and Chuquicamata still finds its way to Antofagasta; short but heavily loaded trains carrying copper ingots trundle along rails laid on the streets of the city, bound for distant markets such as China and Japan. Today, Antofagasta is the largest city in northern Chile, and Bolivia still uses it as a port through the earlier-mentioned agreement between the two countries that was ratified in 1904.

Back in the early 1890s, when San Román's career was reaching its zenith, he also published several important written works. These include the *Reseña Industrial e Historica de la Minería i Metalurjia de Chile*, which when published in 1894 was a comprehensive and detailed guide to the mining industry. San Román began this work by highlighting the legendary, rich, and romantic history of earlier mining that I discussed in Chapters Two and Three, adding that the late nineteenth century was witnessing many modern developments, most imported from Europe and the United States.[34] During this same decade, the commission under San Román's direction also published a massive two-volume work similarly titled *Desierto i Cordilleras de Atacama*. Volume One (1896) consisted of two books (itineraries of the explorations and the mission to the United States). This volume noted the many factors that challenged the commission, including limited funding that made it follow what he called its "curso de zig-zag" or route that had to jog back and forth rather than cover the entire desert systematically.[35]

Volume Two comprised the third book and focused on the region's hydrology, emphasizing the early and ongoing concern about water in relation to the geomorphology. In one section, San Román recounts the trials and tribulations of early travelers who were nearly done in by scarcity of water, rugged slopes, and distances that were very difficult to estimate. In a delightful sentence that characterizes the elation experienced by early expeditions traveling here, San Román playfully alters the spelling of the word agua (water) by lengthening it— "iagua!...iaagua!...iaaagua!"—to mimic their thankful exclamations of relief. His main focus, though, is on how surface and subterranean water resources could support modern mining prospects and agricultural developments.[36] The publication of this fact-filled volume on hydrology coincided with San Román's death in 1902.

Although Chile became a major minerals producer, I would be remiss if I did not also mention the internal politics that constrained development. In one form or another, mining interests played a role in Chile's nineteenth-century revolutions (civil wars), in which mining interests in the north squared off with agrarian and urban interests farther south. By the later nineteenth century, it was becoming abundantly clear that Chile was quickly losing its edge in international copper production. As historians William (Bill) Culver and Cornel Reinhart convincingly demonstrate in their comprehensive article "Capitalist Dreams," copper mining in Chile's desert north had been eclipsed by the mines of Michigan and especially the American West. Why this happened is still open to debate, but Culver and Reinhart make an impressive case for politics.

Chile's copper mining industry was hampered by a system "which had been structured under a code based on Spanish colonial goals of creating revenues for the state in the mining of precious, not industrial, metals." Meanwhile, American mining benefitted from a system that freed it from responsibility to social well-being, limited the liability of corporations, subsidized transportation, and restricted tariffs. As Culver and Reinhart conclude, "Chile had a mining policy of [generating] state revenue; the United States had one of economic growth." By the 1880s, the changes in Chile's laws amounted to too little too late. In reality, the sentiments and wishes of agrarian (Valle Central) and urban (Santiago and Valparaíso) interests helped cripple copper production in the desert north. Interestingly, nitrate and coal interests did not tend to support the laws that copper producers advocated.[37] As we shall soon see, it would take several decades for Chile's copper mining industry to catch up.

In the early twentieth century, the Chilean government slowly got on board with incentives, but transforming the Atacama Desert from wasteland into Chile's treasure house of mineral riches took time. Ultimately, the varied minerals that occur here—gold, silver, copper, borates, iodine, and nitrates—helped position Chile as South America's Golconda. One manifestation of Chile's increased involvement in the development of this region was the subsidization and construction of the Longitudinal Railway in the 1910s, its goal being to connect the railways and communities of the far north with the capital city of Santiago. As the name implies, this railway ran pretty much on a north–south axis, closely paralleling the meridian of 70 degrees west. This was a major, and rather controversial, investment at the time. In 1913, geographer Isaiah Bowman had become familiar with railway development here. Bowman's field notebook from August of that year contains an interesting observation about the value of this then-new railway to Chile. As he observed: "great r'y [i.e., railway] developments in Bolivia is quite unlike that in Chile—In Bol.[ivia] every line is a means of econ. develop[ment]; in Chile the great longitudinal is a tax—parts of it will not have enough traffic to pay." The Longitudinal Railway would be completed in the 1920s, but Bowman was right in that portions of it were little used.

While in this comparative mood, Bowman a few pages later turned his powers of observation to minerals, in particular the nitrate and borate deposits of Antofagasta province. Bowman wrote that nitrate and borax deposits differed in their locations, and he speculated about their origins, observing that although "there are some boraterios [i.e., borate deposits] at lower levels & toward the coast the greater no. are in higher basins & best are in highest where there is rare comb[ination] of pp'n [precipitation] enough to carry away [illegible wording, perhaps "and actively"] deposit the salts of the volc. eruptions that occur at heights in greatest nos." By contrast, Bowman noted that "the nitrate is not in a region pred[dominantly?] volcanic & its greater dryness keeps nitrate intact whereas—if [it had] as much pp'n as in borax basins nitrates would be carried away." Bowman was above all interested in geomorphology, which was the key to where water would be found. In fact, he begins the field notes by mentioning the area's true lifeline, a pipeline built to convey water from the Andes to Antofagasta. This water made possible the Atacama Desert's modern mining industry and urbanization.[38]

The ghostly presence of nitrate is palpable throughout much of the Atacama Desert, even in the heart of its cities. The late-nineteenth-century architecture of Iquique, with its gothic churches and British-style clock tower, reflects this heritage. So too does a similar clock tower in the plaza at Antofagasta. A postcard from the early twentieth century (Figure 4.10) shows Antofagasta's Calle Prat. This major thoroughfare was named after Arturo Prat (1848–1879), the dashing Chilean lawyer and naval officer who died in the

Figure 4.10. Postcard of Calle Prat in Antofagasta, Chile, ca. 1910–1915, reveals a bustling street scene during the nitrate boom. (Author's digital collection)

Figure 4.11. In 2015, a dog slumbers along a portion of Paseo Prat in Antofagasta, oblivious to pedestrians who now stroll and shop along a street devoid of motor vehicles. (Photograph by Damien Francaviglia)

Battle of Iquique at thirty-one years of age. This postcard dates from around 1910, within the first decade that such cards were mass-marketed in Chile. Many postcards depicting places worldwide were actually made in Germany, but this one was produced in Valparaíso. In my collection are several others by this same company, some in black and white. This one is gorgeously tinted. The colorist may never have actually seen Antofagasta, but the selection of colors seems fairly accurate. Brightly colored awnings and vibrant signage contrast with the varied colors of the buildings. The street stretches to the buff-colored arid landscape in the far distant horizon under a light blue sky. In the foreground, a curious boy faces the photographer as a man crosses the street, which is laced by the converging tracks of the local tramway lines. Along the street, shoppers stroll down sidewalks past horses tied to hitching posts. Barely visible in the far distance, amid the horse-drawn traffic, a mule-drawn tram plods along the street. As this card reveals, electric and telephone lines were also run along this major street.

Although there is considerable evidence that the content of photographic postcards like this one were manipulated in the studio—a common technique that could enhance a scene and hence improve sales—overall, this postcard suggests a city on the move and rapidly modernizing even though no motorized vehicles are visible. In fact, automobile traffic would ultimately become so heavy in Antofagasta a century after this postcard was made that a portion of this street was converted into the Paseo Prat—a three-block pedestrian mall that provides a respite from cars and motorcycles, and is safe enough for a dog to sun himself as foot traffic moves along (Figure. 4.11). While on the subject of picture postcards, I should highlight that regardless of how romantic to some and obsolete to others they appear today, they represented yet another improvement in communications technology when first introduced about 1900. I should also note how the content of these cards is selective. All of the early postcards of the Atacama Desert region that I have found focus on developed places; a few show the nitrate oficinas, but the vast majority feature the cities with their prominent buildings, serene plazas, and bustling commercial streets. In fact, of the four hundred or more Atacama region early postcards I discovered, I was not able to find even one showing an empty desert landscape. This is not surprising, as the desert was considered harsh, and hence off-putting and unattractive. In contrast, the cities offered amenities and comfort, and even beauty.

In 2015, walking around otherwise modern Antofagasta in the waning light of early evening, I spied several once-stately Victorian-era homes associated with the nitrate booms. One of them along the Avenida José Miguel

Carrera faced the verdant blocks-long linear park at the south edge of downtown. Once associated with the prosperous nitrate barons but now standing vacant and boarded up, the home seemed out of place. Whereas most of the other buildings around it had the more familiar semi-modern Latin American–style architecture, with flat roofs and masonry construction, these earlier homes were constructed of wood imported from a great distance, including southern Chile and even California and the Pacific Northwest. In the balmy Atacama air, this Victorian-style home seemed to be slowly disintegrating. Still standing pretty much intact, but shuttered and forlorn, the ornate house was almost hidden behind a series of makeshift fences made out of wire and wooden slats. The sun was just about to dip behind the house, so I would have to shoot the photo into the shadows, hoping for the best. The photo's composition would also be a problem because I had no space to peer through the viewfinder. Gingerly sticking my camera through one of the openings in the fence, I pressed the shutter, not knowing how the photo I'd taken would look. When I checked it out on the camera's visual display, I realized the photograph had superbly captured the mood of the scene, its faded pastel look perfectly matching the faded glory of the home (Figure 4.12). Looking at this photograph, some readers might consider the bluish halation on the lens to be an imperfection, but in my mind it adds to the effect of looking into the past through a filter, as does the gloomy, flat quality of the early evening light.

Throughout much of the length of this desert, the evidence of the nitrate booms is easy to spot. About 250 kilometers (ca. 150 miles) south of Antofagasta, the small city of Taltal represents a once-bustling port. Like Antofagasta, it too depended on the nitrate boom in the pampa located on the other side of the coastal mountain range about sixty-five kilometers to the northeast. In the 1870s, the rich deposits of potassium nitrate began to be developed. The British-built Taltal Railway connected the thirty-plus oficinas of the Lautaro Nitrate Company to this port city, which huddles on a low terrace above the harbor. Population wise, as cities go, Taltal has been frozen in time, as there are about as many souls here today as there were a century ago. In contrast to Antofagasta, which continued booming, Taltal has stabilized at about eleven thousand people. This lagging population growth meant that there was less competition for space and a better chance that the buildings associated with the nitrate industry would escape demolition. Moreover, the nitrate mining operations continued in Taltal much later than in Antofagasta, in fact, until the 1960s and early 1970s, albeit at a much-reduced scale. Remarkably, the Taltal railway soldiered on until the early 1970s,

Figure 4.12. Along Antofagasta's Avenida José Miguel Carrera, a once-stately home associated with the nitrate boom now stands forlorn. (Photograph by author)

despite the dwindling amount of nitrate traffic. As a consequence, Taltal remains one of the best-preserved nitrate ports in Chile, and arriving there is like stepping back in time.

When Damien and I arrived in town after the four-hour drive from Antofagasta, we parked the rental car in the shade of a beautiful pepper tree and soon found ourselves eating a delicious lunch in the interior court-yard patio of a restaurant. Adding to Taltal's antiquated charm, the railroad remained steam powered until the end, and today one of the locomotives reposes in the city park near that restaurant. The town recognizes this British

Figure 4.13. View of Taltal, Chile, in 2015 showing the once-busy harbor and town developed on a narrow terrace of land at the base of the arid coastal cordillera. (Photograph by Damien Francaviglia)

Figure 4.14. A view from a hill above the park at Taltal looks eastward up the long canyon of a normally dry stream running through the city center. (Photograph by Damien Francaviglia)

steam locomotive as one of Chile's historic gems.[39] With its many historic buildings in the vicinity of the historic plaza, some by British architects, Taltal's downtown remains one of the most attractive in Chile. It draws tourists from various countries, but Brits seem especially moved when they see a touch of home in this arid subtropical setting. After lunch, Damien and I walked around town, photographing the many historic sites. The old nitrate mining company office building on a hilltop above the park drew our interest, and we hiked up to it to enjoy the views of town from that elevated position. A couple of our photographs from the prominent hills above the harbor area capture the historic quality of the town. One of the photos (Figure 4.13) looks south along the harbor, while the other looks east over the town to the arid hills that embrace it (Figure 4.14). That striking mural in the latter photograph is of course a modern touch underscoring Taltal's pride in its Indigenous population, which outlasted the mining company and is here to stay.

Comparing the 2015 images of Taltal to those taken by geographer Isaiah Bowman about a century earlier confirms that the city retains much of its historic appearance. However, one clue that some things had changed is evident in Bowman's stunning panorama of the city and its port looking due west into the vast Pacific Ocean (Figure 4.15). In Bowman's time, around 1915, a half-dozen large vessels powered either by wind or steam, as well as many smaller fishing vessels and lighters, crowded in the harbor. In 2015, the thriving maritime trade here was long gone, leaving today's harbor virtually empty. Of his visit, Bowman noted that Taltal had much in common with many cities on the Chilean coast, which were also built on marine terraces; however, he added that the terrace at Taltal was unusually wide, "with protecting headlands and islands on either side."[40] Protection may be somewhat illusory, though, for even a cursory glance at the photos just mentioned confirms that Taltal is built on a charming but somewhat precarious site where a stream draining a large area inland enters the Pacific Ocean.

Although this canyon made a natural route for the railroad line and the road into Taltal from the interior, on rare occasions flash floods rip down the channel and inundate parts of the central city. Adding to the drama, official posted signs inform residents and visitors which parts of Taltal are above the reach of tsunamis. When one glances back toward the ocean from one of these signs, it becomes apparent that much of the historic downtown and port area are vulnerable. To those who do not live along the coastline in the Atacama or other hazardous coasts such as in Oregon, this may seem like an edgy way to exist. However, most of the people in Taltal have learned to live with the dangers and thrive in the process. They are prepared to heed the warnings but do

Figure 4.15. In 1915, American geographer Isaiah Bowman captured this panoramic view of the busy harbor at Taltal, Chile. (American Geographical Society Library, University of Wisconsin at Milwaukee)

not obsess about hazards in the way that many Americans do. To me, Taltal epitomizes a small port that is positively serene as it slumbers, while a century ago, of course, it was booming and had a grittier, more industrial feel. After glimpsing the graffiti that taggers had sprayed on some of the historic buildings and objects in Taltal, Damien and I suspected that not all of its population embraces this gentrification. Nevertheless, this small city seems destined to benefit economically from its selective preservation of the past.

In the 1910s when Isaiah Bowman traveled throughout the Atacama, he made special note of mining operations. Bowman was also well aware that Philippi had traversed much of this area about sixty-five years earlier. One of the sites visited was what Bowman called "the famous Dulcinea Mine" and its smelter, which he illustrated in his book (Figure 4.16). This copper-producing operation was located about six miles west of Puquios in the backcountry northeast of Copiapó. The word dulcinea in Spanish means a sweet dream and is sometimes used colloquially to refer to a beautiful woman—a reminder that mina (mine) is a feminine noun. Operating since the 1840s, the Dulcinea was one of many copper mines in a north–south-stretching mineralized belt whose buckle—the city of Copiapó—was a thriving mining center.

The Dulcinea was situated at an elevation of 6,600 (ca. 2,000 m) feet above sea level and had become Chile's deepest mine (ca. 3,600 ft or 1,100 m) by the time Bowman visited. Bowman was impressed with its longevity, concluding that being served by a branch line of the Copiapó railroad since 1852 helped keep it profitable through lower shipping rates. As his photograph shows, despite its exemplary production, the Dulcinea smelter was minimalist in appearance and offered scant protection from the elements. Given the smelter's location in this driest of deserts, its roof was likely meant to keep the sun off the workers rather than repel rain.

To give his readers some idea of the isolated nature of this operation, Bowman took another photograph looking west between Puquios and the Pacific Coast (Figure 4.17).[41] I include this photo for two reasons. First, it beautifully illustrates the topography of the coastal ranges, which are composed of rocks that weather into smooth shapes. The landscape almost conveys the feeling of being covered by short grass, but it is in fact virtually devoid of any vegetation. This was the result of both humans and nature, for mining operations tend to consume anything combustible. My second, and equally important, reason for reproducing this photograph is that it is nearly perfect in the way it depicts a desert landscape. Being about 27 degrees south of the equator when he took this photograph, the sun is to Bowman's right, which is to say north, at just the right angle to bathe the south-facing slopes

Figure 4.16. In the 1910s, Isaiah Bowman photographed what he called "the famous Dulcinea Mine" northeast of Copiapó in Chile's mineral-rich norte chico. (American Geographical Society Library, University of Wisconsin at Milwaukee)

Figure 4.17. A century-old photograph by Isaiah Bowman showing the topography of the coastal range near Puquios beautifully captures the sensual nature of desert landscapes. (American Geographical Society Library, University of Wisconsin at Milwaukee)

in shadow. This lighting brings the topography into full relief, suggesting something else about the landscape. Deserts may be barren places, but as this photograph reveals, their contours can be sensual indeed. In this light, the desert landscape reminds one of a person reposing under only a thin sheet.

Bowman visited the Atacama at just the time that mining operations were again about to increase in scale with new technology. By 1919, when American geologists Benjamin Miller and Joseph Singewald published their definitive study *The Mineral Deposits of South America*, they confirmed that the Atacama Desert was Chile's premier mining region. To describe what was occurring, Miller and Singewald discussed operations in this desert region's four major political provinces (Coquimbo, Tarapacá, Antofagasta, and Atacama) using separate maps.[42] This compartmentalization respected the political geography, but the two geologists also noted that the desert environment itself transcended such political subdivisions. To set the broader scene, they observed: "This rainless belt, the most desert part of the South American continent, long regarded a dreary waste, has become the richest part of the entire country on account of the valuable deposits of sodium nitrate, iodine, borates, and soluble copper minerals that have accumulated there because of the absence of rainfall."[43]

Of the copper deposits, they singled out the Dulcinea mine as an example of how climate affects the deposition of metal ores through deep oxidation. As they put it, "the great variety of minerals found in the oxidized zone and also the depth to which they occur are the unusual features which distinguish the Chilean copper deposits from those of any other part of the world." They noted that minerals such as "brochantite, atacamite, chalcanthite, and kröhnkite have been found in other regions but never in commercial quantities, but in many of the mines of northern Chile these are the chief copper minerals." They also hinted about something else that would soon affect the copper mining industry here. Deeply buried under these mines, a treasure of copper sulfides was lying finely disseminated throughout the igneous rock. These porphyry ores were low in total copper but vast in quantity, and they were now awaiting exploitation as large-scale earth-moving equipment could make the task possible—and profitable.

At this time, however, nitrates were still important and easily exploited. To illustrate their discussion, Miller and Singewald reproduced a *Sketch map of northern Chile showing approximate locations of nitrate lands*, which was distributed throughout several provinces in the Atacama Desert (Figure 4.18).[44] This map, which contains an inset outline of the South American continent, works on two levels to help readers visualize the geographical and historical consequences of modern mineral development. First, it places in

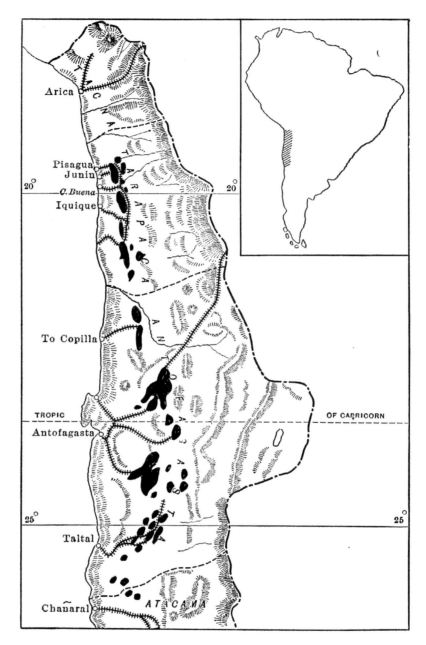

Figure 4.18. This Sketch map of northern Chile showing approximate locations of nitrate lands by Walter Tower (1913) appeared in *Mineral Deposits of South America,* by Benjamin Miller and Joseph Singewald (1919). (Author's digital collection)

geographic context the geopolitical consequences of Chile's victory in the War of the Pacific; the nitrate-rich north now puts the finishing touches, as it were, on the modern map of South America. Second, this map depicts Chile's resolve to fully develop the nitrate industry in partnership with outside capital. More effective production processes, such as that developed by James Shanks, had boosted production region-wide from 1900 to 1920. The Shanks method enabled lower-grade nitrate ores to be worked economically.[45] Aggressive development based on new processing techniques was not confined to nitrates, for as I hinted at and shall further describe later in this chapter, a new type of copper mining was about to transform a portion of this vast region that Chile had wrested from Bolivia. By 1920, many were gambling on the futures of the copper and nitrate industries, but as history would show, one would emerge a winner and the other a loser.

Although economically driven enterprises characterized the Atacama Desert in the 1920s, the region's art and literature began to come of age then. One writer who helped put the region on the literary map was the poet and educator Gabriela Mistral. Born Lucila Godoy Alcayaga in the small village of Vicuña, she soon moved to the even more isolated village of Monte Grande, which lies in a steep-sided canyon at the base of the Andes. In a rags-to-riches story, Lucila grew up in poverty but became one of Chile's most celebrated writers. By around 1909, she changed her name to one with considerably more flair after experimenting with a few pen names. By 1923, she published the evocatively titled memoir *Desolación* (Desolation), which garnered acclaim. Growing up as she did in the southern part of the Atacama, Mistral knew life on the frontier. In her writings, the Elqui Valley metaphorically offers a glimmer of hope as a verdant lifeline in a brooding, inhospitable world. To Mistral, the desert was a place of punishment and despair. In her poem "My Books," she describes "the horrible desert." In another "The Uneasy Shadow," which has a biblical theme and setting, she recounts the trials and suffering of Abraham's wife Agar (Hagar) in "the vast, blazing waste land." In "Hymn to the Tree," Mistral emphasizes how trees provide succor and shade and constitute "the tender womb of a woman."

However, not all trees were characterized as tender. As Mistral notes in the poem "The Thorny Acacia," which also has some biblical references, this desert tree "fastens to a rock" and is "the spirit of the barren land, twisted with anguish and sun." Although other trees such as the oak and myrtle are beautiful, according to Mistral the acacia "was made like Vulcan, the horrible blacksmith god." And yet the acacia is the tree that knows "in the thickness of its million thorns" how badly Mistral's heart has been bruised. In the end, Mistral embraces the acacia "like a sister... in a bond not of tenderness, but more of desperation!"[46] On a spiritual level, Mistral's ambivalence about this desert tree—and the desert itself—is both profound and prophetic. In the twentieth century, an increasing number of people would find themselves attracted to the desert as a place of spiritual renewal, much as pilgrims are drawn to holy sites.

By my reckoning, tourism—that is, noneconomically motivated travel to this desert region—began as early as the 1910s and 1920s. Beginning small and remaining so for decades, tourism here was characterized by a pioneering quality. Intrigued by what they had heard or read about the bleak but fascinating desert, a few hardy souls came to experience the archaeological sites, and others wanted to see firsthand if the world's driest landscape was really as bleak and forbidding as Charles Darwin and others had claimed. This marked an important change, as up until this time most people simply avoided deserts whenever possible. Of these curious souls who thought and acted otherwise, some were American and European cactus aficionados who found their way here in search of specimens to record or collect.

In the United States, particularly California, real estate booms had positioned plenty of eastern wealth in places such as Pasadena and Santa Barbara, where cactus plants could be grown outdoors. The fabulous Huntington Gardens in San Marino developed at this time and epitomized the coupling of wealth and exotic gardening. Personifying this connection was Ysabel Wright (1885–1960). Born in Cuba as Ysabel Suárez Galban, she migrated to the United States and married New York educator and photographer John Dutton Wright (1866–1952). In 1918, the Wrights moved to a twenty-seven-acre estate in Montecito, California, called "Quien Sabe." Inspired by this hilly, semiarid site that faced the Pacific Ocean, and perhaps by their travel to portions of South America (including Peru, Chile, and Argentina) in 1920–1924, the Wrights decided to do something magical with their estate. Ysabel's growing acquaintance with Ralph Hoffman of the Santa Barbara Museum of Natural History was also a major factor influencing her decision to create an extensive collection of cacti at Quien Sabe, which quickly became a site worth visiting. Although the Wrights were never quite as wealthy as their neighbors, and Ysabel was modest indeed as she always considered herself a student of cacti, in reality she became highly knowledgeable and a respected collector within the span of just four years (1928–1932). In the January 1932 issue of the *Cactus and Succulent Journal*, Jacolyn Manning praised Ysabel for having assembled "a botanic garden of cacti from the two Americas, and all the lands and islands in between."

Mysteriously, at the height of her fame, Ysabel and John moved back to New York. Although Ysabel's garden was not maintained, specimens from it formed an important part of the world-renowned Huntington Garden's cactus collection, which was being redeveloped to include "a comprehensive collection of South American cactus—the most difficult to obtain of all desert plants."[47] A recent list by Catherine Phillips, entitled "Plants known to Ysabel Wright and grown in her garden at Quien Sabe" lists nine cacti known to have originated in northern Chile that were acquired by the Huntington collection in the 1930s and early 1940s. It is unknown, however, whether the Wrights actually collected these plants on their trips to South America in the 1920s; more likely, they were provided to Ysabel by intrepid early-twentieth-century botanists such as the famed Czech "cactus hunter" Alberto Vojtech Fric (1882–1944).[48] Also actively collecting at this time, the scientific community was diligently conducting fieldwork and classifying species according to increasingly sophisticated taxonomic systems. The work of these botanists is briefly outlined in a section entitled "Breve reseña histórica del estudio de las Cactáceas," in the varied editions of the popular book *Cactáceas En la flora silvestre de Chile* by Adriana E. Hoffmann J.

In addition to cactus collecting, widespread interest in mineral collecting also lured some collectors to far-off Chile and that country's even more remote Atacama Desert provinces well over a century ago. In the early twentieth century, New York science teacher Harry F. Keller travelled to Paposo and other areas where copper was mined. By this time, the Atacama Desert was well-known for its superb specimens of colorful and unusual ores of silver and copper that had found their way into museum collections worldwide. Keller, though, not only collected some typical less-than-spectacular specimens, but he also wrote about them in a 1908 article, "Notes on Some Chilean Copper Minerals" in the *Proceedings of the American Philosophical Society*. At that time, philosophy and science were much more closely connected, and Keller's article was written for the educated lay public, not just scientists, even though his observations were technically accurate, including his chemical analysis of the ores. Moreover, this took place in the days before journals and magazines for avid rock and mineral collectors appeared, and when collectors and scientists frequently corresponded enthusiastically. In increments, such dedicated collectors added to the body of literature; Keller's article demonstrated that atacamite was present hundreds of miles from its originally recognized location closer to Copiapó.[49]

The process by which major institutions such as the Smithsonian and Harvard University obtained mineral specimens for their collections occurred in steps. As noted earlier in this chapter, it had begun in the mid-1800s. By 1912, when the Harvard University Mineralogical Museum received the collection of American mining engineer Albert F. Holden (1863–1913),

Chilean minerals had become widely available to collectors worldwide. Most prized of all were large, crystallized specimens of jaw-dropping beauty that enhanced both the scientific reputation and the public appeal of these institutions. As Harvard's curator of minerals Carl A. Francis noted, Holden's collection "contained at least two dozen Chilean specimens, mostly silver minerals, particularly Chañarcillo proustites—and every important collection received since then has included Chilean minerals. However," Francis quickly added, "Chilean minerals only became important at Harvard with the addition of a suite that was collected by Mark Bandy in northern Chile in 1935."[50]

Mark Chance Bandy (1900–1963), a well-known and energetic mining engineer and mineralogist, collected museum-quality mineral specimens from the Atacama for both the Smithsonian and Harvard. Bandy was the perfect match for this assignment. A native of Iowa, he had travelled widely and had already done mineralogical work in the nitrate pampas of Chile (1930), where he discovered a new sulfate mineral there. On his 1935 excursion, Bandy left his verdant Redfield, Iowa, home to embark on a three-month odyssey that found him visiting mining properties, scouring local museums, and developing contacts with locals in the Atacama. Bandy's notes reveal that he traveled throughout much of the region from the nitrate mines near Iquique in the north to the metals mines near Copiapó and Chañarcillo as well as Coquimbo in the far south of the Atacama.

Much had changed since Darwin visited these same areas exactly a century earlier in 1835. These changes included the tremendous expansion in mining activities and the development of the transportation system, especially the introduction of railroads and the improvement of the road system. Above all, an urgency characterized Mark Bandy's narrative, because this aggressive development had resulted in the depletion of the best mineral specimens. Like Darwin, Bandy had found the nitrate pampas bleak, but he was also well aware that they were of great mineralogical interest. To him, their salt-encrusted areas "looked not unlike a field of white flowers, the only feature out of character was the entire lack of any green tint in the whole landscape." After connecting with a Dr. E. Lopez, Bandy was able to find "my first real collection of nitrate minerals." These included "a truly fine specimen 10 X 10 X 3 inches, more or less covered by sharp rhombs of soda nitre. A beauty indeed." Bandy's term "rhombs" was shorthand for rhombohedral-shaped crystals, and the specimen was impressive. However, he quickly added that "the rhombs were about 1/10 inch in size and did not compare in size and beauty in any way to the specimen I sold Harvard University, but they were very good indeed." As a mineralogist and entrepreneur, Bandy was somewhat conflicted, adding: "My how I regretted time after time, selling

Figure 4.19. A superb specimen of crystallized native silver from Chañarcillo was collected by Brad van Scriver in 1987. Specimen size: 11.5 cm (ca. 4.5 inches). (Photograph by Wendell Wilson, Tucson, AZ)

Figure 4.20. Called "Drops of the blood of Jesus" by miners for their ruby-red color, crystals of the silver-bearing mineral proustite from Chañarcillo are among the most coveted of all Atacama minerals. Specimen size: 3.8 cm (ca. 1.5 inches). (Photograph by Wendell Wilson, Tucson, AZ)

that specimen for $25.00 to Harvard. They evidently didn't know its value and certainly I didn't, although I knew it was worth more; but I needed the money and needed it badly at the time so took what they offered."

As Mark Bandy visited other oficinas and connected with other company officers, additional specimens began to appear. His luck in the south, though, was not quite as good. In Copiapó, which he observed "is either favorably or unfavorably noted for its dusty streets," he visited the offices of the American Smelting and Refining Company (ASARCO). He then went to the School of Mines, which involved "a long, quiet, nice and dusty ride to the west end of Copiapo." Facing this street were a few impressive residences as well as the main building of the School of Mines, which still stands among the few venerable early buildings on today's otherwise modern campus of the University of Atacama. On his memorable ride, Bandy noted that the city was unusual in several regards, stating that "there are very few buildings of two stories in the city." Moreover, he observed, "it sprawls in all directions and the streets are either unusually broad or exceptionally narrow, and they are at least as dusty as their reputation suggests."

After meeting Mr. Torres there, Bandy noted, "I was disappointed in the collection, as I had expected a much better one." Spectacular silver specimens were on Bandy's mind, and he encountered some interesting examples from the mines near Chañarcillo. Among both institutional and personal collectors, Chilean silver from that part of the Atacama was prized. Some of the more spectacular specimens of native silver consist of twisted ropelike "braids." Then, too, others appear much like woven fabrics of elongated crystals. The specimen from Chañarcillo illustrated here (Figure 4.19) reveals the spectacular, distinctive reticulation that is sought by collectors. In the context of collected specimens, this is a latecomer. It was collected by Brad van Scriver in 1987 and confirms that high-quality silver specimens, though rare, are still being acquired by collectors.

Such is not the case with proustite, the silver sulfarsenide mineral that is now extremely rare at Chañarcillo and other locations in the Atacama region. Although Mark Chance Bandy was not too impressed with the minerals shown to him by Mr. Torres, that quickly changed when he was shown a spectacular proustite specimen that, as he put it, "simply took my breath away." Of all the minerals from the Atacama, proustite was the prize. Although Bandy had heard about other proustite specimens described by Domeyko and numerous mineralogists in the 1800s, of this one he speculated, "My Oh My, just to have a specimen like that to look at each day and a sweet wife to enjoy it with you and life would be worth living, even though you almost starved to death." Wistfully, Bandy concluded: "I'll never own one but I can always remember that specimen, if I never see one finer, *'Gotas de Sangre de Jesus'"* (drops of the blood of Jesus). Alas, even after traveling to Chañarcillo over rutted roads in search of such specimens, Bandy found none as superb.[51]

As a mineralogist, Bandy knew that time had pretty much run out for those seeking the finest proustite specimens, for none could compare to those collected in the 1800s. A case in point is the unrivaled Chañarcillo proustite collected in the 1800s by American mineralogist William S. Vaux (1811–1882). In recent correspondence with me, mineralogist Wendell Wilson noted that Vaux's specimen was "the most beautiful I've ever seen." As evident in this photograph taken by Wilson in 1972 (Figure 4.20), Vaux's specimen is perfectly crystallized and intensely colored. Small wonder it was the envy of collectors worldwide. Vaux's collection was large, but this stunning proustite was among the "Famous 25" specimens retained by his family after his death. The rest of Vaux's large collection went to the Philadelphia Academy of Sciences, but this proustite was held by, and was a personal favorite of, William's nephew, George Vaux, Jr. (1863–1927). Upon George's death,

Figure 4.21. Three tokens from varied oficinas in northern Chile date from the nitrate boom period of a century or more ago when mining companies controlled all aspects of life in the Atacama Desert. (Author's digital collection)

it remained in the Vaux family until the 1970s when they agreed to sell the entire Famous 25 to the Smithsonian Institution, where it resides today.[52] Few mineral specimens have been as coveted by collectors as this one, which makes it easy to understand why a vernacular name for proustite is "ruby silver."[53]

Collecting, of course, is far more than it appears at first sight. Much has been written about the subject, or rather phenomenon, some of it deeply psychological. Some suggest that the need to collect stems from "a subliminal need to deal with a flawed childhood relationship with a parent, especially the mother." As an inveterate collector of geological specimens who lost his biological mother when just one year old, I could plead guilty as charged, but collecting is actually more complex than that—especially when one considers the subject historically and sociologically. Collecting reached a near frenzy in Europe in the late-eighteenth and early-nineteenth centuries when that continent was both industrializing and developing overseas empires that threatened Indigenous peoples with ultimate extinction—by outright subjugation or assimilation.

In a recent book on twentieth-century artist Joseph Cornell, art historians Sarah Lea and Lynda Roscoe Hartigan introduce the elements of time and space into the equation with collecting. Subtitling their book "Wanderlust," they note that Cornell created, or rather re-created, an entire world far from his native New York—a world that he would never actually see but represented in three-dimensional collages, or montages, as he called them. As Lea shows, for Cornell "collecting was the basis of imaginative transport: ecstatic voyaging through endless encounters" with old materials that he had accumulated over the years.[54] Hartigan confirms the chronological dimension as well, observing that one aspect of collecting is "about creating a different, highly personal sense of time." As she argues, "a collector seldom begins the act of collecting with the expectation of closure; the thrill of the hunt for the missing object is more powerful than the need to finish or complete." This, by the way, is reminiscent of the adage that undertaking a journey is more important than reaching its destination, or words to that effect. Ultimately, Hartigan sees collecting as "an effort to control or stop the hands of time, as if to collect is a present- and future-oriented act of salvation from the destructive passage of time."[55]

Like Joseph Cornell, I became fascinated by "found art," although he began creating his works in the early twentieth century while I began mine in the 1950s. Still, there is something admittedly irresistible in finding an object and contextualizing it according to its form, color, texture, or chemical composition. That act might be scientific, but it is also highly aesthetic. To me there is something compelling about the placement of rock and mineral specimens in compartments. Often their only identification is a number that corresponds to the catalogue indicating when and where it was collected, and ideally by whom. In my endeavors, the older the original label and box, the better—another subliminal nod to time. In addition to natural history specimens, which attempt to isolate timeless pieces in time, collectors also cherish cultural objects and exhibit them in a number of ways.

The fascinating mining company tokens associated with the Atacama Desert's nitrate oficinas are a case in point (Figure 4.21). At one time, these tokens served as currency and were exchanged for goods and services at the *pulperías*, as described by Russell. On occasion, these tokens can still be found among the ruins of old oficinas. Some are made of metal, in which case they feel like coins. Over the years, I've collected several of these from the Atacama. My personal favorite is the solid brass 10 centavo token from Oficina Alianza inland from Taltal, which can be distinguished from a coin by its two crescent-shaped holes. Other tokens in my collection are made of compounds such as "vulcanite" (more properly ebonite), a durable vulcanized rubber, or early plastic. The largest of the tokens illustrated is a vulcanite from the Oficina Curico of the Compania El Loa. Cast in the denomination of one dollar, it is about the size of a silver dollar, and each side is a different color (oxide red vs. moss green). The other smaller token is from The Colorado Nitrate Company Ltd. and served as a 10 centavo piece. Being of a lighter-weight material than brass or copper, these tokens feel more like

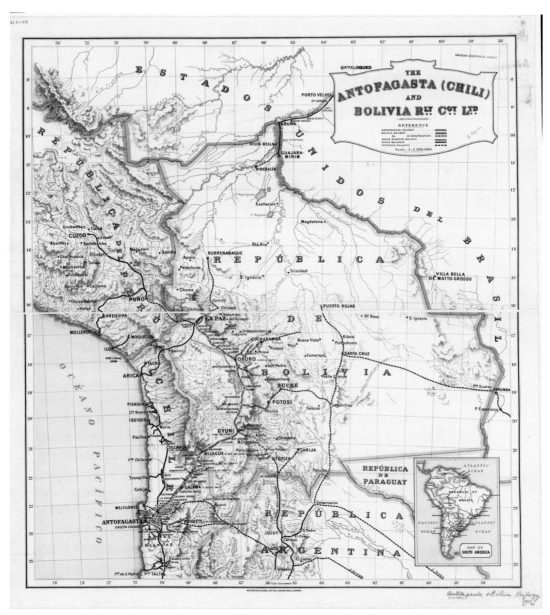

Figure 4.22. Map by the Antofagasta (Chili) and Bolivia Railway (London: Waterlow & Sons, 1916) artfully uses color to encourage travel between the Pacific port city and La Paz in the Andean interior. (American Geographical Society Library, University of Wisconsin at Milwaukee)

poker chips when handled. Although some of these tokens from Atacama oficinas have developed a patina, others protected from the elements appear almost new.

As tangible objects, the tokens stand the test of time, while the companies and individuals that created and circulated them have long since ceased to exist. The tokens illustrated here may seem like simple objects, but each has an intriguing story to tell. The realization that they were circulated locally in the Atacama, passing from the pockets and hands of employees to the *pulperías* and then back into circulation as wages is fascinating, but only part of the story. Tokens also played a role in a more tumultuous part of the region's history, namely, the bloody Santa Maria School Massacre in Iquique more than a century ago. As one of many demands made by miners, thousands of whom converged on Iquique from the outlying oficinas, mining tokens were among the many sore points, and miners demanded that cash payment replace them. This demonstration occurred at a time when the price of nitrates had declined, as did wages. Fearing a complete breakdown of the entire country's economic system, Chilean military troops turned their guns on the men, women, and children who had brought nitrate production to a virtual standstill. The December 21, 1907, retaliation resulted in mass casualties, though the exact count of those gunned down has long been in dispute. One seemingly authoritative source claimed just under two hundred were killed, but other accepted sources claim two thousand, and still others as many as three thousand two hundred. Although hushed up for many years by the Chilean government, the Santa Maria School Massacre remains a bloody stain on the region's history, and was finally commemorated in 2007.[56] In the long and checkered economic history of this region, the massacre figures as yet another indicator that nitrate mining was both labor intensive and unsustainable in light of changing expectations. To many critics, these tokens are essentially counterfeit currency. Although marked as such, the tokens weren't really worth the denominations indicated on them, because the company had purchased goods at a fraction of the price they charged workers, who had no alternatives given the isolation of the oficinas.

Ultimately, these tokens are a reminder that life was cheap but came at a high price to workers, who were often subjected to hazardous working conditions, and who the cost-conscious British companies often pitted against each other by nationality (Peruvians vs. Bolivians vs. Chileans). Regarding working conditions in the early twentieth century, it was often cheaper to replace a killed or injured worker than make the process safer, a condition that led to labor unrest and, in turn, unionization or a closing down of operations, and sometimes both. As economic historian Michael Monteón noted in a 1975 article entitled "The British in the Atacama Desert: The Cultural Bases of Imperialism," the repercussions played out in both time and space. At first they were confined to the pampas and ports of the remote north. As Monteón wrote, "in the Atacama desert the workforce became militantly anti-imperialistic, that is anti-British, in reaction to the poverty and repression in the nitrate factories and ports." Ultimately, though, this disenchantment would spread to the very center of Chile and would have repercussions well into the 1970s. As Monteón concludes, "the formation of the strong working-class consciousness in the north that became an essential element of the socialist and communist movements dates from the treatment Chilean laborers received at the hands of the British merchants and *oficina* owners."[57] I will say more about these political ramifications in the next chapter, for they affected the Atacama and its bleak pampas salitreras (nitrate pampas) in unexpected ways.

In retrospect, the nitrate industry—which Chileans simply call "Salitre"—boomed and went bust well within the timeframe of this chapter (1865–1945), and I would like to again step back into that time period. At the time that the nitrate decline was well underway (ca. 1920), Isaiah Bowman was preparing the informative maps that appeared in his *Desert Trails of Atacama*. To place Bowman in the broader context of this region's historical development, I invite readers to briefly return to his two maps illustrated in Chapter One (see again Figures 1.6 and 1.7). Although drafted about a century ago, Bowman's maps could still be used by travelers today. Of course, the Pan-American Highway was not yet in existence, though it roughly parallels the route of the longitudinal railway; moreover, a number of the railway lines on Bowman's maps have since been abandoned, often showing in the landscape today as slightly elevated earthen lines long stripped of their steel rails and wooden crossties. An example of one such railway line runs from Taltal inland, its demise a result of the collapse of the nitrate industry.

Despite these changes, though, mining was expanding into other areas in Bowman's time. In fact, that industry helped Bowman prepare these maps, as some of the information on them was acquired from mining companies. However, what makes Bowman's work—both narrative and cartographic—different from all others to date was that it represented a new, introspective way of describing the Atacama Desert. By this I mean that Bowman was experiencing this desert not as a place that was being explored for the first time, but rather as one that was now being scoured for what it revealed about how people in the past had used it. Bowman recognized that there was continuity between past and present, for in many cases people were still conducting venerable agrarian and pastoral activities as well as actively mining small

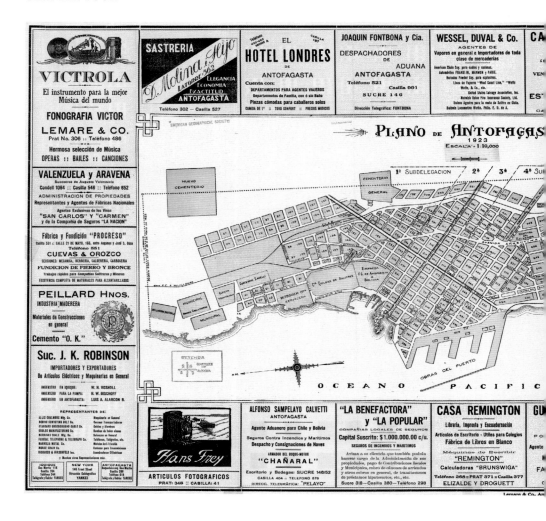

claims. Bowman's *Desert Trails of Atacama* is the single-most important document on life and landscape there in the early twentieth century.

That having been said, I hasten to add that Bowman was just one of many mapmakers depicting the Atacama Desert at this time. For example, in 1916 the Antofagasta (Chili) and Bolivia Railway Company published a superb map that not only showed their line in bright red but also delineated the three countries in the northern part of the Atacama using a color code: yellow for Chile, orange for Bolivia, and green for Peru. This colorful map (Figure 4.22) reveals that the region's cartography had become an art, which is to say, part of the broader developing art of advertising. This map makes it clear that the railroad line facilitates travel, taking less than two days to complete a journey that would have otherwise required upwards of five or six

Figure 4.23. *Plano de Antofagasta* (Antofagasta: Lemare & Co., 1923) shows the city developing along the shore of the Pacific from its central core near the plaza. (American Geographical Society Library, University of Wisconsin at Milwaukee)

weeks. Along the route, its narrow-gauge track traverses varied topography, including broad open spaces of the pampa, and the steep slopes and rugged canyons of the cordilleras, which are shown using a shading technique. The map also indicates that the line serves the many nitrate oficinas, which are indicated in pink. This map simultaneously depicts a region and sends a message: for business or pleasure, the FCAB is the best way to travel from the coast at Antofagasta to the city of La Paz in the highlands.

Of this railroad's western coastal terminal, another beautiful map, *Plano de Antofagasta* (Figure 4.23), shows the prosperous city in the year 1923. The scale is 1:10,000 and the map itself is oriented eastward (that is, east is at the top of the map). Its intense colors are noteworthy: in bright pink, it shows the city filling block after block along the shore of the Pacific, the square

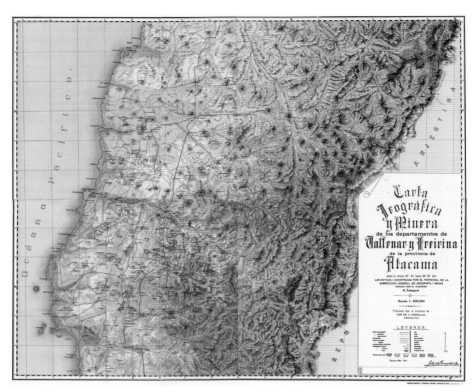

Figure 4.24. *Carta Jeográfica y Minera de los Departamentos de Vallenar y Freirina de la Provincia de Atacama* (Santaigo: José del C. Fuenzalida, 1914) depicts a mineralized portion of Chile's Norte Chico. (American Geographical Society Library, University of Wisconsin at Milwaukee)

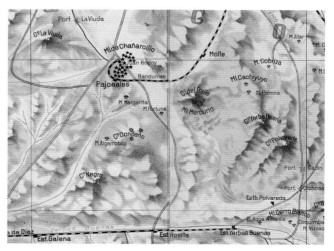

Figure 4.24 detail showing the mines near Chañarcillo.

plaza in the north complemented by the linear Parque de Brazil, which runs in boulevard fashion southward. At just this time, the homes of nitrate company officers and prosperous merchants lined the streets. This map reminds us that some maps address far more than the geographical area they depict and can also serve as vehicles for advertising and promotion. The margins of this map contain a riot of colorful advertisements showcasing varied business establishments that operate in the city. Even at this time, Antofagasta had a tendency to develop a long and lean form not unlike the shape of Chile. Today the city is less than a mile (ca. 1.5 km) wide, hemmed in by the coastal mountains, but stretches about fifteen miles (ca. 21 km) along the adjacent narrow coastal plain that gently tilts toward the Pacific. Given all indications, Antofagasta is likely to continue growing along this north–south axis well into the future.

Meanwhile, maps of mining enterprises also kept pace, cartographically speaking. During this period, namely the 1910s and 1920s, mapmakers often used the latest techniques to depict developing areas and their particular mineral resources. Given the ongoing importance of mining, it is not surprising that most maps addressed it. A growing number of maps appear about this time, and so I shall only introduce a few representative examples, and briefly describe them, here. First, I introduce an important map of the area at the southern margin of the Atacama Desert, namely a portion of the Norte Chico, which normally receives more rainfall than the area farther north but is still considered a desert. Published in 1914 and entitled *Carta Jeográfica y Minera de los Departamentos de Vallenar y Freirina de la Provincia de Atacama*, this colorful map (Figure 4.24) shows the large number of mines clustered in many areas that comprised mining districts. Next, I introduce a 1918 map entitled *Carta Topográfica Departamentos de Antofagasta i Tocopilla* (Figure 4.25) that was prepared by Manuel A. Rojas Rivera. As with many such maps, it indicates mines, railroads, and cañerias (pipelines) in considerable detail. Prospective investors used these maps to help them better understand the complexities of both landscape and industry here. Rojas's map is both beautiful and utilitarian, taking advantage of developments in color printing.

Far less visually appealing but even more informative is the businesslike blueprint map *Carta Minera y Geográfica de la Provincia de Antofagasta* (Figure 4.26). As the abundant text notes, the information compiled on this map was begun in 1888 but is updated to 1920. This map has an unusually comprehensive legend (leyenda) that itemizes the varied minerals mined in this desert region, including metals and salts such as iodine. Compared side by side in this mini gallery of early-twentieth-century mining maps, these three reveal the diversity of cartographic techniques used to frame or position mining

Figure 4.25. *Carta Topográfica Departamentos de Antofagasta i Tocopilla* by Manuel Rojas Rivera (Santiago, 1918) contains a wealth of information on mines and related infrastructure such as railroads and pipelines. (American Geographical Society Library, University of Wisconsin at Milwaukee)

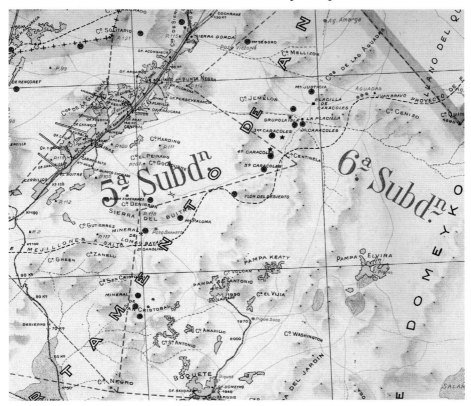

Figure 4.25 detail showing nitrate oficinas and silver mines near Caracoles.

enterprises in the context of the broader region that was now commonly called "el Atacama." That nomenclature dropped the word desert because it was implicitly understood; moreover, although it may seem to downplay the importance of the desert, it does just the opposite—just as the term "the war" is often used for the most important and disruptive event in a peoples' collective memory.

By 1924, when the American Geographical Society published Bowman's book, the Atacama Desert had been completely explored and mapped—or so it seemed. In terms of mining, major developments were still about to take place. Copper, too, would soon shape the fortunes of this desert. Although this red metal had long been mined throughout much of the region, the area around Chuquicamata would become the flagship property for its extraction. As early as 1899, a way had been sought to make use of the finely disseminated copper sulfides in the porphyry rock. However, although Chileans had tried to work the mountain with low-grade methods, imported technology would ultimately prove successful. Developed early in the twentieth century

Figure 4.26. *Carta Minera y Geográfica de la Provincia de Antofagasta* in blueprint format contains information compiled over nearly thirty years (1888 to 1920) and features an especially comprehensive legend. (American Geographical Society Library, University of Wisconsin at Milwaukee)

in the United States, these new processes involved crushing the copper ore into a fine powder and then running it through a flotation process—whereby the metal-rich ores are gathered at the top of the liquid. This may seem counterintuitive as the metal-rich particles are heavier than the fluid and thus should sink, but because they have a molecular affinity for the foam at the top, they are held there in suspension before being driven into nearby vats. This process was pioneered and applied in the American West, where large porphyry copper deposits similar to Chuquicamata existed.

Northern Chile proved the perfect place to expand American interests at just this time. Throughout the 1910s, Americans had taken a strong interest in Chile's copper mines after noting the similarity of Chile's ores to those being successfully mined through the open pit process first used in Bingham Canyon, Utah, and then in Bisbee, Arizona. At this point in time, a bold move by the Guggenheims, one of the world's most wealthy families, was about to transform the Atacama Desert. American born and self-made, the Guggenheim brothers had successfully developed copper interests in the American Southwest and now moved into the Chilean scene. They broadly called their initiative the Chile Exploration Company, but in fact their mission was American and their main focus was in the northern Atacama Desert. Here, on Chile's northern fringe, mines had been exploiting rich copper deposits that consisted partly of native copper and mostly of copper oxides (e.g., cuprite), carbonates (e.g., azurite and malachite), sulfates (e.g., chalcanthite), as well as other copper minerals such as brochantite and the region's namesake mineral—atacamite. The mountain at Chuquicamata was so rich that much of it was ore; therefore, steam shovels of the type that helped open the Panama Canal were used.

As Miller and Singewald observed in 1919, there was another much larger part of the Chuquicamata mine where copper sulfides such as chalcopyrite and bornite were finely disseminated throughout porphyry rock. Although the Guggenheims pioneered mineral extraction procedures at Chuquicamata, they had yet to implement them. As Miller and Singewald reported, "no arrangements have yet been made to take care of the sulphide ores as it will be a number of years before they are reached in the mining operations."[58] Under the ownership of ASARCO (whose name changed to Anaconda in the 1920s), the property at Chuquicamata witnessed the aggressive development of open-pit mining techniques that had proved so successful in the American West. This type of mining involved extracting the small amount of copper in the rock—usually less than two percent—by essentially pulverizing the entire ore body. Where non-mineralized rocks covered those deposits, the topography itself was blasted into rubble and this overburden moved out of the way.

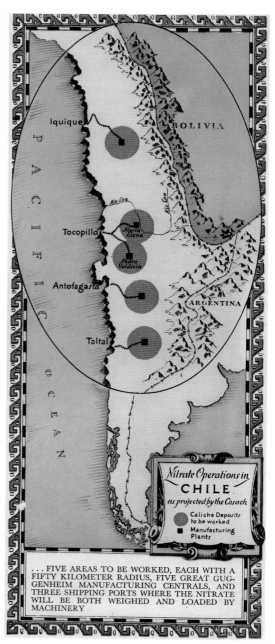

Figure 4.27. This *Fortune* magazine map from 1930 shows existing nitrate mining operations and their oficinas and rail lines in the Atacama Desert. (Author's digital collection)

Figure 4.28. This *Fortune* magazine map from 1930 shows proposed changes that would supposedly make nitrate mining more profitable (compare with Figure 4.27). (Author's digital collection)

In a series of spiraling steps or terraces downward, the once-hidden ore body was dynamited into smaller fragments, scooped up by steam shovels, and hauled away in railcars or dump trucks. The sulfide ores yielded tons of sulfur dioxide from smelter stacks and cast a pall over the desert. With proper treatment, these aggressive processes could also remove other valuable metals, including silver, gold, and molybdenum.

In nitrates, too, the Guggenheims moved to modernize the industry in the 1920s. An example was the electrification of their railroad line from the port city of Tocopilla inland to the nitrate oficinas in 1927. This bold move eliminated the use of steam locomotives, which consumed prodigious amounts of imported fuel (coal and oil) and freshwater, as well as requiring considerable maintenance. The need to radically reform the faltering nitrate industry was quite evident by 1930, the year in which *Fortune* magazine published its first issues. In their October 1930 issue, *Fortune* focused on the Atacama Desert, where it touted the Guggenheim's new schemes to transform the nitrate industry through improved processing technology that was supposedly forty percent more efficient. As predicted in that issue, rationalization and consolidation of holdings would focus on the best properties, in effect abandoning less efficient ones.

Two maps in that *Fortune* article are noteworthy, and are reproduced here. Both reveal how effective cartography could be in advancing an agenda. The first map (Figure 4.27) shows existing nitrate mines in the Atacama Desert. It is beautifully drafted and stunning in the simple but effective way it shows the complex web of railroad lines reaching ports from the many oficinas. Although this map depicted the present condition of nitrate mining, a second map (Figure 4.28) showed the industry as it was planned to be after visionary consolidation overseen by the newly formed Chilean Nitrate Company (COSACH). This included streamlining, as it were, of nitrate production using more efficient methods of processing and closing down inefficient mines. The net effect would reduce costs by shuttering the cost-prohibitive operations, simultaneously abandoning oficinas. A further cost savings would result from the abandonment of railroads to those closed plants. Compared side by side, the maps offer a breathtaking juxtaposition of present and future. However, as economic historian Elisabeth Glaser-Schmidt demonstrated in 1995, government mismanagement and interference, coupled with poor timing, had disastrous consequences. "Guggenheims had put themselves into a precarious position when they opted for an unnecessarily high, yet potentially rewarding stake in Chile's unstable nitrate sector." Alas, a gathering economic depression and government ineptness scuttled the Guggenheim renaissance of the nitrate industry; moreover, the collapse of this industry also negatively impacted copper production.[59]

Thanks to geographer Homer Aschmann (1920–1992), the process by which mines develop and then decline was demystified, though its consequences are no less profound for people and landscape alike. In his seminal 1970 article "The Natural History of a Mine," Aschmann used examples worldwide and a detailed study of mines in northern Chile to conclude that all mines were characterized by stages of development, namely: 1) initial prospecting, in which mineral resources are discovered in a risky, costly, and labor-consuming process that confirms the mine's potential; 2) capitalization, including development of infrastructure and a dependable labor supply as corporations become involved in the process; 3) sustained operation, a phase that begins as profitable (for example, open-pit mining) but ultimately experiences increased costs and reduced profits as wages and taxes rise; 4) cessation of mining activity, often occurring despite cost-cutting measures. This last phase may be averted, as some mines may reopen as metals and nitrates prices increase or technological fixes prolong their lives, but all mines inevitably close down. In the case of nitrate mining, the development of artificial sources dealt the industry a fatal blow. Aschmann further noted that there is an additional challenge facing mining areas in the Atacama Desert. As he put it, "in the mining region of Chile, notably the Norte Grande, the economic bareness of entire regions has long posed a severe physical limitation." Aschmann grimly concluded: "Once a mine closed down, be it a producer of nitrates, copper, or precious metals, it was improbable that any other economic activity in the vicinity could be found."[60]

Although active mining continues in the Atacama Desert and has kept Chile the major mineral-producing nation in South America, this activity resulted in a truly impressive series of relict features such as mining ghost towns, abandoned mineral-processing plants, ore dumps and tailings piles, as well as other above-ground features. Recently, geographer Paul Marr documented these types of relics in an article entitled "Ghosts of the Atacama: The Abandonment of Nitrate Mining in the Tarapacá Region of Chile."[61] As noted by historian and geographer José Antonio González Pizarro of the Universidad Católica del Norte (Antofagasta), nitrate mining played a pivotal role in shaping the historical geography of the Atacama Desert. Ultimately, though, the rigors of the climate, ever-changing technologies, along with the vagaries of labor with its strikes and collective social conflict, led to a total collapse of the oficinas in the nitrate pampas.[62] The economic and social activities associated with mining, though, would soon be romanticized, as another phase of Atacama history was about to begin.

Five

"A Playground for Scientists and Explorers"

The Atacama as Exceptional, ca. 1945 to the Present

The Atacama Desert's rise in the popular mind was already underway before this last (and current) period began, but today the desert is more associated with tourism than resource extraction. In this period, too, we find occasional cartographic and narrative paradoxes: despite this desert's well-earned reputation as a storehouse of mineral riches, it sometimes went missing in atlases and maps. For example, in its text describing the world's deserts, *The International Atlas of the World* noted, "In the Southern Hemisphere are the Kalahari Desert of southern Africa, and the vast internal deserts of the Australian continent. Farther south lies the Patagonian Desert of Argentina, covering some 300,000 square miles." The atlas, originally published in 1947 with subsequent editions into the 1960s, was silent about the Atacama Desert. This might seem inexcusable to those familiar with the Atacama, but the atlas also neglected another important coastal desert—the Namib of southern Africa. Even the maps in *The International Atlas of the World* neglected to show the Atacama Desert, although the Rand McNally map of South America included in it did show "Atacama Prov[ince]" along with the neighboring provinces of Coquimbo and Antofagasta.[1]

Although this atlas seemed to take a step backward—after all, school books such as *Frye's Grammar School Geography* had depicted the "Desert of Atacama" since the early twentieth century—it was countered by the ever-growing number of maps and atlases that did include the Atacama Desert.[2] Among these was Rand McNally's popular *Goode's World Atlas*, which began publication in 1923 and has been subsequently revised in twenty-two editions to the present.[3] Where descriptions accompany maps in atlases, they usually stress the Atacama Desert's aridity and simplify the causes. Typical of these recent descriptions and representations is the *Grand Atlas Gallimard Pour Le XXI Siècle* (1997), which indicated "Désert d' Atacama" on three of its maps of South America; the atlas tersely attributed its location to the presence of the Andes Mountains ("la barriere des Andes, qui prive de pluie tout le Sud du continent").[4]

This burgeoning literature ensured that the Atacama Desert would become fairly well known to students and the general public in the twentieth century. Two aspects of this desert—its bleakness and resource riches—are usually emphasized in publications. In 1966, geographer Peveril Meigs wrote an informative synopsis of "Sector 36," as he classified the Atacama Desert in his UNESCO book *Geography of Coastal Deserts*. Meigs began this essay by noting that "the Atacama Desert of northern Chile is the driest coastal desert in the world," and he concluded that "the desert region of northern Chile is, in a sense, an economic appendage to the centre of Chilean power, and its development will continue to be informed by the power centre outside the desert," namely, the Valley of Santiago.[5] These two concepts—uniqueness and dependence—summarize modern perceptions of the Atacama Desert and undergird a rapidly developing tourist industry.

Before discussing that relatively recent industry, though, I should note that the landscape of the region mirrors a trend that began about two centuries ago and has accelerated in the late twentieth century: subsistence agriculture was being eclipsed by mining. In 1963, when William (Bill) Edward Rudolph published *Vanishing Trails of Atacama* for the American Geographical Society, he observed that "little change has occurred in the agricultural production of Atacama's oases since Isaiah Bowman's book of forty years ago. Indeed, it may be said that a trend of slowly diminishing production continues."[6] Rudolph attributed this to two factors—water and labor. Mining consumes vast amounts of water and provides high-paying jobs. Rudolph's extensive research was based in part on his earlier employment in the mining industry and his experience in diverting water for the copper mining properties in the vicinity of Chuquicamata.

In the late 1960s, geographer Khairul Rasheed conducted considerable fieldwork for his doctoral dissertation in the area where Rudolph had worked—the numerous Atacameño villages northeast of Calama. Rasheed's central theme was "how man has been able to survive and develop a sedentary pattern of livelihood in the arid highlands of Chile," and his work had two major objectives: 1) to identify the processes by which people use and master the desert, and 2) assess the impact that people have had on the way the desert landscape evolved. Rasheed identified four types of communities—urban, agricultural, mining, and railway settlements—and discussed them in considerable detail. His major conclusion that indigenous agriculture had declined in relation to mining was not surprising, but his discussion of why this occurred is noteworthy. Mining not only competes for scarce water resources but also offers higher wages. Farmers no longer maintain the water system and the land terracing needed to produce crops as they migrate toward higher-paying jobs in mines. Their families may stay on the farms for a period, but sooner or later many join the miners in the mining communities. This process may appear to be a one-two punch, but Rasheed concluded that two factors were operating simultaneously: the "pull" of high wages in mining draws labor away from the farms, while the "push" of declining agricultural output from poorly maintained farmlands makes remaining on the land untenable. This, coupled with increasing mechanization, jeopardized the region's indigenous agriculture.[7] The consequences of this are evident in today's landscape, where farm villages subsist but are under constant threat of virtual collapse. However, a growing sense of ethnic awareness, in part supported by the central government, keeps funding flowing to agrarian villages.

I should mention something else about this agrarian landscape. Agricultural abandonment affects both the lives of farmers and the look of the land itself. A case in point is the oasis of San Pedro de Atacama. Readers may recall that a German research team studying this area in 2002 (see again Chapter Three) observed that the thickness of the soil had increased about six feet (180 cm) as farmers diverted water and shaped the land into terraces over the centuries. However, the research team noted that the reverse was now occurring there following abandonment of many fields in the 1950s. As they concluded, "Irrigation water flowing unchecked over abandoned fields without any vegetation caused the cutting of deep gullies." The traveler today can see the consequences of this erosion: the labyrinthine landscape here consists of steep-sided, deeply creased, buff-colored terraces that are being swept away by running water. That this is happening in the world's driest desert may seem ironic, but there is yet another irony to consider; what at first appears to be a natural desert landscape is, on closer inspection, an artifact of agrarian abandonment. Ultimately, the sediment removed from this area will find its way into the vast Salar de Atacama.[8]

Another type of land use—mining—has had an even more profound effect on land and life in the Atacama. As confirmed by virtually all studies, mining remains *the* major economic force in this region. During the period following the Second World War, the continued development of larger and larger earth-moving equipment, along with more aggressive mineral exploration, resulted in increasing mining activity. These met the ever-growing international demand for metals, though mining activities occurred in the context of the inevitable roller coaster of boom times and recession that characterizes the industry. Although nitrate mining was winding down, copper mining was still a major activity, and Chuquicamata was the crown jewel; additionally, silver, molybdenum, and potash mines continued to develop in the postwar period. As noted in Chapter One, an impressive monument commemorating the location of the Tropic of Capricorn lies along the Pan-American Highway northeast of Antofagasta and southwest of Chuquicamata. Erected by the Chile Exploration Company and the Rotary Club of Chuquicamata, this marker symbolizes the power of the mining industry to move mountains and the resolve of the community to support such activities.

It is tempting to think of these mining companies as ruthlessly exploiting resources and labor, but there is another side to the story. As Anita Carrasco notes in an article about American capitalist interests in the Atacama Desert, Indigenous peoples developed a close relationship with the division of the Anaconda Company called "La Chilex," or the Chile Exploration Company. According to Carrasco, local Indians from the village of Toconce were treated well by this company as it fairly rewarded their labor, helped create and improve schools, and developed dependable sources of water. As she observes, "Tococeños' lives have always been intricately connected with the world of mining through employment and because of the effects of the industry on the availability of water." Carrasco identifies two men in particular, the Chilean-born but American-educated Don Rudolfo Michels, and the American-born mining engineer William Edward Rudolph. As noted earlier, the latter was not only a man of action but also one of introspection.

Carrasco notes that Michels and Rudolph helped La Chilex build relationships with the local Indians and were well respected, but that William Rudolph was especially visible and active in these communities. In fact, word of his work in the desert north reached Santiago, and he became something of a luminary throughout Chile. In 1955, Rudolph received the Order of Bernardo O'Higgins and its accompanying gold medal from the Chilean government, "in recognition of his humanitarian work in remote villages of the Atacama Desert, [including] San Pedro de Atacama, Caspana, Toconce, Ayquina, Cupo, Toconao, and Santiago de Rio Grande." As described and mapped in an unpublished report by Rudolph in the American Geographical Society Library in Milwaukee, the water-diverting infrastructure constructed by the Chile Exploration Company included five major pipelines: one 102.8 km (60 miles) long from the Inacaliri River to Chuquicamata; another at Toconce; two at the San Pedro River to the mines; and one at Salado Spring that carried saline water 70 km (ca. 40 miles) to the mine's sulfides plant.[9] Through such pipelines built from the 1920s through the 1950s, the copper company transformed the landscape and the cultural ecology of the Indian communities that cluster in the eastern edge of the Atacama Desert flanking the Andes Mountains.

Focusing on the relationship between companies and local Indigenous communities, Anita Carrasco's research sheds new light on the ethics and values of both protagonists. Rudolph's personal diary entries confirm that this is a fragile ecological zone where people live tenuously but tenaciously. As Rudolph noted in 1954, "These people need only 2 things to support their way of life, their llamas and their water." He quickly elaborated: "The llamas were beasts of burden, besides furnishing weekend meat. The waters supported dry farming—corn, wheat, alfalfa, and some fruits and vegetables." These were traditional close-knit communities, ranging in size to about three hundred people, many seasonally tending their animals grazing in the mountains. Nevertheless, their way of life was changing. Rudolph wrote, "Of late years, the men have been going away for work in the sulfur mines, to bring back money to exchange for tea, sugar, and other articles which they do not have." I should here restate that pure Indian identity in modern-day Chile is rare—only about three percent of the population. The largest category is considered European (about 53 percent), followed by mestizo (about 44 percent). In an attempt to address the importance of Indigenous identity, Chile in 1993 formalized the use of the term Atacameño for any Native person living in the Atacama Desert region. In doing this, the nation, through CONADI (National Corporation of Indigenous Identity), institutionalized a term long used for Indians in this arid region. However, one commonly hears the complaint that it may be too heavy-handed—more of a bureaucratic way of administering Indigenous peoples than a reflection of their actual identities.

With this in mind, I would like to share with readers a remarkable passage from Rudolph's 1954 diary that conveys something of the respect that he felt for the Indigenous peoples. "These Atacameño Indians of northern Chile have one point in common with the Quechua who conquered them, their 3 commandments, 'Do not lie, do not steal, do not be idle.'" Rudolf then provided a rather humorous example of this ethic. In describing how much diligence was required to keep pipelines operating efficiently, he recollected that the company had trained a dog to bark when it detected the rattling sound of a device in a pipe, indicating a blockage. Once identified, that blockage could be cleaned out by a work crew. This method of detecting problems was ingenious, but the Indians from a local village protested. As Rudolph recalled, "they said they could serve us as well as the dogs even though they couldn't bark, and besides their religion commanded them to work which the dog's did not." The company respected this request, and Rudolph tersely noted: "The Indians got the job." This solution led to a long-term relationship that translated into sustained employment. As Rudolph concluded, "Now-a-days the cleaning apparatus is followed by a jeep; an Indian drives the jeep."[10]

Copper mining had not only transformed the interior portion of the Atacama but also helped shape national affairs. Chile's national election of 1970 exemplifies this. Political scientist William Culver sums it up thusly: in that election, the socialist candidate Salvador Allende won the presidency, but not by the popular vote, receiving only a third of the votes cast. By law, the Chilean Congress was duty bound to elect a president from among the top two vote getters, and conservatives threw their support behind Allende after he signed a pledge to respect private property and due process. In 1971, Congress unanimously voted to have the Chilean state corporation take control of the major foreign-owned mines. According to Culver, conservatives supported this popular action, in part as a way of punishing the United States for supporting agrarian reform in the 1960s.[11] The entire situation serves as a reminder that agrarian and mining interests cannot be separated in Chile, and that both the far left and the far right sometimes behave in ways that are counterintuitive, and even downright counterproductive. As a case in point, the outcome of these political decisions in the early 1970s had far-reaching ramifications: the nationalization of Chile's copper companies slowed production and plunged the entire country into an economic crisis as inflation increased 300 percent. Some claim that the United States deliberately manipulated the copper market to damage the economy of this now left-leaning country, but government mismanagement of the economy also helped to worsen the situation.

During this tumultuous period, the government-created copper company called CODELCO (Corporación Nacional del Cobre de Chile) was selected to operate the mines. Today CODELCO is the largest copper company in the world, and it also produces molybdenum and other metals, including the very rare metal rhenium, which is used in jet engines. In the Atacama Desert and elsewhere in Chile, the reputation of CODELCO varies depending on one's political persuasions and memories. In her fieldwork among the Tococeños, Anita Carrasco observes, "Their historical involvement with mining may explain why the people from Toconce make frequent comparisons between the times when the mining company was the property of American capital interests and the times when it became the property of the Chilean state." Carrasco notes that there is considerable ambivalence about CODELCO, partly because the villagers "tend to associate all positive aspects of the company back to the times when it was owned by American capitalists, mainly because the employment opportunities for native people were high—or at least that is how they remember it." Carrasco's study corresponds to findings by Janet Finn, whose book *Tracing the Veins* is a revealing comparison of two Anaconda Mining properties (Chuquicamata and Butte, Montana). Finn writes, "When working men spoke of the days when Anaconda ran the mines in Chile, they spoke well of the operations and of the gringos who recognized and rewarded skilled labor."[12] On a Friday evening return flight from Antofagasta to Santiago, I was sandwiched between two talkative miners headed home to southern Chile for several days. When they mentioned that they had just spent a couple of weeks working in the copper mines and were looking forward to the down time, I innocently asked them if they worked for CODELCO. Their response was immediate. Simultaneously, and almost as if in stereo, they responded with retorts that amounted to: "Of course not! We work for a real company—Antofagasta Mining!"

Both of these miners were mestizos who traced their heritage to Araucania and Los Lagos well south of Santiago, but I also wondered how the Indigenous Atacameños were fairing in places like Chuquicamata, where COLDELCO now reigns, and also in places such as Calama, which have substantial Indigenous ethnic communities from nearby. Curious about how Indigenous people living in villages adjacent to the Andes have fared in more recent times, I consulted anthropologist Anita Carrasco's 2011 dissertation, "One World, Many Ethics." Carrasco observes that the Indians here do not form one single bloc of people but are complex indeed. As she notes, "Atacameños are still in the midst of this process of reconstruction of their many scripts" and are confronting "the world of mining corporations that restrict their horizons and cause damage to their lives." Carrasco confirms that even the name Atacameño is disputed. An Indian woman in Toconce told her that "we are Quechua, not Atacameños," and added that the name Atacameño was "an invention of CONADI," which is to say the Chilean government.

Carrasco's fieldwork confirms that mining has disrupted traditional life in many ways, especially in the smaller villages. As Carrasco demonstrates, "water in the rural villages of Toconce, Cupo, and Turi in the Loa river basin is symbolically constituted through ritual offerings—the cleaning of canals ceremony—that intend to reciprocate with the Earthmother for the extraction of 'her' resources." But this age-old belief may conflict with modernity. "In contrast, the Chilean state has aligned itself in this matter with corporations promoting development of large-scale resource extraction projects that require great quantities of water." This neoliberal government considers the real owners to be those who hold legal titles, rather than those connected to them through tradition.[13]

Carrasco also cautions that "Mining-Community relations in Atacama are not homogeneous," adding that "on one extreme we find communities that have received tangible benefits from their relations with mining companies, and thus, whether willingly or not...have become their strategic partners." The relationship between CODELCO and Toconce is a case in point. However, "in contrast, we find that communities making claims are usually those that have not received tangible benefits from mining corporations." San Pedro Estacion is one such community, and its grievances are not only against the mining companies but also against the government of Chile. In a multipage manifesto to Chilean authorities in 2007, the town's president Sarah Ramos Mendoza demanded that Indigenous people "must be involved in the building of the region's society, and in such a process...have much to contribute." The issue may appear local but is in part driven by a growing international consciousness about the rights of Indigenous people. As Carrasco notes, "Atacameños have two things the mining corporations want—water and moral approval of mining operations." The issue of water rights is central here, as is often the case in desert regions. Carrasco concludes with an astute statement about mutual interdependence that characterizes the entire region: "Without water, no copper can be extracted; without water, Atacameños will perish."[14]

The mining drama will continue into the foreseeable future, but tourism is noticeably on the increase in Chile, and I shall shortly address that in more detail. First, though, I should note that I could find no published histories of tourism in Chile, much less in the Atacama Desert. Consequently, I shall try to construct a basic outline of one based on anecdotal information collected from travel agencies and local hostelries. I should mention at the outset that tourism in Chile is largely oriented toward the spectacular scenery of the southern part of the country, with its glaciers and forests that awed Charles Darwin and other early visitors. Given a choice, I suspect that more people would rather take in a majestic forested mountain view than one as austere as the Atacama. That having been said, I shall also remind readers that the Atacama has long intrigued outsiders, precisely because it is so stark. As with tourism in general, the type of visitation to this region may focus on historical sites as well as those oriented toward understanding the environment, the latter commonly called ecotourism. Of course, ecotourism also conveys a sense that tourists are personally concerned about the sustainability of their footprint on the land. Nevertheless, as noted in Chapter Two, the temptation to leave one's mark on so empty a place, say, in the form of a pile of stones, is likely a natural impulse, too.

By the 1950s, increasing numbers of tourists interested in exotic cacti and beautiful minerals also made their way to this desert. Many acquired specimens for their personal collections, which continues to the present. In 2016, an article by John N. Trager in the *Cactus and Succulent Journal* introduced two cacti out of their yearly selection of ten worldwide that were being sold to collectors for reasonable prices. Removing specimens from the Atacama Desert is prohibited, but these propagated plants were "produced under nursery conditions without detriment to wild populations." Trager observes that *Copiapoa dealbata* "is one of the species of this Chilean genus that cactus enthusiasts yearn to see when traveling to Chile." He further comments that "the species is known only from a 15 km strip along the coast from Carrizal Bajo north, hence the synonym *C*[*opiapoa*]. *carrizalensis*." Its price of seven dollars for a seedling is comparable to what one might expect to spend for a common cactus at a regular nursery, but this is quite a bargain for a rare and exotic plant.

Similarly, collectors were introduced to *Eriosyce taltalensis* (subspecies *pauciostata*), which, as the species name suggests, is found in the vicinity of Taltal. This light green to purplish cactus is also available as cultivated plant specimens for around eight dollars.[15] Through such purchases, the cacti of the Atacama are distributed worldwide as part of the educational mission of the Huntington Gardens. In the process, their native habitat is spared the impact of collectors harvesting plants in the field. Habitats in this desert are both fragile and endangered. My cactus collection includes four specimens of *Copiapoa* cacti purchased from reputable sources. These are rare plants indeed, and their habitats and habits are unique. When my *Copiapoa esmeraldana* flowered, I rejoiced—but by the time friends arrived to see it the next day, that large, impressive white flower had already shriveled up! Out in the field, too, the life of the Atacama operates on its own clock, and lucky indeed are those in the right place at the right time.

Like the cactus collectors, many mineral collectors in the 1950s and 1960s traveled to Chile to see firsthand the "type" locations associated with minerals such as atacamite and coquimbite. Their goal was to return with high-quality museum-type specimens for their personal collections. Some of these found their way into the catalogs and shops of mineral dealers such as Burminco, which I frequented as a teenager when it was located in Monrovia, California. Its owner George Burnham had traveled to many of these locales and always enjoyed swapping stories with customers. As noted by Terry C. Wallace and Michelle K. Hall-Wallace, Chuquicamata was one of the more alluring locations, for its mines had long turned out beautiful specimens of copper ores such as antlerite and atacamite, many of these found in the llamperas, an oxidized zone. Moreover, "the extreme climate of the Atacama Desert contributed to some truly exotic sulfates in the llamperas, which made Chuqui important to mineral collectors." Then, too, the name Chañarcillo has long resonated with mineral collectors as the location where superb ruby-red crystals of the silver-bearing mineral proustite are found. As the Wallaces put it, "for many mineral collectors, a fine proustite is the ultimate mineral specimen. The scarlet-vermillion color of a carefully curated proustite makes the specimen seem to glow, and none compare to the fantastic crystals from Chañarcillo, Chile."

The Descubridora mine associated with the ill fortune of Juan Godoy and the wealth of his former partner Juan Callejos Miguel Gallo is also known for fabulous specimens of silver in calcite, some of the silver appearing as twisted masses of wires and others crystallized in a herringbone pattern. In other locales, atacamite and additional minerals associated with this desert can be found, often with the help of local miners. Nevertheless, the Wallaces warn readers hoping to score on the fabulous ruby-silver here. "The value of the proustite is not lost on the inhabitants; if you travel to Copiapó and inquire about buying a proustite, someone will show up at your hotel room with red-colored rock and an asking price of thousands of dollars." But as they conclude, "Sadly, no new proustite has been recovered in more than 125 years."[16] Despite this warning, fine specimens of other minerals are still being found in the vicinity of Copiapó and many other Atacama Desert mining locales. Expecting very little in the way of luck, I have been amazed at the fine, if modest, specimens I've found there. But I too must offer a caveat: although the old ore dumps of abandoned mining areas can be a bonanza for the historian and mineral collector, extreme caution is needed to avoid injury or worse from one of many hazards, such as open mine shafts, jagged rusty metal, and rickety wooden headframes.

By the early 1960s, when the engaging lecturer at Foothill College that I mentioned in the preface visited this stunning region with camera in hand as a modern-day pioneer explorer, the Atacama had popped up on the radar for those who craved travel to remote areas of stunning scenery. However, Chilean tourism, much of it bus oriented, had been around much longer, as the desert north featured some resort hotels where, during the rainy winter in central and southern Chile, travelers could find an easy escape. La Serena was one such place, and Bahia Inglesa another. Some of this travel involved the families of mining company managers and other executives who had spent time in the north, returning home to central and southern Chile with

Figure 5.1. Chilean postage stamps commemorating the four-hundredth anniversary of the Atacama Desert's discovery (1536–1936). (Author's digital collection)

Figure 5.2. Chilean postage stamps acknowledging the importance of mining depict operations at a port in the country's desert north. (Author's digital collection)

tales of abundant sunshine and pristine beaches. For the more intrepid, the desert interior beckons. The Atacama Desert landscape has become an icon since it was immortalized on a Chilean postage stamp (Figure 5.1) commemorating the four hundredth anniversary of its discovery (1536–1936). The actual location is not specified on the stamp, but the image resembles the rugged, reddish-colored stratigraphy of the Valley of the Moon, a serene if desolate area outside San Pedro de Atacama. One consequence of this postage stamp is that it encouraged people to experience the real thing. Discovery was the operative word here, for modern-day Chileans could visit this unique area through a network of good roads that in part served, and continue to serve, the mining industry. In a fitting tribute to the connection between Atacama scenery and Atacama mining, Chile also created a postage stamp entitled "Mineria" (Figure 5.2). In contrast to the Atacama scenery stamp, this purple-colored stamp emphasizes the huge, capital-intensive features associated with mining, including mill buildings and concentrator tanks. The setting is also noteworthy, for although it too is a desert, the location is a port not unlike Mejillones or Iquique. The steamship in the harbor is symbolic, for much of Chile's mineral production is exported.

These mid-twentieth-century efforts by Chile to raise the aesthetic appreciation of desert and increase awareness of its debt to mining occurred when tourism was still in its infancy. William Rudolph in *Vanishing Trails*

Figure 5.3. In the early 1960s, mining engineer William Rudolph correctly predicted that tourists would someday visit Portada on the desert coast north of Antofagasta, which features what some locals today call the "white cliffs of Chile." (Photograph by author)

of Atacama mentioned tourism in a few sentences, and he rightly noted that Antofagasta could become a major destination. As he observed, "Antofagasta's temperatures are ideal—never too warm or too cool for comfort." He added that "the large modern hotel, built during the 1950's, should have considerable attraction for tourists from abroad, for in addition to climate and comfortable accommodations the city has strikingly odd and scenic seashore formations at the Portada several miles to the north, as well as excellent deep-sea fishing." With its vertical cliffs and natural arches, that desert coastline is spectacular indeed, earning the sobriquet "the white cliffs of Chile" (Figure 5.3).

Turning his thoughts inland, Rudolph also mused that the interior desert could be a real draw. "What an opportunity there is for these travelers to look behind the curtain of the coastal hills upon the mysteries of the Atacama which lie beyond!" Somewhat wistfully, though, he quickly added, "But the time is not as yet. Fifty or a hundred years of taming have brought the Atacama Desert much closer to the world outside, but neither

Figure 5.4. The Cordillera de la Sal in the vicinity of San Pedro de Atacama features spectacular upturned strata and ample evidence of water's role in shaping the desert landscape. (Photograph by author)

transportation nor hotel accommodations in the interior are yet of the quality which the present-day tourist demands." As a parting shot, he concluded: "And the altitude will always be an impediment."[17]

Be that as it may, an increasing number of tourists during the 1960s began tackling the elevated puna country and even the Andes. For those a bit more timid, the sights were still spectacular. These included the Valley of the Moon and the stunning Cordillera de la Sal. Nestled at the base of the Andes, and looming just to the west of the popular San Pedro de Atacama village, the Cordillera de la Sal is comprised of layer upon layer of reddish sandstones contorted into spectacular synclines and anticlines and represents a quintessential desert landscape (Figure 5.4). Then, too, Rudolph mentioned the work of German geomorphologist Walther Penck (1888–1923) whose relatively obscure 1933 book *Puna de Atacama: Bergfahren und Jagen in der Cordillere von Südamerika* featured numerous photographs of the often-forbidding high country punctuated by the striking peaks of the Andes. It was country that required very careful preparation by hikers given the extreme cold, high winds, and high altitude.

People who experienced these sights began to spread the word. In the 1950s, most savvy travelers purchased a copy of the ATACAMA map sheet of the American Geographical Society's series published at a scale of 1:1,000,000. Dating from the early 1940s and subsequently revised, the color and clarity of these maps greatly demystified the region. So, too, did Chile's development of its infrastructure, including paving of roads. In 1970, geographer Khairul Rasheed anticipated the rise of modern tourism, noting that plans were underway for a new highway that would not only assist the local Indigenous people in getting to and from markets, but "will also bring the archaeological sites, thermal springs, geysers, and other scenic attractions closer to the tourists, and encourage the exploitation of sulfur ore deposits of the region."[18] For its part, Chile created El Servício Nacional de Turismo in 1975, suggesting that the momentum was well underway. By the 1980s, selected tours to desert places such as San Pedro de Atacama were gaining in popularity. Some of this was high country where both the scenery and the thin air could leave one breathless—a concern that Rudolph had mentioned earlier. But the highway had indeed played a role in drawing visitors. On the western slopes of the Andes, the iconic Tatio Geysers became a popular attraction, and the stunning Valley of the Moon grew in fame. As tourists in California's Death Valley could claim that they had experienced punishing heat, the traveler to the Atacama's puna could boast that they conquered two extreme challenges—high altitude and hyperaridity—more or less simultaneously.

In a 2013 study of the Pukará de Quitor, legal scholars Ryan Seelau and Laura Seelau note that "as late as the 1970s, the Atacameño communities in the San Pedro de Atacama commune were very isolated from the rest of Chile and the world." However, "starting in the mid-1980s, the San Pedro River Valley slowly transformed from an isolated group of Atacameño communities into one of Chile's most important tourist destinations." This, as they put it, "brought with it many consequences—some good, some bad," but the overall effect was detrimental. Tourists visiting the Pukará de Quitor began overrunning the site, which resulted in looting, vandalism, and damage. Chilean law prohibited local oversight of this important cultural resource, but Indigenous people pushed for and finally got a concession in the 1990s. With their council formed in 1994, the locals gained control of the site, which the entire community has a stake in keeping intact. This strategy does more than simply ensure tourism's future but also guarantees the survival of a site that has deep spiritual and cultural significance. According to the Seelaus, the lessons learned from this experience confirm that organization is important; strategic planning matters; and that building capacity for visitation step by step is a plan that works.[19]

In the last twenty years, international tourism has taken off, literally. International flights into Santiago by the Chilean airline LAN and others such as American Airlines connect with the numerous in-country flights by LAN. Today, cities in the desert north—including Arica, Iquique, Calama, Antofagasta, Copiapó, Coquimbo, and La Serena—are easily reached from the hub in Santiago. It is, however, difficult to fly between the desert cites, as they all connect to Santiago—meaning that a trip to another city 200 miles (ca. 300 km) away may require an expensive and lengthy 1200-mile (2000-km) trip via Santiago. Public transportation users do much better riding the modern buses, which are convenient and comfortable. However, for those who want to explore the desert, there is no better way than to rent a vehicle. A wide range of them, from subcompacts to durable SUVs, is available. Those big all-wheel-drive pickup trucks, many painted bright red, seem to be able to go anywhere, but I have seen even these mired down to their hubs in mud—a reminder that an all-wheel-drive vehicle with a careless driver can get stuck even deeper than a conventional two-wheel-drive vehicle with a careful driver. Although gasoline prices tend to be rather high in Chile (the country has to import all petroleum products), the network of roads and highways is superb. Even on the Pan-American Highway traffic is usually fairly light. Especially given the fuel efficiency of automobiles in the last decade or so, worries about running out of gas have been reduced despite the fact that it may be 140 miles (200 km) between gas stations. The Atacama is truly one of the great travel experiences; not a bargain, as prices in Chile tend to be about like those in the United States, but memorable indeed for its wide-open landscapes and innumerable points of interest.

By 2015, William Rudolph's prediction for Antofagasta had come true: the city now has about a dozen fine hotels, in part due to the thriving mining business but also tourism. Other cities, particularly those along the coast, are now well known for their resort hotels. Rudolph was also correct that the interior desert region would present a greater challenge for the tourism industry. Ironically, the demand has outstripped the facilities, and it is often very difficult to secure reservations for accommodations in places such as Copiapó and San Pedro de Atacama. One does not travel to the Atacama on the spur of the moment unless willing to camp out or risk finding rooms through last-minute cancellations. Typically, reservations are needed months in advance. Making matters more difficult, some hotels and hostels do not have interactive websites or a website at all. The traveler's best bet may be using a site such as hotels.com or booking.com, which appear to buy up blocks of rooms and provide the reservation services for local hotels and hostels. Despite apprehensions about making reservations months in advance without ever contacting the hotels, I was pleasantly surprised by how effectively it all worked, as the places met or exceeded my expectations. Still, I suspect the difficulty in reserving accommodations in the interior is a drawback inhibiting even greater tourism. In the Atacama, there seems to be a reluctance to believe the premise of "build it and they will come." But ironically, by not building facilities in anticipation of increased demand, the hotel industry may be discouraging tourists who may decide to travel elsewhere after finding nothing available. In this regard, the desert north of Chile is unlike the deserts of the American West, where virtually any town of any consequence has rather good accommodations and very easy access to reserving them.

Back in the early 1960s, William Rudolph noted that agriculture's importance was waning in this region. This was the result of higher wages offered in the mining industry and the declining quantity and quality of water, due to high levels of toxicity from mining wastes in areas such as the Rio Loa valley. Agriculture's downward trend continues, but there are some notable exceptions. For example, although central Chile is known for its fine wines, several wine-growing areas in the southern Atacama Desert are gaining some attention. These include wineries scattered along the striking Elqui and Limari Valleys not far from La Serena and Coquimbo. These vineyards in the Norte Chico produce grapes for syrahs and other red wines, as well as those used to make pisco, a light yellow to amber-colored brandy common to the coast of Peru and Chile, and widely claimed to be Chile's national drink. In Chile's Norte Chico, winemaking has venerable roots and even earned Europeans praise two centuries ago. In his book *Viajes por el interior de la América Meridional 1808–1820*, the French traveler Julian Mellet observed that the area adjacent to Huasco abounded in "wheat, superb vineyards and the fruits of Europe; especially figs of exquisite taste."[20] Interestingly, in most grocery stores in cities such as Antofagasta, Calama, and Copiapó, wines from the desert north are hard to find, while wines of central Chile dominate the shelves. Nevertheless, a leisurely drive through those irrigated desert valleys in search of unique wines would likely reward the oenophile with some delightful surprises.

No mention of the Elqui Valley would be complete without again discussing the writer Gabriela Mistral, whose home is open for visitation and draws ever-increasing numbers of tourists. Mistral's being awarded the Nobel Prize in Literature in 1945 ensured that her rustic home east of La Serena would have a place in a developing travel literature, and all tourist guidebooks now include it among the sites to see in the Norte Chico. Perennially ambivalent about Chile, and writing some of her more poignant material there, Mistral also spent a portion of her mature life in other countries. Although she had left Chile, she continued to be haunted by memories, vicariously returning to her native land repeatedly as she wrote while in exile. Interestingly, Mistral herself became a metaphor used by Chile's political regimes. During the dictatorship of Augusto Pinochet, she was construed by the conservative government as a symbol of stalwart sacrifice to nationhood, her image appearing on the highest denomination of the national currency. However, since 1990, with the rise of more liberal government, she has been embraced as an icon of social conscience and humanitarianism.

As hinted at above, a thriving travel literature has developed to further inform prospective visitors headed for the Atacama. Of the half dozen or so guidebooks about Chile that are currently available, perhaps the most definitive—and certainly the hippest—is Lonely Planet's *Chile & Easter Island*. Although the Lonely Planet guidebook series originated in Britain, and its earliest emphasis was on Asia, its appeal is now worldwide. Like most guidebooks in the series, *Chile & Easter Island* bears no date that is easily discernible, perhaps to avoid obsolescence. Interestingly, *Chile & Easter Island* possesses the pulpy two-color feel that has become the hallmark of Lonely Planet, especially given the tendency for most guides to be printed in sumptuous four color. This is deliberate; the Lonely Planet aims for, and reaches, a specific market with its more sustainable and down-to-earth production (and consumption). In keeping with its emphasis on localism, the guide almost self-consciously avoids mentioning the Atacama Desert by name, opting instead for the terms "Norte Grande" and "Norte Chico," which resonate as quintessentially Chilean, despite the fact that Chileans use the Grande/Chico wording and Desierto de Atacama almost interchangeably—the latter especially for a particular place within the broader desert north (for example, Arica or Iquique).

In the Lonely Planet guide's section on Chile's Norte Grande, readers are warned that the town of San Pedro de Atacama "seems hardly big enough to absorb the hordes of travelers that arrive; it's little more than a handful of picturesque adobe streets clustering around a pretty tree-lined plaza and postcard-perfect church." Tempering this statement, the guidebook then adds, "However, the last decade has seen a proliferation of guesthouses, upscale resorts, restaurants, internet cafes and tour agencies wedging their way into its dusty streets, and turning the town into a kind of highland adobe Disneyland." Actually, San Pedro de Atacama is more the Chilean equivalent of Santa Fe, New Mexico, where architectural design and color, some patently contrived, seek to retain the charm of a colonial Spanish city. As Chris Wilson notes in his book *The Myth of Santa Fe*, the purpose here is to create a regional identity that may seem venerable, but which is actually a function of modernism's search for authenticity.[21]

In Chile, the village of San Pedro (as it is often abbreviated) has similarly opted for urban design that harkens back to a colonial past but is guided by the tenets of modernism, as evident in its unpaved streets, restrained signage, and carefully managed preservation activity. The Lonely Planet guide has a warm spot for Chuquicamata, not so much because it is a gritty mining community, but rather because of its association with martyred folk hero Che Guevara (1928–1967), who found an ample amount of capitalist oppression here to immortalize in what would become *The Motorcycle Diaries* (1995). The 2004 film of the same name entranced audiences with its stunning cinematography and potent social message. Shooting locales included the mountains and pampas of the central Atacama, as well as Chuquicamata. These three sources—Guevara's memoir, the movie, and Lonely Planet—agree that Chuquicamata marked a turning point for Guevara. As Lonely Planet concludes, "In a footnote to this much-analyzed encounter, the 'blonde, efficient and arrogant managers' [*sic*] gruffly told the travelers that Chuquicamata isn't a tourist town.' Well, these days it receives around 40,000 visitors per year."

In describing the Norte Chico, the Lonely Planet guidebook also focuses on the lasting impact of mining. Its characterization of Caldera and Copiapó are noteworthy. Of the latter, it recounts with ambivalence that "powerful men wrangle through the bars and strip clubs of this perfectly pleasant mining town. But like most places on the frontier, there's a certain edgy visceral attraction to it all."[22] As we have seen, Copiapó was much like that in Charles Darwin's day, too. I find it ironic that travelers who claim to be in search of the real character of places often find the real characters there a bit too unsavory for their comfort.

Some travelers to the Atacama may seek mind-bending experiences, as portrayed in the 2013 Chilean film *Crystal Fairy and the Magical Cactus and 2012* directed by Sebastián Silva (b. 1979). Crystal Fairy (played by Gaby Hoffmann) is a young American woman on her way to the desert in search of new-age spiritual growth as she anticipates the impending doom predicted by the Mayan calendar, hence the date 2012 in the title. At a party in Santiago the night before she takes a bus north into the desert, Crystal Fairy has a chance meeting with a young, frenetic, judgmental American named Jamie (Michael Cera) and three relatively laid-back Chilean brothers who are about to embark on their own "psycho-active voyage" in a Chevy Suburban—their goal being to find the San Pedro cactus and use it to get high at a beach on the desert coast. The columnar San Pedro cactus (*Echinopsis pachanoi*) is native to the Peruvian Andes but grown as an ornamental and medicinal plant in home gardens throughout the Atacama. It is also a source of mescaline and hence a potent psychotropic drug. The next day, the four men and Crystal Fairy meet again unexpectedly on the plaza in Copiapó. Spirited away by Jamie and his fellow travelers, she now becomes part of their group, and the five seekers travel northwestward toward the coast on a journey that is part road trip and part internal quest. After arriving at a secluded beach near Pan de Azúcar, Crystal Fairy finds a pile of white seashells and—in a long-practiced Atacama tradition—uses them to spell out her name on a barren slope. In the evening, Jamie begins rendering the cactus into a concoction, which they imbibe the next morning and experience some surprising revelations about each other and themselves.

Central in the film's title and plot, the magic-inducing cactus provides yet another example of *mestizaje* in action. Locals note that the Catholic church initially forbade its use, but to no avail. Cleverly, the popular name San Pedro (Saint Peter) recognizes both indigenous and imported traditions and suggests accommodation; like Saint Peter—and the user experiencing its psychotropic qualities—this plant reaches to heaven while still on earth. Thus, the sanctified ascendancy to heaven in Christianity and its spiritual cosmos-seeking counterpart in Indigenous culture come to mind when this plant is mentioned. Filmed on location in Chile's Norte Chico desert country, *Crystal Fairy* deliberately uses an unsteady handheld camera technique that imparts a home movie quality, as does its somewhat soft focus in various shots. Both techniques suggest informality and intimacy while capturing the kaleidoscopic colors of sketches in Crystal Fairy's diary and the more subtle colors of landforms in the Atacama. Filmmaker Silva is also a playwright and artist, and this shows in his work. *Crystal Fairy* personifies Chile's nascent folkloric and underrated film industry, which also produced the 2016 gem *Neruda*.

Filmmakers have joined the long list of those who popularize places and thus draw tourists, but their medium is especially adept at showcasing things that writers have struggled to describe. One aspect of the Atacama that enchants tourists is color, or rather contrasts in color. These contrasts reflect the difference between the natural landscape—which features a palette that is largely in soft pastels such as buff, ocher, pink, mauve, and the like—and the cultural landscape, which often features bright saturated colors. Some writers have been so impressed by the colors used on buildings that the desert seems colorless in comparison, a technique that goes back more than a century. In 1890, for example, William Howard Russell observed that the colorful town of Pisagua stood out from its pale desert setting. As he put it, "the inhabitants atone for the want of color in their surroundings by painting their houses red, yellow, blue, orange, &c., [and] Pisagua, illuminated by the rays of the sun declining in a bank of orange-tinted clouds, looked quite pretty and *coquet* from the sea."[23]

The use of bright colors for exterior building walls is common in Latin America, but nowhere is it more noticeable than in deserts, which have their own subtle coloring. Still, I am impressed that the colors of some townscapes can mirror the muted tones of the desert. A modern photograph appearing in Donald Binns's book *The Anglo-Chilean Nitrate & Railway Company* (1995) provides a case in point. Taken in the city of Tocopilla, at first glance it impressed me as a sepia print. Because everything in the photo is in a brownish tone—the train, the houses clustered on a hillside behind it, even the rocky mountainside that looms in the distance—I figured the publisher opted for a historic quality by printing an otherwise black and white photo that way. Upon looking more closely, though, I was startled to see one bright turquoise-colored house in the lower right-hand corner and realize that it was a color photo![24]

Tourists with whom I've conversed often admit that there is something pleasing, even natural, about houses that are the same color as the landscape. After all, many places in desert Latin America, North Africa, and especially the Middle East are adobe colored, and that helps create part of the vernacular charm of such places. However, as further noted by architectural historian Chris Wilson in his insightful book *The Myth of Santa Fe*, this emphasis on subtle colors has posed a dilemma in that tourist-oriented city in New Mexico. Although Santa Fe's design standards mandate earth tones ranging from light cocoa to milk chocolate on all buildings, some in the Hispanic population are pushing back by demanding the option to paint their houses in bright colors that they prefer, which tourists find so jarring. The irony here, of course, is that Hispanics founded Santa Fe about four hundred years ago, long before Anglo Americans arrived in the nineteenth century to impose those design standards. Meanwhile, in San Pedro de Atacama, the colonial look emphasizes muted colors, but many towns in the Atacama less concerned about historic theme-ing opt for bright colors that tourists might find garish. In reality, the Indian population revered color, as did some Spaniards. For their part, some critics even in the nineteenth century found the Spanish Crown's tendency for uniformly whitish buildings to be monotonous. The takeaway here is that the color of townscapes in the Atacama region is varied indeed and deserves further study. As a starting point, it is

Figure 5.5. In contrast to the muted colors of the desert landscape, the interior of a gift shop in San Pedro de Atacama is a riot of primary colors. (Photograph by Damien Francaviglia)

worth noting that Indigenous Andean culture especially favors intense, saturated colors, believed to have ceremonial powers, that are often derived from the region's mineral pigments such as green malachite, blue azurite, and red cinnabar.[25]

Upon entering the innumerable gift shops that line San Pedro's normally dusty streets flanked by adobe-colored buildings, tourists are amazed by the riot of colors of merchandise lining shelves and hanging from racks. Given his interest in marketing, my son Damien took a photograph (Figure 5.5) of a shop interior that again reminds one of the many contrasts in this desert—between interior and exterior, objects large and small, antiquity and modernity. Interestingly, from what I could determine, most of the goods in the photo are made in the Andean highlands and marketed in San Pedro—a time-honored tradition that tourists unconsciously mimic. Although the last thing readers might expect to see in this book is Atacama kitsch, I include this photograph as a reminder that earlier cultural traditions are embedded in modernity. And although it is tempting to think of this commerce as exploitation of the locals, I remind readers that trade has been an important part of this village for centuries.

An indication of how serious the local merchants are about their trades was revealed to me when I mentioned to one that I was looking for a turtle amulet for my son-in-law Curt Yehnert, who was recovering from cancer treatment—the turtle being a symbol of longevity in his Native American (Cherokee) tradition. Not having a turtle, the shopkeeper and his wife actually escorted me about a block to the shop of a competitor who had one—a richly colored Andean ceramic flute in the shape of the "tortuga" that I sought. In doing this, of course, they earned a future favor from a competitor. However, there was far more to it. They knew what this amulet meant to me and my family, and they were pleased that they had helped me find it. I came away moved by this experience, appreciating these merchants who are not only glad to discuss the folkloric origins of their own merchandise, but also know the vagaries of trade in their community. San Pedro's merchants impressed me as people who are proud of their region's heritage and delighted to share it with others through a lively commerce. One more thing about the merchandise-filled photo is worth noting. As Damien and I were exploring this desert in modern times, we concluded that we were also tourists ourselves at times, despite any loftier goals.

San Pedro de Atacama is a verdant oasis, but in most guidebooks the Atacama Desert's aridity is showcased, as it has become nearly legendary. As suggested in earlier chapters, a fascination with this aspect of Atacama identity has been evident for nearly three hundred years. It reached near perfection as documented myth a century ago. For example, in Iquique, a busy port city in the notoriously arid northern part of the Atacama Desert, geographer Isaiah Bowman noted in 1924 that "the foreigner who comes out on a three- to five-year contract may stay his time and depart without having known a drop of rain to fall; and he may even assert that it never falls and speak as one who knows because he 'has lived there.'" Bowman, though, cited this as a cautionary tale. Based on a fourteen-year absence of rain in Iquique, the British Consul actually advised a would-be visitor to leave his umbrella at home in England. However, as fate (or Mother Nature) would have it, that visitor experienced a full-blown rainstorm the night that he disembarked from his steamship![26] This particular traveler might have learned a lesson, but scientific observation is no match for the exuberance of mythmaking. About a century later in 2014, the New World Encyclopedia online perpetuated the myth of a totally rainless Atacama Desert, authoritatively claiming: "For as long as people have been recording rainfall, none has ever been measured in this area."²⁷ Although this statement is patently false, it does confirm the tendency to emphasize superlatives and hence distinguish this desert as a place unique in the entire world. The point worth remembering, though, is that it can and does rain in the Atacama Desert, although infrequently enough to cause widespread comment and major problems when buildings are damaged and roadways or bridges washed out. I am tempted to add a few historical quotes here about flooding in the Atacama, but I will instead call on personal experience, as related in the next few pages.

In March of 2015, the weather was on my mind as Damien and I met with Professor José Antonio González Pizarro in Antofagasta, who kindly offered to show us the city firsthand. As a historian and geographer at the Universidad Católica del Norte, Professor González knows the city and its environs intimately. After exchanging greetings, we hopped into his SUV for a tour. He knew every street and alley and we traversed many of them as we drove around the dramatic site. From the rugged seashore, the city marches up into the hills, a series of multicolored cubic buildings that seemed nearly piled atop each other. In earlier conversations, Damien and I speculated that Antofagasta, like Arica and Iquique before it, could be destroyed by a strong earthquake—the kind of event Chileans call a *terremoto* but seismologists rank at about eight to nine plus on the Richter scale. In our minds, this quake would be followed by a tsunami that might rush on shore and sweep away everything lying near the coast. After all, we'd seen that horrific footage of one breaching the sea wall and destroying the city of Miyako, Japan, almost exactly four years earlier.

In response, Professor González noted that such a scenario, though certainly possible, should not draw our attention too far from another possibility—the destruction and suffering that might be caused by a hard rain. When conditions are right in the Atacama, such rains fall and create havoc, sweeping homes and automobiles away as if they were matchsticks. That led us to discuss the prospects for rain in the immediate future, for all of the weather websites we had consulted predicted a nearly perfect storm for us and any others who ventured into the Atacama in the days to come. After a few days of glorious sunshine, we could expect five or six days of rain throughout much of the region! This forecast was a little more than ironic, considering that two weeks earlier the London-based *Telegraph* had posted a travel article online about the Atacama predicting "The forecast is very dry and sunny for the next century."²⁸ Like us, Professor González was aware of the forecasts predicting rain. He solemnly observed that such rains were dangerous indeed, though occurring perhaps only twice or three times a century. When

Figure 5.6. Grave markers in the Caldera municipal cemetery reflect the international heritage of people who worked here during the city's time as a major mining export center. (Photograph by author)

we informed him that we'd be travelling throughout much of the area targeted by the ominous forecasts during that time period, he urged us to be careful, concluding that "las lluvias son el enimigo del desierto" (rainstorms are the enemy of the desert). We vowed to keep that in mind as we ventured farther south, hoping that the forecasts were wrong.

Upon reaching Caldera two days later under clear skies, we were impressed by the town's historic architecture. Many of the buildings associated with the mining boom about a century and a half earlier were still in existence and well preserved, including the original railway buildings dating from the early 1850s. We expected this because we had read the tourist literature, but we were especially impressed by what we discovered quite by accident as we decided to visit a place not mentioned in our guidebooks—the municipal cemetery, which was surrounded by a wall and seemingly off-limits. Finding the entrance, we gingerly opened the ornate wrought-iron gate and made our way inside. As noted in an article I wrote forty-plus years ago entitled "The Cemetery as an Evolving Cultural Landscape," the design and funerary architecture of a community's cemetery frequently reflects the broader design of the community through time.[29]

The Caldera cemetery, which was the first secular cemetery in Chile when inaugurated in September 1876, was no exception; in fact, it is one of the best examples of this phenomenon that one can find. It beautifully

Figure 5.7. In a newer section of the Caldera municipal cemetery, mausoleum-style burials predominate. (Photograph by Damien Francaviglia)

reflects the complex socioeconomic stratification and varied ethnicities typical of mining-related communities. In this cemetery, the older part of which has numerous family plots housed beneath shed-like stylized buildings, it was easy to tell a person's country of origin by the type of structure erected over their family's graves. For example, the graves of a Greek family were in a classically Greek-style structure, a Chinese family in a characteristically Chinese-looking building, etc. These structures sequestering the remains of the wealthy citizens were located toward the main street that the cemetery faced; moreover, the gravestones of individuals from places such as Scotland were similar to those of the native country. As time progressed—as evident in the dates on markers—there seemed to be an increasing democratization of the cemetery, as there had been in the town itself. In our almost nonstop photographing of this fascinating cemetery, I focused on the more ornate historical markers (Figure 5.6), while Damien made his way toward the newer section where vaults were stacked upon one another (Figure 5.7). So engrossed, we barely noticed that the weather had begun to change: the blue sky was beginning to fill with altocumulus clouds, and by the time we had finished exploring the cemetery, cumulus clouds began forming.

That evening, we enjoyed dinner on the patio of our hostel overlooking the port of Caldera, witnessing a gorgeous orange-tinted sunset that highlighted every cloud. After sunset, though, this luminous cloudscape began to transform into a leaden sky filled with far-off lightning. The wind still came off the Pacific, but rather than continuously from the southwest as it had for days, it now seemed to shift about from the west as the air cooled slightly and the humidity increased. After a drink or two we decided to turn in early, ready for new adventures in the vicinity of Copiapó the next day. About midnight, however, we were awakened by not only the sound of thunder drawing nearer every minute, but by the din of dozens of dogs barking and howling, and hundreds of birds singing in the dead of night! The rain soon came in fitful drops that only dampened the cement patio and moistened the white crushed-shell streets. In Texas, I wouldn't have thought much of this thunderstorm, but here it had gotten the complete attention of locals, who went outdoors to experience it and take stock of its impact. We had been lucky to witness a rare phenomenon, rain in the Atacama, and that realization made it difficult to fall back asleep. Instead, as I listened to the almost-musical sound of occasional raindrops on the metal roof, a poem—some might call it a parable—came to me in Spanish, as they sometimes do. I call it "Tormenta" (Rainstorm) and have here translated it into English, reminding readers that poems often sound far more musical in Spanish.

TORMENTA (RAINSTORM)

> An old man in the Atacama told me
> anyone who sleeps through a rainstorm
> in this driest of deserts,
> is either very deaf or already dead
> and in any case quite unlucky.

The next morning, under a cloudy sky, the ground was still damp and the streets wet. We drove to Copiapó by the backroads, tracing some of Charles Darwin's historic trek. Rudolf Philippi was also on our mind, for he had noted how lush the area looked after a good rain. When we reached Copiapó by late morning, locals mentioned that they had received a rain shower overnight. Curious, I asked a middle-aged man how long it had been since he'd experienced rain, and he replied "twenty years ago, when I was

Figure 5.8. A surprise in the heart of the Atacama Desert—a mud- and debris-strewn Pan-Am Highway requires very careful driving. (Photograph by author)

fifteen." Upon leaving Copiapó, I figured we'd head back to Antofagasta by the twisting road that led to the copper mining town of Diego de Almagro, and then rejoin the Pan-Am Highway at Chañaral; Damien, however, thought that this wasn't a good idea, given the weather forecast. Looking at the sky, I agreed, and so we set off directly by way of the Pan-American Highway. This would be faster, and we were sure we'd have a nice lunch in Chañaral. It turned out to be a wise decision. After a delightful and inexpensive lunch of chicken, rice, and salad, we headed north under a menacing sky. Within two hours, as we arrived in the heart of the Atacama Desert, the weather caught us by surprise. Storm clouds raced above the rugged hills, dumping torrents and sending mud and debris onto the Pan-Am Highway in places. Still, we kept driving in a line of slowly moving but brave Chileans, who soldiered ahead on roads veneered with mud and rocky debris (Figure 5.8).

By evening the storms had ended, and we arrived late but safe at a bone-dry, and very fortunate, Antofagasta. Turning on the TV in our hotel room, we learned that a heavy rainstorm had ravaged Copiapó. This rain, combined with runoff from a heavier rain upstream, had sent a wall of mud-laden water

Figure 5.9. An abandoned building in Oficina Anita, a normally bone-dry nitrate town southwest of Calama, is seen here after a thorough soaking by heavy rains. (Photograph by author)

charging down the Copiapó River Valley like an express train. Earlier in the morning, the people we had met in Copiapó were commenting on the rare rain event of the evening before; now their city was in the headlines and their lives in shambles. Moreover, the town of Diego de Almagro was especially hard hit and portions had washed away, along with the lightly traveled road I had thought about taking. Glued to the TV news much of the evening, we were stunned that the little Rio Saladito between Diego de Almagro and Chañaral, along which that road runs—the very road that I thought we should originally take—had turned into a roaring torrent of water, mud, and debris. Moreover, the normally dry Rio Saladito, which enters the coast at Chañaral, had effectively torn that small port city in two. Even the small truck-stop-style restaurant in Chañaral—where Damien and I had enjoyed our lunch earlier that afternoon—had been completely inundated just hours later; the newscasts showed water to the top of its windows. Moreover, just a block or two away from the truck stop, the Pan-American Highway bridge

Figure 5.10. On Highway 25 eastbound about 60 kilometers (40 miles) southwest of Calama, heavy downpours slow travel considerably as squalls move through the area during an "Invierno Boliviano," as some locals call such rain events. (Photograph by author)

over the Rio Saladito that we had crossed just before stopping for lunch was now gone after being swept into the Pacific Ocean!

Early the next morning, Antofagasta was partly cloudy but dry, and the sun even shone from time to time. With a sigh of relief that proved only temporary, we headed northeast bound for San Pedro de Atacama via Calama, a city of about 150,000 people who rarely experience rain. At first, the sky was a mix of sun and clouds, but about halfway to Calama soon yielded to the latter. As soon as we left the Pan-American Highway and headed east on Highway 25, it did not take us long to realize that we were now traversing country that had been walloped the day before. As we explored some abandoned nitrate mining towns, including Oficina Anita (Figure 5.9), they presented an unusual picture indeed. Whereas travelers usually describe them as dusty places, the soil was now saturated by the recent rainfall and several shades of darker gray than normal. Moreover, occasional puddles marked places where yesterday's rain had not yet permeated the ground. Looking skyward, we noted that dark clouds were now threatening more heavy rain.

As we walked around shooting pictures of the abandoned main street and office buildings in this brooding light, we noted that our boots soon accumulated clods of mud that made us, or at least made us feel, a couple of inches taller. Back in the rental car, the floorboards were soon covered with a mess of mud and gravel.

As we headed northeastward, the sky became even darker, and the clouds now seemed much closer to the ground. We were almost imperceptibly gaining altitude as we drove toward Calama, and those clouds soon reached the ground as lightning flashed—two pretty good signs that more heavy rain was on the way (Figure 5.10). It soon became apparent that we were directly in the path of this storm, or rather storm system, which arrived in squalls moving out of the northwest and hitting us broadside, with temperamental winds shifting unpredictably from several directions as they buffeted the car. Here in the heart of the Atacama Desert we again had the two-speed windshield wipers on high and there was virtually no visibility. After about ten minutes, the rain would slack off to a drizzle and the sky would lighten a bit. Then, just when we felt that the worst was over, another squall would arrive in a downpour that slowed the traffic down to about five miles per hour on the highway.

With nature in complete control, the traffic on two-lane Highway 25 obediently meandered from one lane to the other at a snail's pace, yielding to opposing traffic one car at a time as drivers maneuvered around standing water on their side of the road. The highway was nearly, but not quite, impassable—almost inundated, with muddy water now lapping its edges and churning along in the roadside ditches and swales. Finally nearing the quintessential desert city of Calama, we found it as rarely seen; lightning flashed from almost every direction as sheets of rain poured down, and the streets were partially flooded. Moreover, the electricity had been knocked out, and this effectively left every intersection in the city a four-way stop. Remarkably, Chileans appeared to have no idea of yielding to the car on the right and then moving across the intersection. Instead, fearing a collision, they waited for others to go first. It was an interesting sight, and yet another manifestation of the safest and most courteous drivers I've ever seen. Rather than being in urban Latin America, with its nearly legendarily aggressive drivers, it seemed as if everyone in Calama was a student driver accompanied by a very strict instructor.

With no prospect for a hot lunch and no chance of getting additional Chilean currency at an ATM (they were all out of commission), we decided to abandon Calama and push on southeastward to San Pedro de Atacama, thankful for the big stash of power bars and generous supply of bottled water

Figure 5.11. More than a century ago, geographer Isaiah Bowman commented on San Pedro de Atacama's tendency to flood, as substantiated here after the heavy rains of March 2015. (Photograph by author)

we had on hand. Luckily, the sky began to clear about half an hour out of Calama, and, despite some flood damage and high water in places, we got to experience the Andes in one of their many stunning moods. To our surprise, fresh snow had fallen to about 10,000 feet, leaving them crowned in white and fleeced by scudding clouds. As I now comprehended, we had experienced a phenomenon that locals call "Invierno Boliviano" (Bolivian winter), or "Invierno Altiplánico," which as William Rudolph dramatically noted occurs in some years and brings with it "spasms of rain, sleet, hail, and snow during the months of January to March."[30] The use of the term winter for this phenomenon, though a misnomer because it occurs at a time of year that is more properly late summer and early fall, nevertheless conveys something of its contrary nature. Some locals claim that the abundant moisture is derived from the humid Amazon Basin east and somewhat north of the Andes Mountains, but from our limited perspective, these clouds seemed to come from all directions. The higher elevation was clearly a factor as the topography in effect lifted and cooled the air, causing the heavy precipitation. It was March all right, and we were in the throes of this rather rare, and truly spectacular, weather phenomenon.

Recalling Isaiah Bowman's description of San Pedro de Atacama as a flood-prone community, we wondered what we'd find when we arrived. As the sun was now in command, and the worst of the weather had passed by a few hours before, we soon discovered the town had weathered the storms and was drying out—but that the streets were rim deep in viscous, adobe-colored mud (Figure 5.11). The town was open for business, but barely. Wielding long-handled brooms, shopkeepers were pushing a few inches of muddy water out of their stores, urging wayward water across the narrow sidewalks and into the streets. Locating and then settling into our soggy but otherwise comfortable hotel, we met a German hiker who had been unable to get outdoors for a few days and was not in a good mood. Our spirits, though, were high. We'd experienced the worst weather that the Atacama had dished out in about a third of a century and lived to tell about it. Besides, the hotel clerk told us that sunny skies were forecast, and that was exactly what we needed to explore the stunning cactus groves of the Andes the next day, which we did without any problems, though some low spots in the gravel road were just barely passable. On our return trip we noted that the railroad line from Calama to Antofagasta had been completely wrecked in some places, with twisted rails protruding from the mud and several trains stranded, but safe, in more fortunate locations that had escaped devastation.

That stormy period went into the record books as one of the periodic inundations that shatter the myth of perennial aridity. The statistics tell part of the story: three regions (Antofagasta, Atacama, and Coquimbo) had been affected; 26 people were dead, most being in the Atacama Province as we had expected, given the TV news footage and our own harrowing experience; 150 people were reported missing; a total of at least 29,741 people had been directly affected or displaced. For the record, at least by the standards of other regions, the amount of rainfall recorded was underwhelming, only about 24.4 mm (ca. 1 in) was recorded, but that weather station was not in the hills, where more rainfall had evidently fallen. Nevertheless, even a small amount of rain here, say about a quarter to half an inch (ca. one centimeter), can cause havoc if it falls quickly, because virtually all of it runs off rather than soaking into the ground. This may seem surprising because the ground is bone dry, but that, paradoxically, is one of the factors that actually *reduces* its ability to absorb moisture that is delivered rapidly. However, some of this rain did soak in, the catastrophe leading Chileans to rebuild what was damaged and await seeing their desert north in bloom. Consulting the historical literature, I determined that a similar event had happened in 1952 but was apparently forgotten by most people and was never even heard of outside of Chile. However, such events are now flashed around the world via the internet, with one web page entitled "Northern Chile Floods, March 2015—Facts,

Figures, and Photos."[31] But it is the exceptional aridity that rules the imagination here, and rain events seem more easily forgotten than the myth that it never rains.

To the many superlatives about the Atacama, we can also add time, or rather age. As geologist Jonathan D. A. Clarke observes, arid conditions have characterized the Atacama Desert for a surprisingly long time (approximately 200 million years), since the late Triassic and early Jurassic. Clarke bases this finding on a study of the region's geological strata, which consist of minerals such as halite (common salt); varied and unusual salts such as chlorates, iodates, and nitrates; as well as dense beds of gypsum and anhydrite. Clarke attributes this sustained aridity to three factors: in contrast to other arid regions of the world, the Atacama Desert has stayed at a similar latitude, maintained its north–south orientation, and has remained at the western margin of the continental land mass. These led Clarke to conclude, "The Atacama Desert is thus almost certainly the oldest continuously arid region on earth."[32] Some sources claim that the closest competitor in this contest is the Namib Desert of Africa, which, at about sixty million years old, is only about a third as old as the Atacama. Nevertheless, as noted in Chapter One, there were exceptions, including the most recent Ice Ages of the Pleistocene, when conditions even here were wetter and cooler than at present.

These fluctuations aside, the concept of sustained aridity is a factor in the Atacama's renown as a place to collect meteorites. Given the sparseness of vegetation and scarcity of rainfall throughout much of the region, rocks on the surface tend to stay put for a long time. A careful eye is needed to differentiate a glassy meteorite from stones on the surface, but the external form of such chondrites is easy for a scientist or experienced collector to detect. Meanwhile, the metallic meteorites are easy for anyone carrying a metal detector to find. Once in hand, their knobby forms help differentiate them from debris such as scrap metal. The pallasites (meteorites featuring a mix of glass and metal) mentioned earlier in this book are especially prized and fetch high prices. Here, though, caution is advised. Whereas the original weight of the Imilac meteorite was about one ton (2,000 pounds or 1,000 kilos), the brisk market online—even at a few grams per sale—gives one pause about the authenticity of the most coveted of "Atacama" meteorites. I will say more about meteorites in the conclusion, for as touchstones to the heavens they seem to echo the theme of the Atacama as the home of present and future outer space exploration.

Relating the Atacama Desert's mid- to late-twentieth-century history gives me an eerie feeling, for my personal memories converge with the historical record. An informative and highly readable book entitled *The Useless Land* is a similar case in point. Written to commemorate the personal experience of four Cambridge scholars, it contains some disarming observations. Published in 1960 by John Aarons and Claudio Vita-Finzi, this book provides an intimate look behind the scenes of an expedition whose goal was "to see what we could find out about the past peoples and climates of the Atacama." The book was published around the time I started working at Rand McNally and candidly notes the strengths and shortcomings of the maps available at the time. As the authors put it, "The existing maps of North Chile are available on [*sic*] one scale only, 1 in 250,000." Aarons and Vita-Finzi begin the discussion diplomatically enough, noting that "at first sight they appear to have all the features of good maps: adequate but uncluttered detail, helpful but unobtrusive colouring, and good representation of relief and communications." That last item—communications—was especially important to the researchers, but they soon discovered a problem. The maps had actually been compiled from "'trimetrogon' air photos, flown, in fact, by a U.S. mapping service." These aerial photos either provided way too much or far too little relevant information. As they concluded, "Only someone very familiar with [a] region at ground level could really be qualified to interpret the photos, and, considering the unlikelihood of any of Atacama's scanty population ever finding their way into the cartographical offices in Santiago or Washington D.C., these departments have done the best that could be done." Their verdict was delivered with characteristic caustic British wit: "The result is a map which has the bland, well-groomed appearance of a door-to-door salesman, with all his sinister deceptions." They also added an astute note that "the Atacama has its uses as a playground for scientists and explorers, as well as a source of wealth for materialistic miners."[33] That certainly characterizes the Atacama as I first came to know it, by reputation, in the early 1960s.

As they traversed this desert and studied numerous Indian villages over a winter, these Brits came in contact with William Rudolph, who helped them make connections and shared his insights about the people and the landscape. Written a few years later, Rudolph's 1963 book *Vanishing Trails of Atacama* not only mentions and quotes the Cambridge team's book but also reflects upon Rudolph's own long friendship with Isaiah Bowman. Rudolph astutely concluded that "there is little to add to what Isaiah Bowman wrote to bring the historical phase of the Atacama story up to date. Singularly enough, that which can be added belongs to the remote past rather than to more recent years."[34] This was Rudolph's way of noting that modern-day exploration into the desert's antiquity—and antiquities—is revealing startling discoveries. This is reconfirmed by the emphasis that contemporary guide-books place on the region's cultural antiquity. The Atacama is perhaps most famous for the world's oldest and best-preserved mummies at Chinchorro. These are generally claimed to date from about 6,000 to 2,000 BCE, though some may date from 7,000 BCE. These mummies from southern Peru and northern Chile are not only old but intriguing, as so many forms of mummification were practiced. These include blackened, bandaged, red colored, and mud covered, as well as mummies whose skin had been removed and earth used to replace it. An entire family was found mummified, as was a fetus.

Although they are pre-Columbian in age, these mummies are included in this chapter because they have generated so much recent attention—despite their antiquity and the fact that serious study began in 1917, when the German archaeologist Max Uhle published the first articles about them. In his introduction to Bernardo T. Arriaza's book *Beyond Death: The Chinchorro Mummies of Ancient Chile* (1995), archaeologist John Verano observes that "the last fifteen years…have witnessed an impressive array of multidisciplinary studies that focus on the physical remains of the Chinchorro people themselves." Throughout his book, Arriaza shows how these mummies are forcing anthropologists to reevaluate their assumptions about the social significance of mummification.

According to Arriaza, most studies conducted in other parts of the world have led anthropologists to conclude that competition and social needs, rather than spiritual beliefs, lie at the root of mummification. Mummification thus seemed to characterize wealthy and power-conscious peoples. However, in the Atacama, a different picture emerges, for "the Chinchorros were a politically simple fishing society with extraordinarily complex mummification practices that could not be based on economic gain." Instead, Arriaza speculates that the mummies were viewed as "'living' entities that used the same space and resources as did the living in a similar manner as the Laymi people of Bolivia, where the dead harvest chile peppers from the same fields." Arriaza argues that Chinchorro mummification practices enabled the dead to be "an extension of the living" rather than a "regeneration of life" as may be the case elsewhere. This places mummification in the realm of spirituality rather than politics. It also hints at a continuity that is remarkable. As Arriaza describes it, the love bond between individuals was so strong that death did not separate the living from the dead. Ultimately, Arriaza notes, "artificial mummification can also be interpreted as an adaptive ideological strategy for group survival to the arid Atacama Desert."[35] Interestingly, the word Chinchorro has political connotations; it is also locally used for an area adjacent to the border of Chile and Peru. In January 2016, proposed Chilean urban expansion north of Arica into the Chinchorro was interpreted as threatening the sovereignty of Peru. As noted in the previous chapter, the border between Chile and Peru, and for that matter Chile and Bolivia, is still a sore point. In the Atacama, the past never remains buried for long, and certainly not forever.

Recently, these Atacama Desert mummies made headlines for another reason. In an article entitled "Atacama's Decaying Mummies," *Archaeology* magazine reported that "since the moment they were excavated, they've been undergoing changes that even millennia in the ground didn't bring about." Despite being kept in climate-controlled conditions at Arica's Archaeological and Anthropological Museum of San Miguel de Azapa, the mummies' skin is increasingly blackening and turning gelatinous, as recorded by archaeologist Marcela Sepulveda. DNA and other testing by microbiologist Ralph Mitchell of Harvard University confirmed that this process of decay, which has accelerated, may be due to a common culprit: climate change. According to Mitchell, "The climate in this region has changed from cool and dry to warm and damp, and the increased humidity has caused microbes that are common to all of our skin, but usually get washed off, to grow and damage the mummies' skin." For her part, Sepulveda also theorizes that "increased agriculture in the region might be affecting the humidity around the museum."[36] Interestingly, some sources suggest that global climate change is having that very effect on the climate of the Atacama Desert, but on the other hand more study appears to be needed as to why a climate-controlled system cannot accommodate changes in exterior humidity which may also occur naturally during varied weather cycles in this region. I will say more about climate change in the concluding chapter, as it could be a major factor in affecting perceptions of what is widely regarded as the world's driest desert.

In addition to pre-Columbian antiquities, the ruins of the more recent mining-related past also have their allure. In this sense, the appeal of the Atacama Desert is both prehistoric—containing some of the New World's most significant ancient archaeological features—and historic, as evident in its portrayal and marketing as a graveyard of industrial artifacts. The nitrate officinas are perfect examples of preservation here during what geographer Homer Aschmann glibly characterized as "a long quiet period of slow decay and rather poor future prospects." Aschmann was concerned with the vitality of mining itself, but he noted that relics long outlive it in arid areas such as the north of Chile. As he put it, "such ghost towns are singularly characteristic of desert areas. Elsewhere, flooding and the more rapid decay of

Figure 5.12. La Copiapó, the oldest extant North American–type (4-4-0) locomotive, reposes in its namesake city at University of Atacama's School of Mines. (Photograph by author)

fixed investments under humid conditions soon erase them from the landscape."[37] The nitrate oficinas stand forlorn, but during the last sixty or more years since their abandonment have gained strong appeal precisely because of their status as ghosts. This, as I noted in a book entitled *Hard Places*, is part of a complex process I call "technostalgia."[38] In addition to ghost towns, the ghosts of vanishing technology, such as steam locomotives, are venerated. In the Atacama Desert, rows of defunct steam locomotives can be found in Chile and Bolivia. Originally too difficult to dispose of when they faltered or simply replaced by diesels, they now develop a burnished patina in the desert climate and become objects of interest to tourists. In Caldera, the old rails of Chile's first railway were turned into a compelling sculpture in front of the still-extant railway station. In the Atacama, objects from times long past often remain visible for the ages and figure in the storytelling of subsequent generations.

I confess to being intrigued by such outdoor preservation, whether intentional or not. As a person with a lifelong love of railroads, the older the better, a visit to see one of the oldest American-built steam locomotives in existence was high on my list. The locomotive in question was built by Norris Brothers of Philadelphia in 1850 and is an international treasure, though little known in the United States. This locomotive helped open the original

line between Caldera and Copiapó and now reposes in a courtyard at the
University of Atacama, surrounded by the buildings of the School of Mines
(Figure 5.12). As seen in this photograph, the stark mountains rise behind the
university, their slopes partially covered in sand dunes—a vivid reminder of
the region's aridity. Closer inspection of the locomotive reveals its original
boiler front featuring the name and location of its manufacturer in raised
letters. As with many early locomotives, this one has a name rather than a
number: Copiapó. In his book *American Locomotives: An Engineering History,
1830–1880*, Smithsonian curator John H. White, Jr. explains that the Copiapó
"is the only surviving 4-4-0 of that period; it is the standard product of a
major builder and, unlike most locomotives preserved in the United States,
it has undergone little reconstruction."[39] The climate has played a role in the
Copiapó's preservation, for an iron locomotive sitting outside for a century
and a half in a wetter location would have badly rusted. The seemingly miraculous preservation of "La Copiapó," as Chileans still proudly call it, helps add
to the impression that time stands still in the Atacama.

The same is often said of the desert landscape itself, which may seem
timeless to some but frequently reveals the passage of time on its wind-blown
and water-scarred surface. Tourists are drawn to these stark landscapes that
appear to be in a state of suspended animation, frozen in stone, as it were.
To find such evocative desert places and sights/sites, the modern-day traveler consults detailed road maps, four of which I will discuss in the next few
pages. I will also illuminate sections of these maps that depict the Atacama
Desert. Generally, these maps depict topography, transportation lines and
roadways, and cities, but they do so using varied techniques. I should also
comment that most modern road maps are in part based on data gleaned
from the impressive set of official topographical maps produced between 1971
and 1975 by the Republic of Chile at a scale of 1:500,000.[40] Naturally, they
are updated to include road construction and new recreational sites such as
national parks.

The first road map I will discuss is the Nelles *Map of Chile [and] Patagonia*,
which is printed at a scale of 1:2,500,000 (Figure 5.13). On it, places of interest are printed in red and highlighted by a red star. These include locales
associated with the prehistoric past—for example, the "Dinosaur Tracks"
near Chacarillas and numerous *geoglifos* (geoglyphs) such as the "Geoglifos
de Tiliviche." As noted in Chapter Two, the rubric geoglyph includes thousands of archaeological sites that consist of petroglyphs as well as freestanding designs made from rocks, or combinations thereof. On the Nelles
map, locales associated with more recent mining that are now abandoned,

such as "Humberstone Saltpeter Works" and "Sta. [Santa] Laura Saltpeter Works," are named and located, while others are simply described as nameless "Abandoned Nitrate Town[s]." Placing all of these features in context, the Nelles map uses the bold name "Desierto de Atacama" *twice*, as that region runs for such a great distance latitudinally. Moreover, because the name of the desert is placed north–south, that is, perpendicular to most of the other place names on the map, it has greater visual impact, making the text and the desert it describes all the more visible. By size and placement of its name, then, the Atacama Desert is emphasized as one of Chile's great natural features along with the adjacent Andes Mountains.

This is understandable, as tourists now visit the Atacama Desert in increasing numbers to experience sites of historic interest and scenic beauty. Many are bound for the spectacular sights of the Altiplano in adjacent Bolivia, but the Atacama is a must-see on the way. Some organized tours focus solely on the Atacama, and a combination of stark desert scenery and historical sites represent the major draw. About one million people live in the Atacama Desert today, mostly concentrated in cities, but one would not know that from the tourist literature. It is solitude, rather than company, that many of these tourists crave. And yet, people being people, even these travelers feel the need to be connected to both place and fellow humans. The communication technique called geocaching is a case in point. Geocaching involves finding a place designated by GPS, the place being a cache of objects hidden in a container at selected coordinates. In this type of pursuit, a traditional map is virtually useless, especially in helping one find the container at the end of a search. By the year 2000, geocaching began to be popular in Chile, no doubt brought there by Americans and Europeans.

A prime example of a geocache in the Atacama is one established, or rather hidden, by "acargill" on September 26, 2004. As acargill explains: "The cache is not very far from San Pedro [de Atacama], it's in a hill where you can see the town, a valley and the archaeological ruins of an ancient place." The creator of the cache continues, "it's a good natural viewpoint, so take your camera!" The reference to archeology is noteworthy, because a cache is likewise in a location where artifacts can be found. The cache itself is a prize of sorts, and acargill appears to have a deep interest in this desert. "Inside the metallic container, you will find some seeds (chañar), a piece of an aromatic plant (Rica Rica), a local map, some coins and a CD with 10 digital photos I took of some beautiful places." The creator reminds fellow geocachers about etiquette, adding, "If you take the CD, go ahead, but please replace it with your own photos and log yourself in the book." Online, one can trace

Figure 5.13. On the Nelles map, the name "Desierto de Atacama" is positioned at two inland locales—one between Copiapó and Antofagasta, and the other between Antofagasta and Iquique. (Author's digital collection)

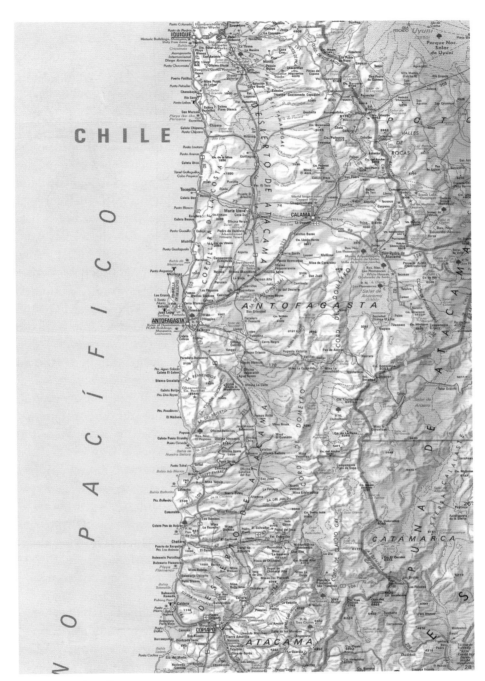

Figure 5.13 detail.

the history of this particular cache, which was found 338 times between its establishment in 2004 and the website's posting in July 2015. After confirming "Found It," most of the responses feature statements of appreciation, such as "Quickly found by the kids as we went for a long walk around the top of the valley above San Pedro," or "Our first cache in Chile! Great View!"[41] Like those mysterious *signos del camino* described in Chapter Two, it is apparent that markers are still left by people traversing this desert, if not in stone, then hidden in containers that can only be found using technology.

Although people are increasingly relying on GPS for travel, many still use commercial printed travel maps. Continuing my discussion of such maps, I should note that most are in synch with broader travel literature, which portrays the Atacama as a place of antiquity and mystery. On the Nelles map, for example, one of the region's oldest communities, San Pedro de Atacama, is touted for its archaeological museum, which was established by Father Gustavo Le Paige in 1963. Similarly, in Copiapó, the words "Churches, Museums, Mining Centre, Vineyards" promote the community where the previously discussed locomotive La Copiapó reposes on the campus of the Universidad de Atacama.[42] Also in harmony with the narratives in travel guides, the Nelles map makes note of a "Picturesque Village" near the Parque Nacional Volcán Isluga, and the *DK Eyewitness Travel Guide: Chile & Easter Island* refers to Isluga as "an Aymara [Indian] village with a beautiful 17th century church as well as the Pukará de Isluga, a ruined fortress dating from pre-Columbian times."[43] In a nod to modernity, north of the city of Calama is shown as the location of the "World's largest open [pit] Copper Mine" at Chuquicamata, a spectacular human-made topographic feature that is invariably also mentioned in travel guides.[44] However, the Lonely Planet guidebook may be most current in reference to this site, as it notes that the world's largest mine in terms of actual production is now the Mina Escondida, which is located about one hundred miles (170 km) southeast of Antofagasta. Mina Escondida (hidden mine) is an apt name, as the ore was hidden under overburden and only recently discovered (ca. 1981). It is operated as two properties by the parent company, BHP Billiton, an Anglo-Australian mining enterprise.

Modern road maps and guides can barely keep up with the rapidly changing mining industry. Chile's dependency on this industry, in particular copper, is inseparable from its recent political history. Although the copper industry was nationalized by the Allende government in 1971, the political coup in 1973 returned portions of it to private ownership. Still, the reality is that copper is such a cash cow to Chile that even the conservative Pinochet government kept it in the grip of public control. Anaconda properties continue to operate today under the banner of the national copper company CODELCO, and hence nominally not-for-profit. Anaconda's other properties worldwide are now part of BP, which is no stranger to the controversies generated by mineral extraction. Despite copper price fluctuations and massive damage inflicted by the March 2015 rainstorms, Chile remains the world's largest copper producer. Virtually all of Chile's copper is mined in the Atacama Desert's porphyry copper belt, which stretches almost due north–south from well north of Chuquicamata (at Rosario and Quebrada Blanca), through the belt buckle at Mina Escondida, and then southward to El Salvador, the latter at Diego de Almagro in Chañaral Province. As in all porphyry ore bodies, the mining operations are open pit and far-flung, spreading their waste rock and tailings out across the desert in topographic features that will be visible for thousands of years—provided they are not re-mined for their metals content at a future date. Their sulfide ores such as chalcopyrite, bornite, and covellite are crushed and concentrated, then smelted, to meet the ever-growing quantities of copper that the electrical age demands. These spectacular open pit mines, huge earth-moving equipment, and company town-type settlements are reminiscent of copper mining in the western United States; the similarities are more than coincidental, as many of the original companies such as Anaconda had mines in Chile and other countries, including the United States.

That mining is a fact of life here is apparent from even modern road maps. On most, but not all, tourism maps of the Atacama Desert, the mine at Chuquicamata is shown. So, too, is the name "Atacama Desert." However, at least one popular modern travel map—*Chile* by International Travel Maps, or ITM—restricts the name "Desierto de Atacama" to a relatively small area about a hundred kilometers (ca. 60 miles) south of Antofagasta (Figure 5.14). This conservative positioning, I should note, is different from that used by Isaiah Bowman in *Desert Trails of Atacama* (see again Figure 1.7). Judging by the almost completely barren landscape in that part of the Atacama, ITM's placement may reflect the mapmaker's belief that this part of the Atacama is the quintessential desert; it is, after all, along the stretch of the Pan-American Highway (and the highway running parallel to it west of the Sierra Vicuña Mackenna), where one may drive for a full hour at the posted speed limit and see absolutely no vegetation for sixty or more miles (ca. 100 km). Prepared at a scale of 1:1,750,000 (nearly twice the scale of the Nelles map), the ITM map does indicate "Region III Atacama" as an administrative subdivision lying northeast of Copiapó. Then, too, this compression or reduction in size of the Atacama Desert may be the result of this map's different emphasis.

In fact, the ITM map is one of the most detailed travel maps available showing topography; the various cordilleras (mountain ranges) of the

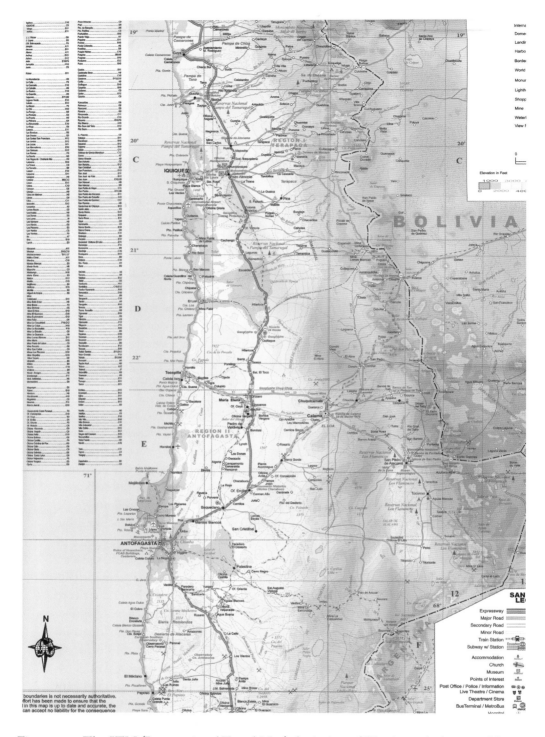

Figure 5.14. The ITM (International Travel Map) depiction of "Desierto de Atacama" by name is conservative, limiting it to a relatively small area about 100 kilometers (ca. 60 miles) south of Antofagasta near Cerro Paranal. (Author's digital collection)

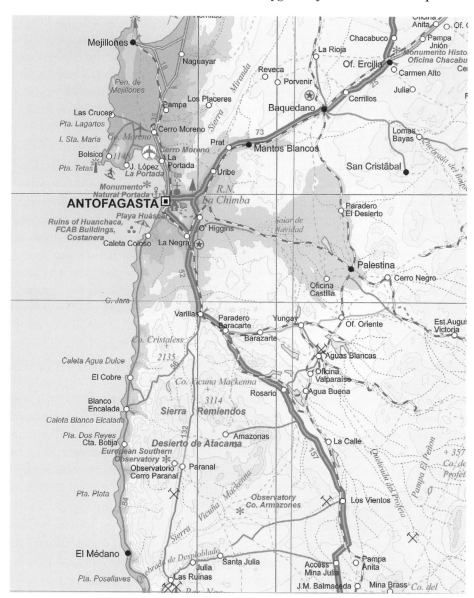

Figure 5.14 detail.

Andes—for example, Cordillera de los Andes and Cordillera de Darwin—are shown. So, too, are various features named cerro (hill) such as Co. Parada and nearby Co. de la Posada, the peaks of which are both shown as 1922 [meters], about 5,000 feet. Clearly, then, the ITM map of Chile is more concerned with physiography than climate. Interestingly, because it uses a standard

color scheme based on elevation (i.e., it is a hypsometric map), the lowest
areas—regardless of climate—are shown in green while the higher elevations
grade upward from buffs and browns to purples and grays at the highest elevation. Although such bright green coloring may subliminally downplay the
Atacama Desert as bone dry, the desert's name printed across portions of
maps uphold the tradition that this is hostile country deserving considerable respect.

Regarding how well maps depict topography, I have long suspected that
some older styles of rendering landforms that now seem dated, such as those
on San Román's 1892 map, might actually do a better job in certain situa-
tions. With this in mind, I had my original copy of the San Román map digi-
tized and printed at the same size as the original map. I then took that copy
to the Atacama to see for myself which maps—the modern road maps or
San Román's—better depicted the topography. Out in the wide-open spaces
west of the Sierra Domeyko, I was surprised that San Román's map was actually more effective than the latest road maps in depicting some topographic
features; in particular, a landscape that in reality was punctuated by several
distinctive mountains might be melded into one shaded blob on the modern maps. San Román's map not only showed but individually labeled Cerro
Blanco, Cerro Pan de Azúcar, and Cerro del Arbol. Still, this should not imply
that San Román's map is perfect, for as Professor José Antonio González
Pizarro noted, San Román's depiction of the topography farther south was
not particularly accurate. Still, the manner in which San Román had depicted
the topography in the area of my comparison was better than the Nelles map
and at least as effective as the more detailed ITM map. In retrospect, the San
Román map was the best in at least one regard: for the person in the field, it
seemed better designed to reveal the salient features of the topography than
even the modern topographic detail on the ITM map. I found myself reaching an almost implausible conclusion: I felt confident enough to use the San
Román map, despite its age at 120 years, because it showed the landforms so
graphically, as veritable landmarks, rather than with contour lines, which the
eye cannot discern in the landscape as readily.

Lest readers think that the name Atacama Desert is too downplayed on
the ITM map, it should be noted that many scientists, particularly those
concerned with habitats, also limit this desert geographically. Consider,
for example, the definitive book *Ecosystems of the World: Hot Deserts and Arid
Shrublands*, in which W. Rauh describes "The Peruvian-Chilean Deserts" in
considerable detail. Only toward the end of his chapter on this long coastal
desert region does Rauh introduce the Atacama Desert by name. "Driving

from Tocopilla through Chuquicamata to Calama and further [*sic*] inland to San Pedro de Atacama one enters the most desolate waterless desert of the world, the Atacama." In this portion of Chile, Rauh notes, "very extended tracts are completely devoid of any vegetation." Even as an ecologist familiar with aridity, Rauh was clearly in awe, observing, "Altogether the Atacama is not only one of the driest but also one of the most grandiose deserts of the world." Rauh also confirmed that the Atacama Desert is among the least known. "In contrast to other deserts of the world there are no investigations concerning the ecosystems of the Atacama." Rauh's concluding sentence that "the ecologists have here a wide-open field for their research" confirms that the Atacama Desert can still stimulate the scientific as well as popular imagination.[45] In recent years, research has indeed increased in parts of arid Chile, as evidenced by the stunningly illustrated and informative 2009 book by Chilean ecologists Raquel Pinto and Arturo Kirberg entitled *Cactus del Extremo Norte de Chile*, and Fred Kattermann's studies of *Eriosyce* cacti.[46]

Cultural features are also important aspects of the Atacama Desert on modern maps. The increased detail on the International Travel Map of Chile extends to mines. In the vicinity of Calama, like other travel maps it shows the Chuquicamata copper mine. However, the ITM map omits reference to this mine's largest-in-the-world status. The ITM map also shows many other mines that are not indicated on the Nelles map. This further suggests that the intended user of the ITM map may be more experienced, and thus already knows this information. Overall, then, the ITM map appears to be tailored to the more seasoned tourist—likely the business traveler with economic interests (especially mining) in the area. However, the map also indicates a number of historical and natural sites. The thought of having this ITM map in hand near a normally bone-dry place such as Calama may seem somewhat ironic to many travelers, for it is made of a durable plasticized material and on both sides features a logo of raindrops and the bold lettering "WATERPROOF."[47]

The next road map I would like to discuss is the Borch map of Chile (scale 1:2,000,000), which was prepared and printed in Germany in 2011 (Figure 5.15). As part of the venerable tradition of Germans mapping this part of the world, it lists seven separate cartographers, including its namesake, K. Borch. The map illustrated here is the fifth edition and is dated 2011. Although the Borch map includes the disclaimer "We accept no liability for information or accuracy of content appearing on this product," it also adds, "If you notice a discrepancy then please let us know, your comments are highly appreciated." This map is "water repellent," and it depicts

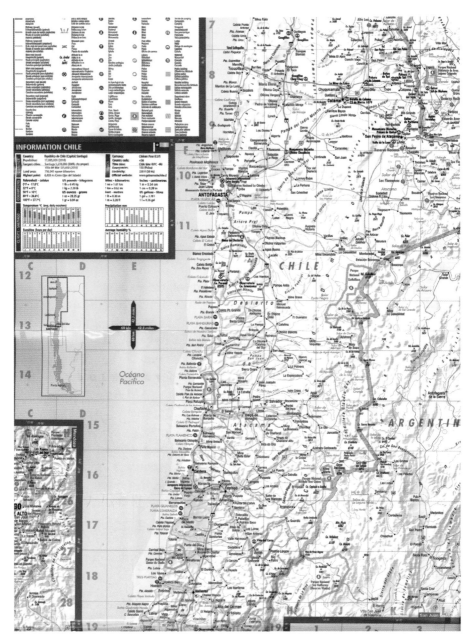

Figure 5.15. The German-made Borch map of Chile applies the name "Desierto de Atacama" to the area between Antofagasta and Copiapó. (Author's digital collection)

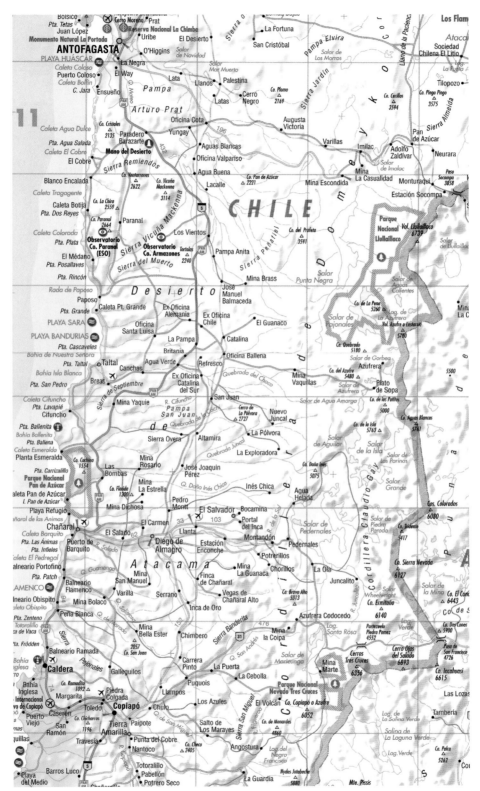

Figure 5.15 detail.

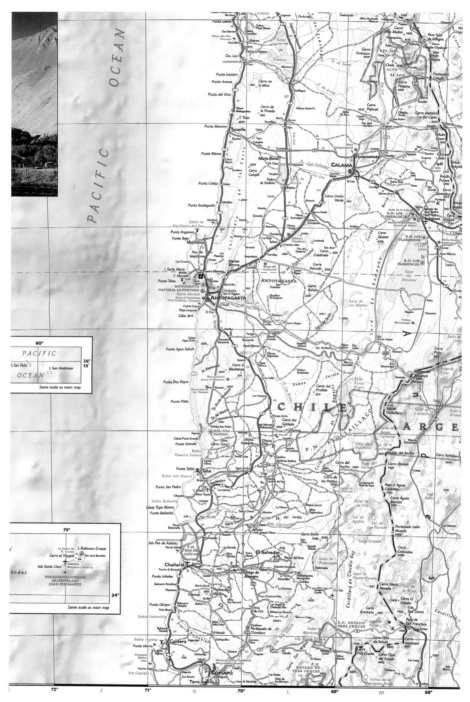

Figure 5.16. The National Geographic map of Chile (2010) features the name "Desierto de Atacama" twice—in large letters stretching north of Copiapó, and in much smaller letters southeast of Iquique. (Author's digital collection)

Figure 5.16 detail.

the desert portion in a separate section. The Borch map features a soft shading to emphasize topographic relief. Of special interest is its coloring: compared to the green color used in the wetter part of the country, the desert north grades off into a very pale, almost cream color. However, the name "Desierto de Atacama" appears in only one part of the arid north—the large area between Copiapó and Antofagasta. Yet it does a fine job of identifying the desert-like pampa regions north of Antofagasta, including the large Pampa del Tamarugal, which is named twice.

The last road map that I will highlight is the National Geographic map of Chile, which bills itself as an "Adventure Travel Map" and claims to be "regularly updated" and "waterproof" (Figure 5.16). My copy illustrated here is dated 2010, and its scale is 1:1,750,000. One expects high-quality graphics and accurate information from National Geographic, and this map does not disappoint. The National Geographic map is as detailed as the others in many regards, and it shows the topography using contour lines. It features the name Desierto de Atacama in two places: in large letters stretching north of Copiapó to near Antofagasta, and in much smaller letters southeast of Iquique adjacent to the Salar de Llamara, which lies to the northwest of Calama. Readers may be surprised that I do not favor any one of the four road maps regarding how the Atacama is named and positioned; after all, each reflects the preferences of the mapmaker, and besides, this desert, like all others, is impossible to depict with precision, as periodic variations in rainfall patterns change things on the ground in remarkable ways. Moreover, aesthetically speaking, these road maps may appear "busy" or cluttered compared to those from two or more centuries ago, but it should be recalled that these recent maps are called upon to show an increasing amount of information travelers may need to make critical decisions regarding routes to take and sights to see. I will say, though, that one aspect of the four maps is worth contemplating before selecting one over the others. As noted, the Atacama Desert can throw some weather-related surprises at the traveler, and I found myself glad to have those weatherproof maps in hand on the way to Calama as the skies opened up in March 2015.

The name Calama brings to mind another, much darker, side of this desert's past. There is a growing awareness that the Atacama played a sinister role in Chile's infamous political coup of September 1973, wherein the socialist government of Salvador Allende was overthrown in a military coup by dictator Augusto Pinochet, who ruthlessly eradicated his regime's opposition. The most poignant references to the Atacama Desert in connection with political violence are found in books and websites, and are not [yet] on maps and tourist guides. For example, in her novel *Bone and Dream: Into the World's Driest Desert*, Canadian-Chilean writer Lake Sagaris recounts two imaginative journeys in the Atacama Desert, the first as a modern-day writer haunted by events of 1973, and the other as a character from the distant past, a high-spirited Indian woman who fell victim to the predations of Diego de Almagro in the 1500s. Sagaris loosely based this character on *La Tirana* (literally, the female tyrant), also known as Ñusta *Huillac*, the warrior princess who is the focus of Chile's largest religious festival every July fifteenth.

Maps as well as narratives play an important part of this story; however, these maps are not the kind printed on paper, but are rather mental or cognitive in nature. For example, Sagaris begins her modern literary journey with a cartographic metaphor: "My book is my map of memory, but also a search for those who left the desert scarred with their trails, and roads seared by their voices." In this journalistically styled novel, Sagaris seamlessly links the ancient Indians brutalized by Almagro (and by extension, Spain) almost five centuries earlier with the much more recent victims of Pinochet's U.S.-supported takeover. Several types of perspectives—Indigenous and European, modern-day and past, and map and place—merge into one story

Figure 5.17. The eerie *Hand of the Desert* (*Mano del Desierto*) by Chilean sculptor Mario Irarrázabal commemorates those who paid the ultimate price for dissenting against the Pinochet regime and were buried in the Atacama. (Photograph by author)

bounded by this desert place. Of her airliner's descent into Antofagasta, Sagaris notes, "As we swerved and dived toward the earth, the whole desert unfolded below us for a moment, a crumpled map spread out on a table." Just as her plane was about to touch down, Sagaris writes that "it sank into the heart of the map, which rose up around us and disappeared." Disappearance is the operative word here. In a manner of speaking, the map swallowed everything, and her drama began to be enacted on, and in, both the map and the desert landscape. Throughout *Bone and Dream*, the past and the present, and the map and the place, become one. In a later passage, Sagaris pens, "I've become obsessed with maps, and now I travel across a landscape as flat and apparently simple as a map, criss-crossed with old roads and ancient borders, seeing from so close and passing so quickly that I can make out only the shadows of the lost territories they once portrayed."[48] The Atacama Desert as mental map and actual place brings Sagaris face-to-face with the oppression that characterized both 1530s and 1970s Chile. Few books have been more explicit about the power of cartography in personal discovery and collective memory. Sagaris shows how maps can work metaphorically to blur the distinction between explorer and chronicler, writer and reader, real and imaginary, and past and present.

Using this literary (and also historical and geographical) technique, the Atacama Desert—or Desierto de Atacama, as it is called on modern maps of Chile—became the graveyard for many of "the Disappeared," shorthand for those people who were murdered, and their bodies disposed of, in the vast desert waste. Their loved ones recorded and commemorated this loss in many ways, one of the more moving being *arpilleras* or testimonial tapestries by mostly working-class women. In discussing these tapestries, Ericka Kim Verba notes that *arpilleras*, while not technically maps in the traditional sense, are nonetheless "'graphic texts...that can be analyzed and interpreted to reveal something about the spaces and times they portray.'" Many of these *arpilleras* (literally, burlaps sacks) are pictorial representations of the settings in which the Disappeared lived, including recognizable features such as buildings, streets, and the backdrop of the Andes Mountains. Verba reveals that "other arpilleras are best read as 'maps of the mind,' where buildings, objects, and figures are arranged in abstract space, devoid of scale or topography, in order to convey a specific concept or emotion." Although many of these *arpilleras* served a therapeutic purpose, they were also evidentiary, being smuggled out of Chile to human rights groups.[49]

Such monumental loss calls for monumental commemoration. Situated within sight of the Pan-American Highway in a desolate, sweeping valley southeast of Antofagasta, a cluster of upright forms catches the attention of travelers. At first sight, it might be mistaken for an outcrop of vertical rock spires, but its smooth contours suggest something more vulnerable. As it comes into closer view, this outcrop reveals itself to be a partially buried giant hand reaching skyward, as if trying to simultaneously draw the travelers' attention and escape entombment under the sandy surface (Figure 5.17). Called *Mano del Desierto* (*Hand of the Desert*), this eleven-meter-tall (ca. 36 ft) surreal sculpture was designed by Chilean sculptor Mario Irarrázabal and erected with the support of the civic group Corporación por Antofagasta. Irarrázabal selected masonry to give form to his vision; constructed of iron-reinforced concrete, the sculpture calls attention to the injustices of the Pinochet regime. The message here is one of hope that the Disappeared will not be forgotten.

Since its dedication in 1992, *Mano del Desierto* has been seen by hundreds of thousands of passersby. A website calls it "even more alien than the desert itself," but it is meant to evoke a common humanity that rises from oppression despite the cruelties inflicted upon it. Although situated about 350 meters (1000 ft) off the highway, it sparks curiosity. Every day, a number of motorists pull off the highway and onto the heavily rutted road that leads to its base. Most, no doubt, are deeply moved, but inevitably this monument also invites personalized commemoration. It is often tagged with colorful painted graffiti, though periodically cleaned by volunteers from the Corporación por Antofagasta. Mano del Desierto is impressive in any light, but when Damien and I visited it at twilight, the immensity of the desert and the immensity of its message seemed especially poignant with evening coming on.

I had read a lot about how Pinochet's regime silenced his opposition, and was familiar with the film *Missing,* starring Jack Lemmon, but it became far more real to me when related by Professor José Antonio González Pizarro, who vividly described Pinochet's tactics. Recalling this conversation, a poem came to me on the plane back to the United States.

THE DISAPPEARED

What was Pinochet thinking
when he turned the Atacama Desert
into a graveyard for dissenters?

Even in those dark days
every Chilean school kid knew
archaeologists were discovering
the world's oldest mummies here.

Today people sifting through
ghostly white nitrate beds
find fragments of shattered bone
not knowing whether they've discovered
a historic site or a crime scene.

In this thought-provoking desert
more than anywhere else on earth
those who die by whatever means
refuse to surrender to the inevitable
without leaving a trace.

Ultimately, everyone knows

that amnesty can protect perpetrators
but the limitless pages of time
will condemn them forever.

Upon returning home to Oregon after that trip, I happened to run into Professor Ken Nolley of the Film Studies department at Willamette University, from which Ken was just retiring. Comparing notes, Ken told me that he was planning to travel extensively, and I casually mentioned that I had just returned from a trip into the Atacama Desert, much moved by the experience. What transpired next was a perfect example of serendipity and the role that it plays in research. Ken, who has a long history of social activism, mentioned that he had yet to visit Chile but that it was high on his list. Remarking that "the Disappeared" was a subject that continued to haunt him, Ken asked me if I had seen the 2010 documentary film by Patricio Guzmán entitled *Nostalgia de la Luz* (Nostalgia for the Light). I had not heard about it and was delighted to find a copy in our university library. At home, after putting it into the DVD player, I was immediately transported in both time and place. *Nostalgia for the Light* begins with close-ups of a beautiful nineteenth-century German telescope, the kind that has long captivated Chileans. Guzmán uses the telescope to reveal his childhood love of astronomy, which he never outgrew. Of an important period in the desert's past, Guzman states that "science fell in love with the night sky," adding that astronomers built observatories in the north because "in the Atacama, one can touch the stars." At about the same time though, he showed how despite Chile's relative isolation and slow pace, it became swept up in the affairs of the world. This is a reference to the ill-fated 1971 socialist revolution which he proudly joined in hopes of moving Chile toward what promised to be social and economic equality.

Of the Atacama Desert, Guzmán has much to say. He has a special love for it, even though he claims it is barren and devoid of plant and animal life—"a resemblance to a faraway world." An astronomer interviewed in the film agrees that the Atacama is like outer space. He also explains how science and religion have parted company in the modern world—a situation he regards as a shame because *both* ultimately ask the same penetrating questions about origins. For his part, Guzmán believes that "a man may soon walk on Mars." The footage of the desert in this film is stunning—beautiful, serene, and impassive—but Guzmán notes that it "is a condemned land permeated by salt." The desert, as he puts it, is "the vast open book of memory." In this place, which yields ancient mummies as well as more recent human remains, "we cannot forget our dead." He here refers to the Disappeared, many of whom were taken to a former nitrate town named Chacabuco that served as a concentration camp from 1973 to 1974. Chacabuco lies not far from other nitrate towns such as Humberstone and is now widely visited. In the film, an interview with one of the prisoners held there reveals how he paced off the interior buildings so that he would be able to identify the place in the future—as evidence that atrocities had been committed. In appearance, Chacabuco bears some resemblance to Manzanar in California and other hastily constructed camps where Japanese were interned throughout the arid and semiarid American West during the Second World War. Wistful and poignant, *Nostalgia for the Light* is both a tribute to and a condemnation of Chile's twentieth century past.[50] Even more ambitious in scope is Guzman's 2015 film *The Pearl Button* (*El Butón de Nácar*), a sweeping look at memories of loss throughout all of Chile, from Tierra del Fuego to the desert north, and from the earliest settlement to the present.

This loss resulting from political excess is a palpable theme in Chile today. In a website entitled *Grief in the Desert*, Paula Allen documents the impact of the Pinochet regime on the Atacama Desert, where women from Calama comb the desert for remains of their disappeared loved ones. Allen presents photos of these somber searches and also cites the works of Isabel Allende Llona, whose uncle, Salvador Allende, fell victim to Pinochet's regime.[51] Another website, entitled Flowers in the Desert, documents this desert search, which is illustrated in a series of stark black and white photographs. This website also posts the words of Isabel Allende, who pens of the Atacama, "That arid land is the perfect metaphor for the unremitting pain of the women of the disappeared."[52] However, although Allende's creative writing was forged by that loss, she can look ahead as well as back into the desert's past. In her book *My Invented Country*, Allende writes that "This elongated country [Chile] is like an island, separated on the north from the rest of the continent by the Atacama Desert, the driest in the world, its inhabitants like to say, although that is not true, because in springtime part of that lunar rubble tends to be covered with a mantle of flowers, like a wondrous painting by Monet."[53] In this one complex sentence, Allende sustains and yet challenges the myth of the Atacama, subverting the desert's extremes by portraying it as a land of paradoxes. Simultaneously, it is a space-age, monochromatic, lunar-like landscape that can morph into a florid, colorful wonderland like that created by a long-dead French impressionist painter. Using the brief flowering of the desert as a metaphor, Allende sees hope amidst bleakness. Writings by Isabel Allende, Lake Sagaris, and Paula

Allen remind us that women are also among the modern-day discoverers of the Atacama. Their works underscore this desert's importance in the collective memory and its prospective cathartic role for Chileans hoping to move beyond an era of genocide that refuses to stay buried.

We can add music to the vehicles that foster these memories of the Disappeared. In 1993, the multicultural group Inkuyo released *The Double Headed Serpent* CD album, which contains a song entitled "Callejon de las Viudas" (Alley of the Widows). This haunting instrumental is about the women of Calama who continue their lifelong search in the Atacama Desert, hoping to bring closure to lives disrupted by political violence. In 2003, on the thirtieth anniversary of Salvador Allende's death, the Chilean group Quilapayún reprised their epic 1970 folk song "Cantata Santa María de Iquique." Originally written during the time of Chile's Nueva Canción (new folk song) movement of the late 1960s and early 1970s, this song has found a new following on YouTube. As the title suggests, this song commemorates the 1907 massacre in that port city, which I discussed in Chapter Four. I use the term epic to describe Quilapayún's song because the poignant, fact-filled lyrics alone fill four single-spaced pages, and the song itself lasts for about thirteen minutes when performed. The mournful lyrics dramatize the plight of the miners who were "killed like dogs." Quilapayún places the death toll at 3,600, but as I noted, that figure is disputed by other sources. Listening to "Cantata Santa Maria de Iquique" reveals how seamlessly place and event can be merged, for it links the desperation of the region's workers to the desolation of the desert pampa landscape and the indifference of an unjust social system. Filled with references to the pampa with its dry silences, rocky narrow valleys, and forlorn nitrate towns once occupied by workers who toiled and barely survived during sun-parched daytimes and lonely fog-shrouded nights, this song indicts both the society and the desert in the tragedy. Like most historically themed folk music penned in the last five decades, "Cantata Santa Maria de Iquique" is ultimately a protest song demanding *justicia para todos*—justice for all.[54]

While on the subject of music, I would be remiss if I did not mention two other albums with an Atacama Desert theme. Both are instrumentals and feature Atacama Desert scenes on their CD covers. The first is simply called *Atacama* by an electronic music group called The Smiling Buddhas. Its seven songs suggest an autobiographical recollection of experiences in the deserts of Chile and Peru. "Walking Through the Desert near Palpa (Hitched a ride in a Battered Collectívo in the end)"; "On the moon on Earth (Valle de la Luna)"; "Ghost Towns of the Coast of Guanera"; "Plateau Pulse on Laguna

Miscanti"; "Gold Mines and Quicksilver"; "Arid Lands Sink into the Pacific"; and "Flying on a Bus (Night Bus from Tacna to Nasca)."[55] The starkness of the Atacama lends itself to such edgy, sometimes discordant electronic music filled with intriguing, unfamiliar sounds. Some of the tracks have an almost industrial quality, but then again so does this desert. Thinking that Damien might find this CD of interest, I slipped it into the rental car's CD player as we rolled south toward Chañaral one sparklingly clear afternoon. His conclusion—"Sorry, Dad, not my taste"—reminded me that music is in the ear of the beholder.

With that CD rejected and ejected, I inserted the second CD, which turned out to be somewhat more to his liking. Entitled *The Acatama* [sic] *Experience*, this album by jazz musician Jean Luc Ponty is rather more traditional. According to Ponty, "this album was produced 'on and off' between January 2006 and February 2007, due to the fact that I also performed with my band on different continents during that same period." As Ponty adds, this travel was actually advantageous, as it "influenced the development of musical ideas, such as sketches of improvisations with electronic effects that came to maturity as a collage on the title track *The Acatama* [sic] *Experience*." Ponty further explains that "these sounds evoke perfectly what I felt in the vast canyons of the Acatama [sic] Desert in Northern Chile, which I visited after a show in Santiago." In addition to the title track, another called "Desert Crossing" was also inspired by this desert. As Ponty noted, "*Desert Crossing* is the very first unaccompanied acoustic violin piece that I ever recorded."[56] His point that the desert is inspirational has long been claimed by travelers: the desert is so different from one's normal habitat that it can encourage one to break from convention and experiment with new forms of expression.

Returning to literature, the Atacama Desert theme thrives in modern fiction, ranging from the crime-solving detective novel to mind-bending magical realism. In the former, Chilean author Roberto Ampuero (b. 1953) uses the desert as a stage to set much of the action in his 1996 novel about detective Cayetano Brulé. Entitled *El Alemán de Atacama* (The German of the Atacama), this novel finds Cayetano trying to solve the murder of a German national. Cayetano is Cuban and has had a wide range of experiences throughout the world, but he now finds himself in new—and very different—territory, facing new challenges. Although far from civilization and seemingly serene, the tourist-centered town of San Pedro de Atacama and the largely mining-dependent city of Calama offer many possible reasons why someone might want to kill the German. These include ongoing disputes about water, archaeological sites, and industrial and mining developments. In this page-turning story, the Atacama is fraught with tension from interests, some local and internal, others external and originating far from the region. As the story develops, Ampuero casts the desert itself as a protagonist capable of guarding mineral and archeological treasures, and even forcing Cayetano to question his sanity. In one scene suggesting a true existential crisis with a maritime metaphor to characterize it, Cayetano wonders if he has lost touch with the reality and rationality of the cities, and he hopelessly runs aground on the sandy bottoms of the desert oasis.

In literature and popular culture, the vastness of the desert is often equated with the immensity of the ocean. Ampuero's writings also make use of the earth, including metals, as when Antofagasta is called "that city located between the Pacific and the copper cliffs of the Atacama Desert." The geographical descriptions in this detective novel border on the ethereal, and Ampuero is a master at characterizing light, especially in the desert. In one passage, he writes that "outside, the cold mantle of the night continued to flutter over the desert which glowed under the perfect disk of the moon." In another, he describes how "the moon was dropping its wreaths of silver on the chañar trees and covered, with a feather dusting almost like quick-silver, the tops of the ramparts and the prehistoric summits of the Cordillera de la Sal." In that sentence, which I have translated as literally as possible, Ampuero plumbs the subtle differences between the appearance of two metallic elements—solid native silver and the liquid metal mercury. This novel is set during the month of May, when the sun is low in the sky and the nights coolest, if not downright cold at times. One of my favorite passages occurs when the detective experiences the desert light of late afternoon and compares it to what he has seen elsewhere in Chile. "It was an ocher-colored light, fresh and thick, that Cayetano thought looked like the color of aged rum. The wintery light of Valparaíso, by contrast, seemed pale and transparent, like the color of a good Russian vodka."[57] Ampuero knows these subjects well, for he lived in Valparaíso for his first twenty years then fled to communist Cuba to escape the heavy-handed right-wing Pinochet regime. He ultimately became disenchanted with much of the same political oppression tactics in Cuba (used this time by the left) and moved to Germany. The peripatetic Ampuero then moved to Sweden and Iowa but more recently (2011) served as the Chilean ambassador to Mexico.

Whereas Roberto Ampuero's detective novel gives the Atacama Desert a European connection and modern setting, another Chilean author—Hernán Rivera Letelier (b. 1950)—casts the desert in light of a seemingly timeless period, using hints of Latin American magical realism and dark satire. To date, Rivera has written five novels dealing with the Atacama. His first—*La Reina Isabel cantaba rancheras* (Queen Isabel Sings Mexican Folk Songs)—was published in 1994 and became an immediate sensation in Chile. Few writers have better portrayed life in the nitrate oficinas than Rivera, who grew up in one until age eleven, when he moved to Antofagasta where he lived alone and earned his way selling newspapers. Rivera's first novel relates the life, passion, and death of a woman named La Reina Isabel, a prostitute who reigns over the town, bringing joy to lonely miners. Prostitution is a fact of life in mining towns, and a prostitute with a heart of gold is a common trope, but Rivera creates such a memorable time and place for her that the story seems fresh. The timeframe is the early twentieth century when the nitrate industry was falling on hard times and dying a slow death. The fact that this unnamed but archetypal nitrate mining community is fading away adds to the feeling of decadence, if not hopelessness. The constant references to Mexico are noteworthy, as that country's music and popular culture have found their way to South America; to many, Mexico symbolizes the boisterous and expansive Latin American frontier, where a cowboy past meets the future. Funeral services always serve well in introductions, and Isabel's death provides a springboard into revelations, both about her many amorous relationships as well as her social position in town.

Rivera is said to admire the masterful storytelling of Gabriel Garcia Márquez, and that is apparent. Rivera also alludes to the somber, soul-searching poems of Gabriela Mistral, with their quest for redemption amidst desolation, and that is evident, too. Of special interest to me, though, is how deftly Rivera describes and interprets this Chilean nitrate town and its landscape. The pampa's punishing sun, immense landscape, and omnipresent dust make it one of the characters in this story; it is pervasive, appearing in almost every chapter. In one passage Rivera claims that when winter occasionally touches the pampa with freezing temperatures, the landscape is transformed into the white steppe of Siberia. Of the dust, which coats almost everything, Rivera writes that during a time without even the miserable breath of warm wind to stir the air in the plaza, "the dust covered trees resemble wrinkled sculptures cast in plaster." In one of the most memorable passages in the book, he describes "the cloud of earth that day and night emanates from the caliche mills, which makes a slow forty-five degree turn and, silent and dense, settles on the dormant camp." This banner-like dust cloud is, as Rivera puts it, "like an immense, weightless dragon." As other scenes take place elsewhere in town, Rivera reminds readers of the "overwhelming grandeur of the pampa," or, less kindly, the "infinite tedium of the pampa."[58]

In 2008, Rivera published another Atacama novel, *Mi Nombre es Malarrosa*. It too portrays life in the desert and uses a female character to reveal the eccentricities and universalities of people whose fortunes are orchestrated by a distant company. Rivera's *El Arte de La Resurrección* (2010) is a remarkable story about a man living in the Elqui Valley who claims to be Jesus and is obsessed with finding Mary Magdalene, a prostitute by that name in a mining town. Like Rivera's other novels, it is thought-provoking and eye-opening. Like many stories set in Chile's desert north, it is also a combination of fact and fiction, based on the real-life story of Domingo Zárate Vega, the self-proclaimed social revolutionary "Cristo del Elqui" (1898–1971). It is, however, Rivera's most recent book, *El Vendedor de Pájaros* (*The Bird Seller*), which was published in 2014, that I would like to further comment on. The town is named Desolación, which is appropriate, as there are no other towns nearby. In the middle of nowhere, Desolación is a stop on the Longitudinal [Railway] where locomotives take on coal and water. Throughout this novel, Rivera uses the train as a storytelling device, for it has different meanings and holds special memories for the varied and often colorful characters who live in Desolación. The story begins on a particular Wednesday when only one passenger—the unnamed seller of birds—alights from the last coach of the train, bringing with him cages of birds for sale. The birds are metaphors for freedom but they can be caged, as readers soon learn of the inhabitants of Desolación.

El Vendedor de Pájaros is a very rich and complex story, but in this discussion I shall highlight how Rivera characterizes the Atacama Desert. In contrast to the brightly colored plumage of the birds for sale, the landscape of the desert is monochromatic, described as "the color of the rock, which is the color of the hills, which is the color of the puna, which is the prehistoric color of the little lizards." Of Desolación itself, Rivera notes that it consists of only five streets that are not named but rather simply numbered. Much of the housing is built of imported "Oregon pine." In the town's bleak, dusty plaza, there is only one algarrobo tree to provide any shade from the relentless sun in February. When the breezes stop blowing the place is stifling, or as Rivera richly characterizes it, the atmosphere feels like it has no air at all. In a couple of passages, Rivera highlights how people fondly remember the only time when rain moistened the parched earth and the air smelled fragrant.

In a passage bowing to the region's literary heritage, Rivera makes reference to Gabriela Mistral, who similarly characterized the desert as unforgiving in her book-length compilation of poems of the same name: *Desolación*. Rivera also notes, as have many other observers, that many nitrate oficinas are named after women. He reminds readers that the name Desolación can only be seen in one place in town—on the wooden sign at the railway station. The bird seller says of this place that "the oficina is the cage, and the desert is the bars of a cage," a cage in which we deceive ourselves with false promises. The last sentence of Rivera's book is a stunning epitaph informing the reader that the only thing remaining of the oficina is the railway station, "where trains replenish their water thanks to the existence of the well at this place, a station that appears on the railroad maps (as it reads in worm eaten letters of the sign)"—a name that confirms "a place so wild, sad, and helpless, that even the buzzards fly by it."[59]

In addition to fictional accounts of life in the Atacama Desert, current events can also become part of the folklore and popular culture. In 2010, the theme of isolation reinforcing the dramatic and tragic was center stage in the successful recovery of "Los Treinta y Tres," as Chileans call the thirty-three Chilean miners who were trapped underground in the San José Mine north of Copiapó for sixty-nine days. As the world watched in suspense, television cameras occasionally panned out across the adjacent austere desert. Many viewers had seen the Atacama Desert before; just two years earlier, it was the background for an epic battle at the end of the James Bond film *Quantum of Solace*. In that 2008 film, the ultra-modern hotel-like accommodations for the Cerro Paranal space observatory about eighty miles (ca. 130 km) south of Antofagasta seem as much a forlorn outpost on some distant planet as a place on planet earth. The otherworldly quality of mining operations in this vast desert region is also apparent in the 2015 German film *Surire* by Bettina Perut and Iván Osnovikoff. Surire is the name of a small community located in the high desert near the border between Chile and Bolivia, and in the film a mining company wreaks havoc on both the environment and the Indigenous society as they strip mine the resources of the dry borax lakebeds.

Underground mining is also a compelling subject, dramatically speaking, and a gifted writer can bring this to life in remarkable detail. Within a year or two after the 2010 mine disaster near Copiapó, books began to be published about this high-tech race against time to rescue the stranded miners. Of these books, Héctor Tobar's *Deep Down Dark* (2014) effectively uses the wide-open desert landscape to contrast with the claustrophobia-inducing interior of the mine. Tobar begins by observing that "the San José Mine is located inside a round, rocky, and lifeless mountain in the Atacama Desert in Chile." After providing the locale, he then adds the element of immense time. "Wind is slowly eroding the surface of the mountain into a fine, grayish orange powder that flows downhill and gathers in pools and dunes." Tobar next orients the reader vertically, writing that "the sky above the mine is azure and empty, allowing an unimpeded sun to bake the moisture from the soil." He then unites earth and sky with a comment about the weather and climate: "Only once every dozen years or so does a storm system worthy of the name sweep across the desert to drop rain on the San José property. The dust is then transformed into mud as thick as freshly poured concrete." A few pages later, having established that nature is not beneficent here, he adds another element of timelessness. "In the Atacama, which may be the globe's oldest desert in addition to being its driest, there are weather stations that have never received a drop of rain." As with all such pronouncements, Tobar does not name those legendary weather stations (no one to my knowledge ever has). In a reference to the region's history that readers will recall from Chapter Three, Tobar notes that Charles Darwin described this desert as "a barrier far worse than the most turbulent ocean." To further underscore the desert's timelessness, Tobar pens that "a man entering this landscape today still sees the relentless Atacama emptiness that Darwin saw." Adding a touch of the extraterrestrial, he writes that when the sun is low in the sky, "the mountains turn maroon and orange, and resemble photographs of the surface of Mars."[60]

After characterizing the exterior scene as limitless and timeless, Tobar then contrasts it with the spatially confined and time-dependent effort to recover the miners after the collapse of a huge diorite rock mass weighing twice as much as the Empire State Building trapped them deep underground. This was a venerable mine. It opened in 1889 and had long produced ores bearing gold and copper. At the time of the collapse, though, the property had fallen on hard times and upkeep lagged. Tobar makes it clear that geology is in control here, and he documents the character of the rocks and minerals that first lured, and then trapped, the miners. He paints a vivid picture of miners' work underground, stating that they had just "removed a few hundred tons of ore-bearing rock, with copper sulfide in fingernail-size specks that glimmer with the same marbled pastels of art nouveau paintings: crimson, forest green, maroon, and the brassy yellow, tetragonal crystals of the copper ore known as chalcopyrite." His description of the mine conditions are spot on, for I have experienced them in geologically similar porphyry copper mines in Arizona at the same depth—2,230 feet (710 m) below the surface—where the temperature and humidity are oppressive. Far below the surface on a dry sunny day, the air is stifling, although mine ventilation can take the edge off. When the collapse occurred, it sent that huge section of diorite downward, blocking off all escape.

Miraculously, all of the miners survived the cascading rock, but then they awaited word from above for help. As hours passed and night fell, the miners' "families gathered around bonfires…cones of weak light swallowed by the immense blackness of the Atacama night." In a situation like this, mine maps are essential for directing help, and yet the maps were reportedly out of date. Shift supervisor Luis Urzúa was "a trained topographer, and thus carries a map of the mine in his head; that map tells him there's nothing to be done." As he put it, "I knew we were screwed." To one of the trapped miners, the huge diorite block looked "like the stone they put over Jesus's tomb." Others called it a "guillotine of stone." Its size was overwhelming—550 feet tall, weighing 700 million Kg or 770,000 tons. To many of the more religious or superstitious miners, their number, thirty-three, also had significance as Jesus's age when crucified.[61]

The rescue was both dramatic and unprecedented, and it took place at a makeshift location soon dubbed Camp Esperanza (Hope). In times past these men entombed below would have perished. When driller Eduardo Hurtado arrived in hopes of boring holes to locate the miners, the situation was dire. Hurtado found it ironic that he had always drilled to find minerals, not *viejos* (miners)! He needed a topographer to tell him where to start drilling, but one couldn't be found. Finally, Hurtado found one as well as blueprints of the mine, and thus started a hit-and-miss procedure that lasted days. Miners on the surface became frustrated as the mine blueprints proved inaccurate; some began chanting "*La planificación de la mina es una hueva*" (the mine's blueprints are shit) when the drilling would encounter nothing.

Interestingly, Carmen Urzúa (wife of shift supervisor Luís) brought a small plaster statue of the Virgin of Candelaria, which was a replica of one in a church in Copiapó. Carmen and other wives decided to build a shrine, placing a few stones around the plaster figure. As she related, "we made a small place where the people could go and let go of their pain, where they could pray for the miners, and start to forget that they might be dead." Soon, more shrines were built for individual miners, their stones covered in candle wax as time progressed. Finally, after sixteen agonizing days, the miners were located, all alive but most in bad shape. They needed nourishment that could be sent down a pipe but only received water and glucose, as NASA had warned that anything heavier might prove fatal. In essence, people from one of the world's oldest terrestrial professions were being advised by authorities representing space-age astronauts.

Now that the miners had been found, an even more daunting task presented itself: how to get them out. For the next fifty-three days, the process of drilling to free the miners proceeded, Camp Esperanza on edge but hopeful for a miracle. This was the most elaborate and carefully watched mine rescue in history, and it was an international effort. Mine rescue technology and expertise arrived in one of the remotest spots on the globe. After sixty-nine days, the miners were finally freed one by one, as a hole just big enough for each had been drilled and they were carefully lifted through it in a capsule painted blue, white, and red—the colors of the Chilean flag. In reality, these miners had been pulled from a dank and smelly tomb, emerging into the clear, dry air of the Atacama Desert. Metaphorically, though, Tobar observes that they were "like men on a mission inside a stone space station, or castaways on a lifeless planet" who helped humanity "learn how men can endure such confinement and isolation."[62]

In 2015, *The 33* arrived on the big screen. This much-anticipated movie version featured several well-known stars, including Antonio Banderas, James Brolin, and Chilean-American actress Cote de Pablo of NCIS TV fame. As anticipated, *The 33* was a big hit in Chile, though ticket sales in the United States were anemic. American critics, too, distanced themselves from the movie, the general consensus being that the film seemed unable to pull off the difficult task of being both a biopic and a docudrama. I often disagree with critics, and this is a case in point. For me, *The 33* succeeded in conveying the plight faced by the trapped miners and their dedicated but sometimes conflicted rescuers. *The 33* emphasizes the political stakes faced by the Chilean government, for the world was watching. The film does an especially good job of portraying the passion of the miners' families, particularly the women who refused to give up hope, and who put pressure on the authorities to keep searching. This film also nicely portrays the dilemma faced by those who used inaccurate maps to try to locate the miners. One scene of a hologram-like map to show the situation underground was stunning. So too was the Atacama Desert scenery in the vicinity of Copiapó—stark, crinkled mountains covered halfway to their summits in a mantle of sand, a characteristic landscape in large portions of Copiapó province. From the ground, and in sweeping aerial views, this film gives audiences a memorable introduction to the Atacama.

The village of Tierra Amarilla, with its picturesque plaza, was also featured. In one moving scene, a light rain falls in the middle of the night, an obviously rare event in this driest of deserts—and perhaps a hint at the miraculous events transpiring. Noteworthy, for Hollywood at least, was this film's frequent reference to the miners' religious faith. This periodic foreshadowing that a full-blown miracle was needed to effect the miners' rescue may have been a bit too much for skeptical audiences. Then, too, the action in the film was limited to the first major rock-fall collapse and a few aftershocks, rather than the nonstop *Road Warrior*–type action American audiences are increasingly provided by studios. In retrospect, a story set largely underground in a desert embodies considerable cognitive dissonance, as does a story about a disaster that claims no lives. As the film's narrator concludes, more than a billion people worldwide witnessed "this miracle in the Atacama Desert." However, given the fact that audiences already knew the outcome, the element of suspense was, to use an intentional pun, undermined. That having been said, *The 33*'s almost science-fictional NASA equipment employed to rescue the miners was impressive, and it further reinforced the association between the Atacama Desert and space exploration. In fact, in several pivotal scenes, the film seems as much a mission to outer space as it does a tunneling through inner space to free the miners.

More prosaically, the Atacama Desert's otherworldliness is also showcased in a thirty-minute educational documentary film entitled *Discover the World: Chile*, which was released in 2011. This film's unidentified British narrator begins by stating that "When God created the world, he quickly swept the remaining dirt and threw it behind great mountains; that's how Chile came into existence, according to a popular version of its history. But," he quickly adds, "there is an impressive country behind all the desert dust." After introducing Santiago, the thriving wine industry of central Chile, and the seacoast resort of Viña del Mar, the film moves "to the north, to the Atacama Desert," which, in fact, is the focus of slightly more than half the footage. The narrator describes the huge, open-pit copper mine at Chuquicamata as a vital organ. "The heart of Atacama beats far from the coast in the middle of the desert. The largest copper mine in the world is Chile's treasury." After describing the copper mine at "Chuqui," as it is affectionately known, the narrator introduces the surrounding desert as "frightening and at the same time very attractive." In showcasing the Atacama Desert, this documentary woos would-be travelers. "If you are looking for freedom, and solitude, you can find it here in the Atacama Desert of north [*sic*] Chile." Over film footage of stark desert landscapes, the narrator voices a common refrain linking tourism and discovery. "The largest desert on the [South American] continent is like a moon, barren and empty; then there is

Figure 5.18. The author's son Damien on Mars, or rather in the Martian-like landscape of the Atacama Desert northeast of Cerro Paranal, Chile. (Photograph by author)

the feeling that even this apparent nothingness isn't totally empty because there are some lovely events waiting for you on the horizon."[63] Anticipation is the key word here, as travelers are about to discover the colorful cultures of the Atacama Desert puna adjacent to the Andes.

If life sometimes imitates art, it should come as no surprise that the future of the Atacama now involves linking this desert with otherworlds in both a literary and literal sense. In the popular press, the discovery of two six-inch-tall "humanoid" skeletons in the Atacama Desert, one cleverly called "Ata-Boy," have puzzled scientists since first publicized by *Ripley's Believe it or Not* in the 1930s. Imaginative UFOlogists claim they are otherworldly aliens, but scientific studies suggest a terrestrial origin for these admittedly enigmatic skeletons. The Atacama humanoids may stimulate the imagination to look skyward for answers to this desert's relationship to the heavens, but science itself plays a role. Two quotes by NASA that connect the Atacama with space exploration will suffice as a final example of where the popular conceptualization of this desert may be headed. The first matter-of-factly states that "A group of scientists, including researchers from NASA…in 2005 announced that they had identified habitats and microbial life using a mobile robot, or 'rover,' in Chile's Atacama Desert." The second quote reveals the motive for such exploration and suggests a connection between this desert and other desert worlds beyond those of the earth, noting of these scientists that "They studied the scarce life that exists there and, in the process, helped NASA learn more about how primitive life forms could exist on Mars."[64] In this scenario, all those earlier descriptions of the Atacama as otherworldly fall into place, as this desert is positioned as a stepping-stone in helping humankind explore the final frontier of outer space. Students of discovery and exploration should note the irony here however: the Atacama Desert, which has for so long been considered "lifeless," may provide the key to understanding the tenacity of life elsewhere in the universe.

Scientists confirm the Atacama Desert's similarities to the Martian landscape. In May 2013, a team of ten geophysicists analyzed a layered outcrop in Chile's Valley de los Dinosaurios (Valley of the Dinosaurs) near San Pedro de Atacama using laser-induced breakdown spectroscopy (LIBS). Their goal was to determine the similarity here with strata data collected by the rover *Curiosity* in Gale Crater on Mars. "With this work," the team recorded, "we demonstrate, for the first time, that semiquantitative chemical stratigraphy can be very rapidly obtained by performing LIBS measurements on visually distinct layers within an outcrop at a terrestrial Mars analogue located in the Atacama Desert of Chile." This conclusion was reached after eight years of research in which other rovers investigated outcrops elsewhere on Mars, including Meridiani Planum and Gusev Crater. These near-vertical outcrops visually resembled topography in the Valley of Dinosaurs. The geoscientists noted that "Even if there are geochemical differences between our study site in the Atacama and similar exposures on Mars, the obvious structural similarities make the VD outcrop an appropriate morphological analogue, and provide a possible case-study for improving Curiosity's tactical operations." Their conclusion—that "the VD outcrop may therefore be a reasonable morphological, geochemical, and environmental analog similar to outcrops in Gale Crater and elsewhere on Mars"[65]—is a reminder that those who imagined that the Atacama was Martian-like were not deluding themselves. It is a basic human trait that we seek, and sometimes find, the exotic in what we personally experience. This is especially true of landscape analogues: just as travelers to the prairies of the American West and the pampa of Argentina two centuries ago imagined they were traversing the Asian steppe,[66] it is not surprising that we can experience Mars in the Atacama Desert.

With that prospect in mind, Damien and I set out to visit the Cerro Paranal observatory. I planned that part of the trip because of Damien's lifelong interest in the red planet. On the way southward along the Pan-American Highway, we took the highway B-170 cutoff to the observatory. About halfway there, we entered a broad valley veneered with red rock and finely pulverized dust. After a few miles of this, Damien mentioned that he had always heard that parts of the Atacama Desert look just like Mars, and now he exclaimed, "it's true!" There was absolutely no traffic here, and it looked as if this jet-black road was stretching across the rust-red Martian landscape. Damien knew all the terms, noting that this looked like the "regolith" from early Mars missions. I volunteered that we use this term on earth for such a desert landscape upon which rocks sit because the forces of erosion cannot remove them, leaving a residual, rock-studded surface. After pulling the car to the side of the road, we turned off the engine, and savored the silence that enveloped us. This was Mars at its most hospitable; the temperature was a perfect 72 degrees and there was just the slightest of breezes. This temperature is about the hottest one can ever experience on Mars, but the nighttime temperatures there can plunge to minus 195 degrees Fahrenheit (-125°C). After walking around this scene with his arms moving slowly to simulate being in a spacesuit while exploring the red planet, Damien sat down on one of the bigger Martian rocks and asked me to take his picture with his iPhone. Looking at his image in the view finder, I pressed the shutter button and recorded for posterity the realization of his long-held dream (Figure 5.18).

I think about this image when I recall a sentence in a book Damien has on his shelf back home, *Postcards from Mars*, which I gave to him about ten years ago. In that superbly illustrated book, Jim Bell describes the richness of images beamed back from Mars by the first Viking Landers in the 1970s and 1980s with the more recent images from the Pathfinder and Sojourner (1997). In Bell's words, "The difference between the views of Mars from the Vikings and Pathfinder and the views from Spirit & Opportunity is the difference between 'acquiring images' and 'taking photographs.'" The scenes of the Santa Anita panorama in the Plains near the Columbia Hills on Mars could easily pass for the locale in Chile we were now exploring on foot. The images have a warmth that suggest the richness of photography rather than the icy quality of some remotely sensed images. This reminds me that no matter how far we advance photographically, there is still a place for the more time-honored artistic traditions. In some circumstances, even a hand-drawn sketch can bring out aspects of a natural history specimen that state-of-the-art photography cannot. As historian Tony Rice noted in *Voyages of Discovery* (1999), despite advances in image-making, "photography did not make the artist redundant, however. Artists continued to accompany expeditions well into the present century and were employed to produce illustrations for scientific publications long after they were no longer routinely taken in the field." As Rice concluded: "Notwithstanding the amazing developments in image technology, including holography, digital photography and the wonders of computer enhancement and virtual reality, no-one has been able to improve on the hand and the eye of a superb natural history artist when a particular morphological feature or the subtlest nuance of colour needs to be shown to maximum effect."[67]

This realization gave me pause, for time and place converge in the Atacama Desert in a remarkable way. Damien and I felt strangely at home here, yet the scene was undeniably extraterrestrial in appearance. I envisioned supplementing my photography with pastels and paper in order to tease out more of the subtleties that even the finest camera might miss. At that moment, time seemed to be standing still, but I was altogether too aware of how quickly it passes. When Damien was young, and I helped him learn how to take photographs, our favorite films to shoot were Kodachrome and Ektachrome slides. We also loved black and white film on occasion, as it could create moods that color film could not. Now, these obsolete films are virtually impossible to buy and we shoot digital photographs with impunity, finally conceding that they are just about as good as the best we shot using "real" film. Besides, Damien experiments with his camera and iPhone and can tweak the color, even make images into black and white at will.

With our Martian-like pictures taken, we jumped back into the car and rolled south toward the observatory. On the way, I realized that I had fulfilled a dream of experiencing the Atacama Desert in its Martian guise, but I was secretly saddened to realize that Damien will likely never actually get to Mars, despite the fact that plans are underway for future manned missions there. However, as we neared the observatory I was quickly disabused of such sentimentality, as my own son was soon expounding on the big Chilean telescopes and the technology being developed to make that journey. In a sense, those first unmanned spacecraft made it possible to experience Mars before people ever get there, which is not unlike the way most people have come to know the Atacama Desert. This occurs through a complex interplay of texts such as narratives, maps, and landscape images. As in my own journey

Figure 5.19. The colorful Puzzle de Piso (floor puzzle) of Chile is an educational tool that depicts the Atacama Desert using cartoon-like pictures of features found there. (Photograph by author)

to the Atacama, this encounter began as remote sensing, so to speak, but was ultimately followed by experiencing the real thing. Like Mars, we continue to need places like the Atacama here on earth. Small wonder that when I ask American audiences when I lecture on maps if they have heard of the Atacama Desert, about half of them raise their hands. By contrast, when I ask those same people if they have ever visited the Atacama, fewer than two percent raise their hands. However, when I ask if they would like to travel there someday, enthusiastically most hands go skyward, as if reaching for a goal they may never realize, but still hope to achieve.

Some of those who raised their hands may not have to travel very far. The armchair travelers I introduced in Chapter Two are very much with us today, thanks in part to the Internet. This is especially true for those whose physical or psychological impairments prohibit such journeying. There is even hope for those afflicted by agoraphobia, the fear of open spaces (or even venturing outside of their homes), as evident in the case of Jacqui Kenny. According to *New Yorker* magazine writer Andrea DenHoed, Ms. Kenny longed "to visit places she could never go herself—the more remote the better."[68] However, Kenny's agoraphobia prohibited her from experiencing firsthand the bleak but evocative places she found so alluring. To empower herself and others, Kenny downloaded Google Street View to capture images from desert places worldwide, including the Sahara, Gobi, and Sonoran Deserts as well as the coastal desert regions of Peru and Chile. She then used Instagram to share these collected views with others, and it became an instant hit. The images Kenny posts are all the more stunning because they were originally taken for one purpose—to matter-of-factly depict street scenes—but they serve another much more profound purpose: they reveal the power of the desert to frame and even dwarf the works of humankind.

What Kenny calls her "favorite image from the Atacama Desert of Chile" is also mine, as it is so evocative.[69] In it, a colorful Spanish-style church in the "Arica y Parinacota Region of Chile" reposes in front of a stark landscape of sand dunes that sweep towards bare hills—a perfect contrast between the inspirational works of humankind and the awe-inspiring desolation of the desert. This picture reminds me that what attracts me most to the Atacama is a quality I call *desiertísimo*—a shamelessly hybridized adjective I coined for the most hyperarid, and hence ultimate, of desert landscapes and conditions, much as the word "muchísimo" means very much (or colloquially "the most") of something. A literary friend from South America balked when she heard me use this term—after all, there is no such word in Spanish. However, she quickly conceded that it made perfect sense to her, and opined that words not yet in existence can, and perhaps even should, be invented. With a mischievous twinkle in her eye, she admitted, "poets do that sometimes—and so do scientists."[70]

In closing this chapter, I would like to introduce the last map that I will discuss in this book. Damien and I discovered this map quite accidentally on the last full day of our trip in March of 2015. We happened upon it in the children's section of a department store at the Calama Mall where Damien was shopping for gifts for his son and daughter. As opposed to the other maps discussed above, this one is not on paper but rather made out of wood. Entitled simply *Mapa de Chile*, this map is packaged in a translucent plastic box containing forty-eight pieces (Figure 5.19). The box calls it a "Puzzle de Piso" (a floor puzzle), and when completed it is large—190x30 centimeters. Though narrow (about a foot wide), this jigsaw puzzle is so long (74 inches, or 6 feet, 2 inches) that it won't fit on most tables. The premise behind this format is that kids can learn and have fun simultaneously. Also, a map such as this emphasizes the unique shape of Chile and hence can build nationalistic pride. Using this map is a challenge, for it involves learning as a tactile as well as intellectual experience. The picture on the box shows a child assembling it on the floor, his position similar to that of conquistador Diego de Almagro, who fist glimpsed what would become Chile in 1536, while an inset shows a part of the map—not just any part, but the area reaching from Antofagasta eastward to Calama and thence into the Andes. This, of course, is much the same area covered on many of the other maps in this book. However, this map is more pictographic than any other I've discussed. As the wording on the box notes, the puzzle "contains illustrations of animals, places and typical objects arranged geographically." Accordingly, in the Atacama Desert portion of the map, important features such as the open pit mine at Chuquicamata and numerous villages are indicated. The colorful houses common to the area are shown, as are chinchillas, miners, and so forth. The overall color scheme is ocher and tawny, as that suggests this part of Chile's aridity compared to the green color in the lush, forested south.

Given the size of the box containing the puzzle map, and our vow to travel light, we decided not to buy it because it would be too cumbersome to carry on board the plane. I snapped a quick picture of the label on the box and resolved that was sufficient; I would order one online when I returned to the United States. During the night, however, I relented. Rethinking my earlier decision, I now vowed to buy one in Antofagasta the next day before we boarded the plane. However, nature had other plans, for when we arrived in that city and headed to the downtown mall, we found much of the complex closed and people mopping up water-damaged stores whose floors had been flooded. Upon arriving at the airport, we realized that it was also impacted, as many flights had been cancelled and diverted. After hours of waiting, we boarded our plane, which finally taxied onto the runway, but then came to a dead stop. In Spanish, the pilot somberly informed us that the entire crew's time had elapsed, and so we had to return to the gate. By sheer luck, I talked a compassionate LAN Airlines representative into getting us on *any* plane going to Santiago, which we did. Arriving there, we were hoping that all flights had been delayed so that we could catch one, but all had departed. Because most U.S.-bound flights leave Santiago in the evening, we would have to wait a full day for the next one. Santiago is a wonderful city, but Damien had to get back home in time to go to work, so we talked our way onto a red-eye flight headed for Panama City, then early the next morning connected to one going to Houston. Leaving Houston in the late afternoon, we caught a flight to Portland that got us there by late evening. When we added it all up, it amounted to a grueling 36-hour trip, but Damien was able to make it to work.

A couple of days after arriving home, I decided to go online to purchase the puzzle. After all, anything can easily be purchased this way, or so I thought. However, typing in the name and other information drew no responses—even when I typed it in Spanish. I then recalled that this puzzle was "Hecho en Chile," and moreover solely about Chile; hence, there would be no market for it elsewhere. My decision not to buy the puzzle turned out to be my only regret on an otherwise perfect—and perfectly fulfilling—trip.[71] My remorse about this dilemma lasted only so long. Turning this disappointment into an opportunity, I've put it first on a rapidly growing list of things to do on my next trip to this irresistible desert, so far away from home but always in my thoughts and dreams.

Conclusion

"KNOWLEDGE OF THE TERRITORY"

Atacama Discoveries in Personal Perspective

It may seem peculiar to some readers that I have put off defining the word "discovery" until this concluding chapter. After all, I use it in the title of this book and many times in its pages. Although discovery is what occurs when one learns about something that they have found, even that definition is nowhere near as simple or straightforward as it seems. The Merriam-Webster dictionary defines the verb discover as "to make known or visible." In the previous chapters I illustrate many ways explorers and scientists made something known or visible—sometimes the latter quite literally, in the form of maps and other visual images as records of what was encountered. With good reason, students of the history of discoveries distinguish between discovery and exploration. As the dictionary makes clear in its discussion of synonyms, "discover presupposes exploration, investigation, or chance encounter and always implies the previous existence of what becomes known." In a sense, much of what I have written about involves both exploration—the act of seeking out or searching for something or someplace—and discovery, the act of subsequently coming to understand in more detail that which was encountered. We tend to confuse these two words and often use them interchangeably, but there is a solid rationale for maintaining the distinction between them. Doing so can help us better recognize that discovery includes some degree of further refinement of that which was first encountered through exploration. The key concept here is that further refinement enables us to more fully understand something, in this case a place, and the varied elements of which it is comprised. Moreover, we can subsequently understand it increasingly better with each successive inquiry we make. In this concluding chapter, I shall show how what has been, and is being, discovered about the Atacama can effectively be placed in the context of other discoveries in other deserts worldwide. Having made that point, I shall note that discovery is also key to understanding how unique a particular place such as the Atacama truly is—as well as how it is rediscovered through time.

Permit me to cite three specific examples. The first relates to the region's fabulous mineralogy. As curator Carl A. Francis notes, "Harvard's involvement

Figure C.1. The recently discovered copper molybdenum mineral named szenicsite is so rare that it is only known to occur in one small portion of one mine in the vast Atacama Desert. Specimen size: 3.8 cm (ca. 1.5 inches). (Photograph by Wendell Wilson, Tucson, AZ)

with Chilean minerals was revived in the early 1990s when Terry 'Skip' Szenics—who had relocated to Santiago to produce specimens and lapidary materials marketed through Aurora Mineral Corporation—began submitting unidentified specimens for study." Serendipitously, Szenics had purchased an interesting mineral specimen at a gasoline service station in the Atacama Desert. This piece of Powellite from the Jardinera No. 1 Mine led Szenics to explore for more like it. In the process, he discovered specimens of a mineral that he was unable to identify for a very good reason: it was heretofore unknown to science. This green crystallized copper molybdenum mineral turned out to be so rare that its only known location was a one-square-meter sector of the mine. Discoveries are still being made, but they take considerable effort and perseverance. Whereas curator Francis observes that the mineral atacamite "is so common in the region that it should be designated Chile's national mineral," the new mineral that Szenics discovered is among the rarest on earth. Appropriately, it bears the name szenicsite in honor of its discoverer (Figure C.1). More recently, Szenics sold his entire Chilean collection to Harvard's Mineralogical and Geological Museum, where Francis notes that it serves numerous scientific and educational purposes, including adding "several world-class display pieces." It joins other notable specimens obtained from Ignacio Domeyko and other luminaries in Chilean geology.[1]

Even though rare minerals like szenicsite are intriguing and help push the boundaries of science, I confess to being equally intrigued by atacamite, which occurs throughout the Atacama as well as in other places such as Australia and China. After all, it was atacamite that helped put this desert on the map. Although the other mineral specimens illustrated in this book

Figure C.2. A prized item in the author's personal mineral collection is this specimen featuring clusters of radiating green atacamite crystals from the Atacama Desert near Copiapó, Chile, seen here positioned on the 1892 map by Francisco J. San Román. Specimen size: 9 cm (ca. 3.5 inches). (Photograph by author)

were photographed from nationally recognized collections, I was surprised that I could not find a high-quality digital image of atacamite, even though showy specimens are exhibited in the prestigious collections. Opening the drawer in my mineral cabinet entitled CHILE, I photographed one of my favorite specimens of this green mineral that had changed so many fortunes, my premise being that no illustrated book on the Atacama Desert should be without one (Figure C.2). This specimen consists of sprays of emerald green atacamite crystals on an iron-rich, grayish brown matrix. Another of my specimens has more pronounced resplendent green crystals on a matrix of dull yellow limonite (iron oxide). The more I study both under a high-powered hand lens, the more I discover about their patterns of crystallization and their relationship to the other adjacent minerals. At an earlier time, mineralogists were intrigued by atacamite, but miners in the Atacama thought little of such green-crusted rock, which they tossed into waiting ore cars. Even though miners have always tended to hold onto really showy specimens, these in my collection were once so common they would have gone to the ore concentrator or smelter. To me, though, this part of my mineral collection is sacred ground, and I've marked "RARE" on the two labels. I take some solace in knowing that my specimens would cause a Chilean miner to pause nowadays, for these enterprising workers realize that ore samples such as these are worth far more to mineral collectors—say about fifty dollars each—than the value of the copper they contain, which would amount to about fifty cents. Miners and collectors alike have discovered that the zones of enrichment full of colorful oxides and chlorides of copper and silver are now found only infrequently.

The second example of rediscovery concerns meteorites. Readers will recall that Fletcher's 1889 study referenced earlier did not reveal anything exceptional about Atacama Desert meteorites. However, given recent revelations about the antiquity of this desert, current researchers suspected that the subject warranted another look. In 2008 and 2009, an interdisciplinary international team of thirteen scientists set out with a premise: because "hyper-aridity and associated very low erosion rates have been ongoing for about 25 Ma [million years]...[T]his area is potentially a very favorable place for the recovery of meteorites." Observing that a surprisingly small number of meteorites had been found in the Atacama Desert since the late-nineteenth century, they embarked on an extensive search, locating forty-eight meteorites. Their fieldwork involved combing the desert on foot using ten-meter grids as coordinates. One area in particular—San Juan in the Central Depression—provided a veritable bonanza of meteorites; moreover, many were relatively unchanged chemically despite the fact that they had fallen about twenty to forty thousand years ago. Comparing these meteorites to those found in other hot deserts worldwide (e.g., the Sahara and Australia), they concluded that "time is not the primary factor controlling chemical weathering in the San Juan DCA [collecting area], or to put this in a different way, that chemical weathering is not removing meteorites in the [same] time scale...as observed in all other desert meteorite areas." This led the researchers to a surprising conclusion: "The way meteorites disappear in the San Juan area may not be so much linked to chemical weathering, but more to eolian [i.e., wind-related] erosion, a conclusion also supported by the flat shape of the meteorites." This corresponded with their earlier observation that "faceting by the wind generated flat shapes that are not typical of meteorites encountered in other hot deserts." In a nod to the exceptional nature of this desert, they entitled their article "The densest meteorite collection area: The San Juan meteorite field (Atacama Desert, Chile)."[2] This study is yet another example of how the Atacama Desert is constantly being rediscovered, and its uniqueness underscored in the process. My colleagues and friends who study and collect meteorites are well aware of this, for they have seen prices skyrocket in recent years as the public has discovered the romance of finding the world's oldest unchanged meteorite specimens.

A third example is even closer to my field of geography. In 1961, geographer Glenn Trewartha (1896–1984) published a book entitled *The Earth's Problem Climates*, which was subsequently revised in 1981. In both editions, Trewartha used the adjective "problem" not in the traditional sense of the word (that is, something that might pose difficulties for human settlement), but rather for conditions that are anomalous or unusual enough to demand greater scientific scrutiny. One such locale is western coastal South America where, he noted, "an intensity of aridity prevails that, so far as is known, is not equaled in any other desert on earth." Interestingly, nowhere in his analysis of this region did Trewartha use the name Atacama Desert, which had long been recognized as a portion of this larger part of arid western South America. Instead, he elected to use the names of the two countries, Chile and Peru, in which this coastal strip lies. Although conventional wisdom pointed to the general factors of a large high-pressure zone and the position of the desert between cool ocean and tall mountains as causes, Trewartha was the kind of geographer who thought outside the box. Based on field observations as well as a collection of data, he first described several factors that influence climate in similar locales such as South Africa. After analyzing all other factors, Trewartha postulated that a condition unique to South America—its sustained southerly winds that run along the nearly straight parallel shoreline—created pressure differentials that increase aridity by causing additional subsidence. As Trewartha further observed, "the coastline itself becomes a climatic control, which intensifies the aridity along the littoral of northern Chile and Peru by (1) producing (in conjunction with the Andes) a closed anticyclonic circulation with paralleling winds, a cool ocean current, and a strengthened upwelling of cool water, and (2) generating coastal divergence and subsidence in the surface winds which approximately parallel the coast." This led Trewartha to conclude that these important factors, along with "a strong and positionally stable anticyclonic cell, abnormally cool coastal waters...and an unusually strong sea breeze—these conjoin to produce maximum aridity."[3]

I could easily cite other examples of how rediscovery works, including the informative 1999 scientific article "Vegetation changes and sequential flowering after rain in the southern Atacama Desert," which clearly shows how plants in this region respond to infrequent rainfall—a subject that was first speculated upon by Charles Darwin in 1835, then subsequently by J. M. Gilliss and Rudolph Philippi who actually witnessed it in the 1850s, and thereafter pondered by generations of visitors awed by the periodic "bloom-ing of the desert."[4] In Chile, this *desierto florido* (flowering desert) draws tourists from the central and southern part of the country, as in August 2017, but it also lures scientists from much farther afield.

However, I would now like to place the aforementioned discoveries in a broader context of Atacama Desert history by reviewing the main stages of its development. First, Atacama was the name that the Indigenous peo-ple used for some aspect of the place—not the place itself. That name, spir-ited from the Indigenous culture by Spaniards in 1536, soon characterized a challenging and exotic political province. This earliest Atacama was simulta-neously *despoblado* and *terra incognita* to the western mind, but above all home to varied Native peoples adept at survival in the face of environmen-tal and territorial challenges. Tellingly, during this early phase, the Atacama Desert was known to Europeans by what was said about it (skimpy narratives hinting at the difficulties it presented to would-be colonizers) as well as what was *not* said about it; these silences, as it were, added to its sense of mystery and heightened curiosity. Those same silences are also replicated on many maps of the period, which often feature blank space as if nothing were there.

However, with the European Enlightenment and the second phase of exploration that accompanied it, the Atacama became one of many deserts that beckoned as frontiers by those handmaids of progress: scientists (though that term was not widely used until the 1830s). Ideally, the Atacama could now be appreciated on its own terms for what it revealed about nature, but those who bankrolled such expeditions almost always had ulterior motives, especially the economic development that sustained colonial expansion. In the next phase, extensive mineral resource exploitation began, as riches underground were converted into working capital by corporate enterprises that coordinated their efforts with recently created independent nations. Toward the end of that period, riches on or close to the earth's surface, in the form of nitrates, lay waiting to be scooped up. Later in this period, under-ground copper mines yielded to huge open-pit mines that, in effect, turned the earth inside out. In the last and most recent phase, the Atacama Desert developed a dual identity as a stage in the drama of human disappearance resulting from political brutality and evermore aggressive, and energy con-suming, mineral extraction. The image of the desert as a place of hidden secrets thus built on earlier phases, but it now exploited them in not only a modern but also a postmodern sense; the latter finds it occupying a neth-erworld between fact and [science] fiction. This edgy quality is closely con-nected to the Atacama Desert's discovery as an otherworldly place: a strange locale that will serve as a stepping-stone for outer-space exploration. That experience might also draw tourists, as hinted at in the Atacama Desert's number two spot on the *New York Times* list of "52 Places to Go in 2017." In a superlative, that list promises "New Ways to Explore the World's Highest [*sic*] Desert"[5]—a reference to elevation and altitude being an increasingly easier claim to make now that the link between the star-filled heavens and the sun-parched earth is so well established in the collective imagination. In this type of superlative branding, the Atacama is not alone. Also in early 2017, the Death Valley Natural History Association summed up Death Valley in three words: "Hottest. Driest. Lowest."[6]

In addition to space, the element of time is also palpable in the Atacama, both for travelers and scholars alike. Important lessons can be learned from the Atacama Desert's timeline of discovery. Foremost, the presence of Indigenous peoples helped outsiders conceptualize and appropriate the region through the adoption of names that had been long used and are now becoming long-lasting. This happened in the case of the Atacama but also characterizes other deserts, for example, the Mojave, Namib, and Sahara. However, at the same time, deserts acquire personalities that are difficult to separate from the seemingly heroic outsiders who "discovered" them for a multitude of reasons; for example, Almagro is an enduring name in the Atacama, as Coronado and Oñate are in the Sonoran and Chihuahuan deserts. Typically, these newcomers applied (or misapplied) Indigenous names to different features than the locals did, and these soon filled their maps. Next, deserts may take on romanticized meanings as they are explored. This is somewhat paradoxical, for at the same time that a culture begins to demystify a desert, popular culture may increasingly portray it as mysterious and beyond human comprehension. This paradox is likely a place-related manifestation of imperial nostalgia, the romantic sentiment in which a conquering civilization casts as noble the very Indigenous peoples that are now endangered by that same civilization.[7]

Another important lesson is related to chronology. Readers may notice that the phases discussed above seem familiar, because other desert places have experienced colonial exploration, mining booms, and the like. Those observations may lead one to ask a reasonable question: do these phases also neatly characterize the other arid lands as well? The answer is both yes and no. For example, at first glance the phases that apply to the Atacama Desert appear to be similar to those that transformed the American Southwest, in particular the Mojave and Sonoran Deserts with which I am very familiar. Both of those North American deserts are admittedly not as arid as, and their vegetation more luxuriant than, the Atacama. However, in the Sonoran Desert, which extends southward into Mexico, initial Spanish military expeditions also worked hand in hand with Church and Crown in the 1500s and 1600s, much as they did in the Atacama. Although the distribution of Indigenous peoples differed according to the types of habitats, missionary activity perpetuated some of these early patterns in both locales. Interestingly, in both the Atacama and Sonora, Jesuit priests drafted maps that emphasized the locations of churches and indicated topography and river systems. In the Atacama, the work of Alonso de Ovalle in the earlier 1600s can be seen, and in the Sonoran Desert, the gifted Jesuit cartographer and missionary Eusebio Kino drafted maps circa 1680.[8] In retrospect, Kino appears to be the more masterful cartographer of the two, but then again, he was operating in the later-seventeenth century, when mapmaking was improving overall.

About sixty years later (ca. 1745), the Croatian Jesuit cartographer Ferdinand Konscak (aka Fernando Consag) mapped portions of Baja California in concert with Spanish naval cartographer Rafael Villar del Val. This, as cartographic historian Mirela Altic notes, was an unusual alliance indeed, as Jesuits and the Spanish military were often at odds elsewhere in Latin America.[9] In the desert north of Chile (and for that matter portions of the area within Peru and Bolivia in earlier times), I could find no record of such intimate cartographic alliances. Of course, these religious cartographers were not mapping deserts per se, but rather documenting missionary activity. Nevertheless, a fruitful area of cartographic history research would be in archives containing records of Franciscan and Jesuit missionaries who may have mapped portions of the Atacama, but whose maps have not been seen, much less studied, since they were submitted to church authorities two or more centuries ago.

Although later mining booms would transform portions of the Atacama and Sonoran Deserts, these were different in key respects. Modern corporate mining developed earlier in the Atacama Desert than in the Sonoran Desert. This fifty-plus-year difference (1830s vs. 1880s) is explained in part by the aggressive encouragement of outside-owned mining by the newly created Republic of Chile, which successfully attracted foreign (especially British) capital and expertise, and which in turn played a major role in developing Chile's desert north. Nevertheless, in both Chile's norte and el norte of Mexico, maps served to bring the territory into the nation-state. As Raymond Craib highlights in his study of Mexico, "exploration, mapping, and surveying became the means by which to identify and assume control over resources, to reconfigure property relations, and to generate knowledge of the territory." This cannot be overstated. As Craib further notes, "These activities were instrumental to the process of territorial integration, to the degree that one could plausibly argue that the state and cartography are reciprocally constitutive." In keeping with many other studies of the relationship between mapping and placemaking, Craib concludes that maps enabled the landscape itself to be "translated into a stable, lined, and punctuated text." Of course, the landscape is constantly changing, but a map can provide the aura of stability. "Fixed on the surfaces of two-dimensional maps, what appeared as practical chaos could be over-written with discursive order." This made cartography an epistemological as well as an empirical endeavor.[10]

Mining was—and remains—key in the development of Atacama Desert identity. The Atacama appears to be the only place in the Americas where Native populations used smelting techniques to treat copper ores in pre-Columbian times.[11] Significantly, the Atacama remains the most productive desert mining region on earth, accounting for much of Chile's economic prosperity—and Chile's enviable position as a Latin American economic powerhouse. The Atacama is perhaps the quintessential desert, in that it embodies so many themes inherent in all deserts, including stark differences between densely settled irrigated areas and vast areas of wilderness, the presence of resource exploitation such as mining that further transforms the landscape into barren moonscapes through huge piles of debris, and the recognition that much of life here so seemingly far from civilization is ultimately dependent on decisions made in distant centers of power such as Santiago, New York, or London. Although there are dangers in such generalizations, which might lead one to paint all of the world's deserts with too broad a brush, this study of the Atacama Desert suggests that a comparative timeline of developments in all arid regions worldwide would be valuable and eye-opening. Doing so would enable a better understanding of the role that the world's deserts play in relation to geopolitics and the global economy. Such a comparison, however, should in no way detract from the uniqueness of each of the world's deserts.

Ultimately, the Atacama Desert has become unique through a well-defined process of emphasizing the exceptional. These superlatives include the driest and oldest arid climate, earliest types of human activity such as mummification and pre-Columbian metals smelting in the New World, as well as the earth's most otherworldly scenery. Other deserts, such as the Mojave and Sahara, have earned their own reputations; for example, the former has the highest-recorded temperature, and the latter the largest expanse of sand dunes, etc. Interestingly, some of these reputations build on one another, as in the fabled Dakar off-road desert races which began in Africa's Sahara but were recently (in the 1990s and again in January 2015) expanded to run through the Atacama Desert. The premise was that the Sahara might be the biggest desert, but the Atacama is the driest, so the Dakar racers proved their mettle by tackling exceptional places with exceptional characteristics. A quieter and gentler premise is involved with the rapidly growing ecotourism industry, which emphasizes the Atacama Desert's unique and endangered habitats—rather than how quickly one can blast through them in an off-road racing vehicle.

In common with other deserts, the Atacama is now being rediscovered as a place that can help an ailing planet regain some measure of sustainability. I note with much interest that in almost the exact place where Bowman described and illustrated the wind-powered railcar with its prominent sail, one now finds batteries of wind turbines. On slightly elevated land at the edges of Calama, they stand like tall sentinels, their huge propellers generating power into the grid. At that time, Damien's wife Sarah served as a consultant to alternative energy companies, and so these giants were a mandatory stop. With his iCamera, Damien recorded them in motion-picture footage and frozen in time (Figure C.3). To me, these wind turbines are above all symbols of how internationalized the world has become, for their components originate in far-flung countries. They remind me of the forests of wind turbines in California and Texas, where the movement of air is predictable enough to warrant a vast investment in infrastructure. The comparison does not end there, for in other parts of the Atacama, such as the Amanecer Solar CAP project near Copiapó, photovoltaic solar panels are generating electricity. On a smaller scale, road signs throughout the Atacama Desert are illuminated at night by power harvested during the daytime from solar panels atop them, and stored in batteries awaiting nightfall.

Meanwhile, a second type of solar power is also thriving in the Atacama. Developed about ten years ago, the Atacama 1 Solar Complex at Maria Elena Segunda generates electric power from heat collected in solar panels that in turn power turbines. California's Mojave Desert has several similar solar-power-generating facilities that provide an increasing amount of electricity to the state's urban areas such as Los Angeles. That comparison is apt, for the Atacama Desert is likewise ultimately the backyard of a vast urban empire that lies in the fertile valleys to the south, including Valparaíso, and along the Pacific Coast. It is tempting to think of solar energy as cutting edge, but as William Rudolph noted in 1963, "the first industrial solar heater in the

Figure C.3. With the western peaks of the snow-capped Andes as a backdrop, huge propeller-driven turbines in the vicinity of Calama, Chile, harness energy from one of the desert's most dependable renewable resources—the southerly winds. (Photograph by Damien Francaviglia)

world was installed in 1874 in Las Salinas, in the very heart of the Atacama, and was used successfully for more than forty years for distilling water."[12] In 1980, near the southeastern Arizona town of Willcox, I explored the ruins of a small turn-of-the-century solar-powered steam-generating plant that had been forgotten long ago by locals after a hailstorm destroyed its special parabolic-shaped glass panels and dashed the hopes of investors. As with wind power, we should view our current emphasis on solar power as a renewal of—not the birth of—interest in sustainable energy sources.

In their isolated and ancient Atacama settings, these recently constructed facilities seem surreal, for following their energy transmission lines takes one into the heart of the most populated areas of the nations that subsidize them. For its part, Chile plans to increase the use of renewable energy by 2050 to 19 percent solar; 23 percent wind power; and 29 percent hydropower. The former are largely Atacama Desert–based, while the latter will be achieved through hydropower stations positioned to capture the normally ample runoff from the Andes of central and southern Chile. Pondering this, I also note another irony: two of the factors that travelers have long found so noteworthy here—the wind and the abundant sunshine—are now essentially renewable resources that connect them to the outside world, seemingly harmlessly, though not without their own environmental impacts.

Even the fog here has been turned into a resource, as locals long ago realized that water condenses on hard, inclined surfaces. When directed into small plots, this water serves to irrigate plants. This technology is now being used to develop hydroponic farming, and crops such as baby carrots, cherry tomatoes, peppers, and cucumbers are growing from nutrient-rich waters in facilities such as those at La Chima. The Chilean Foundation for Agricultural Innovation began pilot programs in 2011 and can help meet the needs of cities such as Antofagasta.[13] Such innovations are a reminder that discovery coupled with the need to address and solve problems can lead to one of the greatest human gifts—the propensity to invent. Through inventions, the desert's seemingly inhospitable environment can be turned into opportunity by human ingenuity. A pre-Columbian example of this is the Nazca Band of Holes in the desert of southern Peru. Locally called *puquios*, these spiral-shaped masonry-lined holes have long intrigued observers. In 2016, archaeologists provided a possible answer, hypothesizing that the *puquios* may have been used to draw irrigation water from underground by funneling the southerly winds into these ingeniously designed human-made vortexes.[14] Ingenuity is a never-ending process; in the future, skills being learned in the Atacama Desert today are likely to play a role in humankind's colonizing of an otherwise inhospitable outer space.

Another aspect of Atacama Desert identity is well worth pondering in this conclusion. After studying maps of the Atacama covering several centuries, one must deduce that this desert is not as fixed or stationary as is generally believed. This is true of all deserts, whose margins may change with human activity or climate; for example, they may famously grow through "desertification." This term is used for a seemingly insidious process by which deserts expand in size to consume arable land, but its causes can be natural as well as human induced, namely, decreasing rainfall and removal of vegetation, causing severe erosion and soil loss.

Inevitably, these deserts will continue to change in the collective human imagination as perceptions and expectations change. The only real constant here is that people will likely use a combination of words and images (narratives and illustrations, including maps) to portray these changes, fixing them momentarily. As we discover more about these deserts, we will create new narratives and images. This last point is worth restating, for it suggests that the history of discoveries is far more relevant than commonly assumed. Just as geographer Isaiah Bowman discovered new connections to the Atacama Desert's past in corporate archives a century ago, we can also learn new things about and hence [re]discover these same places anew. Although many students of the history of discoveries emphasize the earliest-recorded contacts between European explorers and the lands they encountered, it is worth restating that discovery continues to the present. Many new students of discovery are now taking a second, and far more penetrating, look at how Indigenous peoples conceptualized places before those places were appropriated by explorers. As D. Graham Burnett shows in *Masters of All They Surveyed*, the exploration of South America, and by extension all continents, is full of unexpected twists and turns. Most of these lie less in the actual geography than in the mindset of the explorers.[15] Therefore, understanding exploration and discovery requires that we not only examine geographic space but also probe deeply into mental territories. Is the desert ever really separable from the mind and soul that conceive it? The answer, of course, is a resounding no.

Although this book is nominally about the Atacama Desert, it has hopefully taught the reader and prospective traveler about the people who discovered this desert as well as about the place itself. If readers have gained a better understanding of this desert that is good, but my main purpose is to enable them to comprehend how this desert impressed other people who experienced it in the past, and especially how they acted on that information. Studied collectively, they reveal that a sometimes shared and sometimes disputed understanding of it emerged; moreover, by studying the past we can see how and why that understanding continues to evolve. I am but one of many outsiders who hopes to tell the story of this desert through time, but my abiding appreciation of both history and geography helps confirm how inseparable those two disciplines are.

Travelers have had much to say about this desert over time, but the voices of desert insiders who have lived here are of particular interest to me, because they are more durably connected to this unique place. One of these residents of the Atacama is José Antonio González Pizarro, the Chilean professor whom I mentioned in preceding chapters. Like me, Professor González wears two hats, one as a historian and the other as a geographer. However, unlike me, he is a consummate insider here who knows the Atacama Desert in ways I will never be able to fully comprehend. José González and I had corresponded (in Spanish) before my recent trip, and I was pleased to learn that he too had a deep interest in how the Atacama has been encountered through time. When we finally met in 2015, he kindly provided me a copy of a book that his university had recently published on the huge and geographically diverse Antofagasta region. One of his chapters in this beautifully produced book is entitled "El Conocimiento del territorio por Viajeros y Exploradores" (Knowledge of the territory by travelers and explorers), and in it he tells a story of fascination and engagement with the landscape by newcomers over several centuries.[16] As an academician who has spent much of his life in the Antofagasta region, Professor González tells this story in a linear, almost genealogical, manner, occasionally using maps to help clarify points. Although I had written an article for the Society for the History of Discoveries that treated the subject in a very similar manner, Professor González's chapter convinced me that I had not missed any of the major players in this historical/geographical drama. However, recalling that there were many more voices that needed to be heard, and their stories further contextualized as international as well as local history, that chapter inspired me to expand my article into the book you have now nearly finished reading.

Believing that all books are ultimately autobiographical regardless of subject matter, I would like to further insert myself into this conclusion as I have done from time to time throughout this book. In the preface, I hinted that although my doctorate is in geography, I have found myself invariably drawn to other fields and their practitioners. Geography has offered me a bedrock on which to stand, but other disciplines have inevitably influenced my trajectory. However, I had mentors in my profession such as Homer Aschmann who provided insights to me even as an undergraduate student. In the 1970s, Aschmann mentored geographer Conrad J. Bahre (1942–2017), whose dissertation *Destruction of the Natural Vegetation in North-Central Chile* sheds light on the transitional zone between forest and desert in portions of the Norte Chico, specifically Coquimbo Province. Aschmann, of course, had been influenced by the writings of generations of geographers before him, including Isaiah Bowman and Carl Sauer. Each of these geographers had specialties but could apply them more broadly (worldwide) as well as to specific geographic regions. A few of these broadly trained geographers also mentored students who in turn went forth into the Atacama to break new ground.

A case in point is University of Arizona geographer Carl Bauer (not to be confused with Carl Sauer), who is both a legal expert and geographer whose specialty is Chile. Calling upon his legal expertise, Bauer has urged Chile to reform its water laws by moving away from free-market models. However, instead of pushing for overnight reform, he has advocated building on conventional neoclassical principles rather than jettisoning them outright. As Bauer put it, the most effective way this can be accomplished is "to revive older approaches to institutional economics—approaches that are rooted in qualitative and historical analyses and draw on politics, law, and other social sciences in order to understand 'economic' issues."[17]

That approach would have appealed to geographer Homer Aschmann, who advised me and my fellow students to use such sources in his Economic Geography and Geography of Latin America classes. One of Bauer's former students, Manuel Prieto, now teaches at the Instituto de Investigaciones Arqueológicas y Museo R. P. Gustavo Le Paige de la Universidad Católica del Norte, which is located in San Pedro de Atacama. Prieto's 2014 dissertation, "Privatizing Water and Articulating Indigeneity: The Chilean Water Reforms and the Atacameño People (Likan Antai)", reveals how intimately linked identity is to both the local water resources and the dominant Chilean neoliberal state.[18] Prieto's recent (2015) research epitomizes the fusion of serious scholarship with hands-on environmental awareness, and it resulted in an edited book featuring chapters by scholars in varied disciplines. In a chapter entitled "La Ecologia (a)Politica del modelo de aguas chileno," which focuses on the Rio Loa, Prieto urges decision makers to consider alternative strategies that take into account both the future needs of the population and the sustainability of the environment. The irony here, as Prieto and his coauthors note throughout their anthology *Ecología política en Chile* (Political Ecology in Chile), is that depoliticizing decision-making is in itself inescapably a political endeavor. Their book's subtitle highlights four key words or concepts—nature, property, knowledge, and power—that are elements in an enduring drama taking place not only throughout the Atacama but also throughout Chile, since the country's return to a more liberal government in 1990.

It is essential that geographers and others influencing public policy take into account not only the academic literature but also the perceptions and interests of stakeholders. For example, when Homer Aschmann studied mining's impact on the desert north of Chile, he relied in part on Leland Pederson's seminal study *The Mining Industry of the Norte Chico, Chile*—a geography doctoral dissertation published in 1966.[19] Pederson's work is comprehensive, bringing together many facets of an economic activity—in this case, the mining industry—in one place. Moreover, it does so cognizant of time in that it clearly recognizes that the past played, and plays, a role. When geographers and others rub shoulders with miners and appreciate the challenges and satisfaction of extracting minerals using underground and surface mining techniques, it gives their writing a sense of *costumbrista* not unlike that of the Chilean writer and publisher Jotabeche (Vallejo) a century and a half ago. Most of all, though, Pederson's work has a leisurely pace that reflects how the author learned, and encourages the reader to do likewise. I bring up the issue of time here because a study like Pederson's would be hard to duplicate today, not because there isn't a vast base of material that needs to be synthesized, but because a sense of urgency now seems to play a greater role in driving scholarship. On the one hand, this sense of urgency is understandable, as some problems (global climate change, for example) are pressing, indeed. However, jumping in to solve a problem without fully evaluating the situation beforehand is fraught with danger, too. In an important example from the United States, the rush to provide alternatives to fossil fuels resulted in the ethanol industry that now brings with it unforeseen environmental impacts such as massive siltation downstream from agricultural lands. True, we have avoided burning fossil fuels, but in the process we are now depleting soils and damaging the ecology of the Gulf of Mexico.

A closer look at how geographer Homer Aschmann addressed the issue of water use in relation to mining is instructive. In addition to identifying the four phases of mining, which characterized as well as transcended the Atacama Desert, he identified key challenges. Noting that the process of mining may contaminate a water supply, as it did in Chañaral, Aschmann observed that "it is in society's interest that an effort be made to integrate the mine's water requirements and provision with present and potential future agriculture and urban developments in the basin, maximizing opportunities for multiple use of the same water, and minimizing damage from contamination either by purification or segregation." Aschmann did not make this recommendation hastily, for he recognized it as a profound choice with immense consequences. As he concluded, "The mine is inherently transient, but it may have the option of [either] promoting future development or creating a permanent desert."[20] Although Aschmann was describing land that was undeniably desert in character, he used the term desert in that sentence to portray a land that would be dismal indeed—unable to sustain any life—and hence entirely useless. This, as I showed in Chapter Two, was among the definitions of desert used by Felix Fabri about five centuries ago. In the mid-1800s, Charles Darwin was of a similar mind. It is noteworthy, though, that something had occurred in the interim to permit Aschmann to recognize that humankind could actually *create* this kind of desert. That realization had begun to materialize around 1860 with George Perkins Marsh (1801–1882), and was well established by 1956, when the classic eleven-hundred-plus-page anthology *Man's Role in Changing the Face of the Earth* was published—and dedicated to Marsh in recognition of his vision.[21]

That ability to so thoroughly degrade a place is humbling, and yet it seems an unavoidable point on the trajectory of development. Like all places, the Atacama has a timeline of development based in part on the occupation of the land by varied peoples, some wielding more power than others. Whereas the earliest Atacameños were herders and subsistence farmers with a few miners, they managed to make transitions into the colonial economy. The pace of change increased in the nineteenth century; when Darwin visited this desert at just the time the scales had tipped, he was awed by the experience. A century later, it was apparent that mining had completely triumphed, if temporarily. Today, we recognize that Indigenous peoples and landscapes are endangered and feel obliged to intercede on their behalf. Similarly, whereas early scientists sought to better understand the land with extraction in mind—the mining companies handily exploiting its treasures—the modern Atacama is viewed by many as a land in need of remediation and reclamation *from* mining. In recent years, mining's privileged status has been challenged on environmental grounds, as evident in two recent multi-million-dollar lawsuits won by the Chilean government against companies operating in the Atacama. The first lawsuit in 2013 was a 16 million-dollar claim against the Pascua Lama mine owned by Barrick Gold Corporation, a Canadian firm. The Pascua Lama mine is situated near the crest of the Andes close to the Argentine border in the southern Atacama in a highly sensitive zone; as of this writing, the mine may move forward, but with a new alliance of business partners. The second lawsuit involved the Caserones Mine run by the Japanese-owned JX Nippon Mining & Metals, which was ordered to pay 10.8 million dollars in damages.[22] These disputes make headlines, but the mining industry is aware of its nearly insatiable demand for water, and its impacts on the environment, not to mention negative public relations. In another example of ingenuity in the desert, important strides have been made in the Atacama, including the use of seawater for ore processing by Barrick and CODELCO—a technique that is costly but which frees up water for agriculture and other uses, including water supplies for cities such as Antofagasta.[23]

The Internet and scientific literature today reveal an Atacama Desert in crisis, or rather crises (economic, social, and environmental), largely human-induced and interdependent. I have emphasized cacti in this book not only because they are fascinating and beautiful indigenous flora, but also because they can tell us much about how habitats are changing. Cacti are especially sensitive to pollution. Most of the sources I consulted warn that the cacti here are endangered. In many areas dead plants abound, but then again, the rates of decomposition here are slow indeed. Still, those who take censuses of cacti are justifiably alarmed that the dead are beginning to out-number the living, whose numbers are dwindling. Climate change and min-ing pollution are often suggested as culprits. In their prologue to *Cactus del Extremo Norte de Chile*, botanists Raquel Pinto and Arturo Kirberg state that their book represents ten years of study, and they hope that it will encour-age authorities and business enterprises to develop conservation strategies to protect the native flora.[24] Although their book covers a remote corner of the world, it has broader implications; far from public consciousness, and con-science, these arid lands are ultimately important to the overall health of the entire planet.

That having been said, I think we need to be careful not to overreact based on limited information. Without the historical record close by, it is easy to conclude that changes in the environment are not only human-made but also modern rather than endemic. Whereas Isaiah Bowman care-fully plumbed Copiapó Mining Company records compiled beginning in the 1800s, today there is a temptation to disregard what those records revealed, namely, an environment subject to periodic or episodic natural calamities. In addition to human-induced climate change, we should also recall that such storms and droughts may be part of the punctuated hydrologic history of the region that predates the enormous carbon emissions of the recent industrial society. Wisely, some scientists are factoring into their climate change mod-eling the kind of historical data that Bowman found in the musty mining company records. This is not to discount modern climate change, which the United Nations' *Global Deserts Outlook* predicts will increase temperatures by one or two degrees and decrease rainfall by five to ten percent in the Atacama Desert.[25] Rather, I encourage a more cautious and comprehensive look at not only the historical data but also other indicators in the field. I think if we do so, we will recognize that climate change may represent a combination of natural cycles *and* human-made conditions—a realization, by the way, that might help climate change jeremiads and climate change deniers find com-mon ground. While on the subject of problem solving, I should note that Bowman himself was tasked with the pressing need to help the world create more effective and lasting political boundaries, and he readily stepped into that role; however, despite this commitment, he still made the time to share his insights on how humankind related to the environment in the Atacama Desert. One of the discoveries I made in writing this book is that letting others set our agendas can be a slippery slope. It may offer economic incentives in the form of grants, and academic connections aplenty, but hopefully not at the expense of independent observation.

It is a truism, albeit a somewhat postmodern one, that the disciplinary approaches we use to study places (and peoples' relationships to them) have an impact on both the real world and the books written about it. When I began writing this book, I sought the names of reviewers in three major fields (geography, history, and anthropology) who could help my publisher determine how this book met its goals. As I made several phone calls and sent out emails, I had little trouble identifying anthropologists and historians at universities in the United States who were familiar with the Atacama. However, I had a surprise in store when I contacted fellow geographers to inquire about those studying the desert today. Although they were excited about my project, most had to think deeply before coming up with names. To my complete surprise, only a handful of American geographers seemed to be studying the Atacama with the character of place in mind. Among the exceptions is the previously mentioned geographer Manuel Prieto, who now studies this desert as part of a department that weds museum studies and anthropology. My own career had a roughly similar trajectory, for I found positions in varied allied disciplines to sustain my limitless curiosity, including museum-related directorships as a public historian, and teaching stints in varied university departments. I find it relevant and reassuring that Prieto has found a niche in not only academe, but in the very desert that he studies. I did much the same when I directed environmental planning programs and a mining museum in Arizona, my premise being that there is no better place to be than among the people—and in the places—one feels most passionate about. Then again, as Isaiah Bowman noted of the desert outposts of San Pedro de Atacama (Chile) and Salta (Argentina) a century ago, these out-of-the-way places also attract revolutionaries seeking refuge from oppressive national politics.[26] Those revolutionaries no doubt found time and space to reflect, as one can in desert regions. The type of position held by Prieto is ideally suited for reflection and action—much like that held by his Center's namesake, Gustavo Le Paige (1903–1980), the Belgian-born Jesuit who shepherded the Atacama Desert's archaeology while simultaneously shepherding souls as a padre in the desert outpost of San Pedro de Atacama.

As a person who has always worked on the margins of my core discipline, I feel delighted and vindicated that academe has become more welcoming toward ways of thinking that challenge disciplinary boundaries. As a scholar who prides himself on being "undisciplined"—as I recently half-jokingly characterized my mindset to one of my Willamette University colleagues—I have come to see disciplines as administrative conveniences. Rather than despairing for the discipline of geography, which used to lament losing prestige in academe, I welcome the realization that understanding place is more important than any particular discipline (even geography) that claims to do so. Over a nearly fifty-year career, I not only witnessed the discipline of geography diffusing into other fields—sometimes openly and sometimes covertly—but personally participated in it. At the time, this infiltration felt delightfully subversive, but it was part of a larger trend. Through such transference, a geographer such as Karl Offen (PhD, University of Texas) now teaches in the Adam Joseph Lewis Center for Environmental Studies at Oberlin College in Ohio. In 1972, with the ink barely dry on my University of Oregon PhD diploma, I left the University of Minnesota Geography Department to teach in the Environmental Studies program at Antioch College in Ohio. In retrospect, this interdisciplinary move resulted in new opportunities and expanded horizons, but much the same can be said about other interdisciplinary moves, for example, the flourishing "place studies" programs at colleges and universities. These developments have helped resolve the issue that used to plague me back in the early 1970s. At that time, as noted in the Preface, when I would tell fellow geographers about what I was studying, many would respond that it was interesting but asked whether it was *really* geography. I now understand that it was *place*—and especially *place in time*—that has always interested me, along with many others. Unsurprisingly, the colleagues I have come to respect most over the years are in varied disciplines; some in geography but many others are in diverse fields such as anthropology, literature, history, folklore, film studies, even philosophy. On this frontier, happily, there is no shortage of scholars who can help quench one's thirst for knowledge about place, including the meaning of places to the people who experience them as residents and visitors.

Nevertheless, as I wrote this book, I still found it surprising that relatively few professional geographers have studied the Atacama Desert since the 1960s. Their scarcity seems somewhat ironic given the American Geographical Society's a long-standing interest in this region. After all, AGS published Bowman's classic *Desert Trails of Atacama* in 1924, and even published William Rudolph's follow-up study, *Vanishing Trails of Atacama* in 1963 as a tribute to Bowman's original work. However, as hinted at above, several things have changed since the early- to mid-twentieth century, when geographers were luminaries and helped inform other closely related disciplines, including anthropology. In retrospect, geographer Khairul Rasheed's 1960 dissertation, "Man and the Desert: Northern Chile," was written at a pivotal time. Even back then, fragmentation was well underway; although Rasheed was mentored by geographers and others, his dissertation was officially in the field of political science, not geography. As I continued to write this book and recognize that I was unusual as an American geographer studying the Atacama Desert broadly, I took some solace in realizing that geography has *always* been an interdisciplinary endeavor. Some old-school geographers may stubbornly claim to be the real geographers, but this type of inquiry is boundless and, like the polymath Alexander von Humboldt himself that each discipline claims, really belongs to us all.

Although it is obvious that Isaiah Bowman's and Carl Sauer's humanistic style of geographic interpretation is increasingly less popular today than in the twentieth century, it still has its merits. Writing this book about the changing character of Chile's north reminded me of a favorite impressionistic essay by Sauer, which is eloquently entitled "The Personality of Mexico." In this pithy essay, Sauer claimed that the character of Mexico's desert north, as personified by the frontier states of Sonora and Chihuahua, was distinctly different from that of the south. According to Sauer, there was a more adventurous spirit in the northern desert frontier than in the more "colonial" states to the south. In the former, risk-taking and adoption of new ideas was the rule, whereas staid tradition ruled the south of Mexico.[27] Like the Norteños he described, Sauer himself had taken a risk in writing it. One of the highlights of my graduate education at the University of Oregon was meeting Sauer, who visited our campus, gave a few spellbinding lectures, and went with us on several field excursions. At that time, the University of Oregon's geography department was jokingly considered to be "UC Berkeley north," and its dozen or so professors emphasized cultural and historical geography along with many others. The program included study of the physical environment as well, though we nicknamed it humanistic geography. In January 2016, nearly fifty years later, I consulted that same department's promotional material, now on its website, which showcases the activities of "16 full-time faculty whose expertise tackles issues of environmental change, globalization, and geo-spatial technologies through teaching innovation and research." In this regard, Oregon is much like other universities, whose research agendas are closely linked to major grant funding as well as venerable humanistic traditions of inquiry.

In retrospect, Carl Sauer's essay on the personality of Mexico, and for that matter Isaiah Bowman's *Desert Trails of Atacama*, represent a time when geographers' abilities at generalization were not only well honed but highly respected, both within and outside the discipline. I thought such power-ful geographically themed prose was pretty much a thing of the past, edged aside by a heightened caution that comes with the compartmentalization of the social sciences and a pressing need to solve environmental problems. However, in 2015, when Professor José Antonio González Pizarro provided me with a copy of the stunning anthology *Región de Antofagasta: Pasado, pre-sente, y futuro* (The Antofagasta Region: Past, Present, and Future), I learned otherwise. When reading González's descriptive essay on "La formación de la nortinidad" (roughly, "The Creation of Northern Identity"), I realized that Sauer's and Bowman's kind of geography was alive and well—in Latin America. González calls upon the earlier observations of literary geniuses such as Mario Bahamonde (1910–1979), who wrote passionately and author-itatively about the people of the north and their environment. Channeling the works of these writers, and fusing them with his own astute observations, González remarks that the term "northerner" is shorthand for "Atacaman." In his essay brimming with poetic prose, González begins by deftly iden-tifying a number of traits that comprise such identity; the most salient to me is the constant tension between the region's tumultuous history and the seemingly eternal and omnipresent desert landscape and its characteristics of solitude, silence, and ruggedness. In the Atacama, as in northern Mexico, the theme of the desert as exploitable opportunity also prevails, shaping the character of the inhabitants.

According to González, city people, especially in Santiago, may tend to be unaware of or forget such ways; however, Chile's north—with its com-bination of elements such as land, salt, water, oases, and desert—encloses and sometimes even disguises a formidable history of human development. González astutely notes that the northerner embraces all of the varied expe-riences with the land of the desert, from the margins of the Andes to the sea-shore of the Pacific. One of his quotes excerpted from Bahamonde's essay on Pampinos y Salitreros is especially noteworthy: "The northerner, as a regional type of person, exists and has always existed, although to the rest of the country the concept is not very clearly differentiated." I shall paraphrase the rest of this passage about the character of this truly regional inhabitant, who is neither the son of the Diaguita, nor Atacameño, nor Quechua, nor Aymara, all of whom travelled through these mountains in the painful nostalgia of ancient times; rather, the northerner is the son of adventure and not a pris-oner of tradition. Nevertheless, this northerner is a legitimate son from the remote ancestors of earliest Spaniards to the most recent transients. Perhaps the most profound point that González makes in this essay is that one of the most enduring character traits in the Atacama Desert is not memory, but forgetfulness.[28]

As a modern-day chronicler of the Atacama, González is both inspirational and insightful. Using the past as a palette, he paints a picture that captures the character of both the people and the place. This is a kind of literary and artistic alchemy that I hope American geographers can rediscover in the writings of their earlier colleagues. Like the observations of American geographers Isaiah Bowman, Carl Sauer, and Homer Aschmann, these characterizations by Professor González meld two types of sources: firsthand field experience and a deep respect for literature. The latter should not only include the early written accounts, but also modern and even postmodern literature. I was not surprised when Professor González also gave me another book he had recently written, a beautifully illustrated biography of Andrés Sabella, one of the early- to mid-twentieth century literary and artistic geniuses of the Atacama Desert. González's latest book (2016) is a definitive study of the Atacama Desert's nitrate industry from 1880 to 1967 that emphasizes work, technology, and everyday life—and which appropriately enough begins with a chapter on the region's poetry.[29] Whenever I read González's poetic prose, I lament how much American geography might lose in its transition from ecumenical humanities-based scholarship to specialization and its obsession with problem-solving. Please do not misunderstand my intent here. In the United States as in Latin America, there is a place for technical and policy-driven research and publication. As already shown, I have personally spent about half of my career working on such issues, including environmental protection of arid and semiarid lands in the American Southwest as an environmental and community development planning program director (1979–1983). However, I believe that we need to strike a balance. I am concerned that, in the United States at least, the incorporation of poetic writing into geographical interpretation has become every bit as endangered as the endangered species we seek to protect.

Writing this book has made realize something about myself that I had long suspected but could never quite put my finger on. Throughout much of my professional life, there has always been a tug-of-war raging between science and artistic expression. It began in 1963, when I was asked by Foothill College to declare a major, and I declared two: geology and art. Never for a moment did I think these were separate, and I soon discovered in my art classes that Leonardo da Vinci used his artistic talent to represent the positions of fossil-rich strata that so fascinated him—and proved, to his satisfaction, that portions of land now high and dry were once underwater.

Yet, my discipline of geography commonly separated these two impulses—careful observation and creative expression—as if they were oil and water. Stubbornly, throughout my professional career, I have tried to use both to decipher and celebrate landscapes of all kinds. Moreover, I will use anything at my disposal—sketching and poetry on the one hand and scientific tools such as classification systems on the other—to capture their essence. There is a passion in the way I go about this, and that enthusiasm has earned me friends and detractors.

Back in the 1960s when I first became fascinated by the Atacama Desert and learned about its geography from the likes of Homer Aschmann, there was still a place in the discipline of geography for such passion about the character of place. It was, after all, Aschmann's mentor Sauer "who encouraged those interested in landscapes to take the time to 'sit on vantage points and stop at question marks.'" Like many of Aschmann's former students, including geographer Martin "Mike" Pasqualetti, I also remember Aschmann as a man who could, "with little prompting, recite at length from a Gilbert and Sullivan operetta, quote obscure Spanish or German poetry, or tell earthy limericks late into the night at a bar or at his home."[30] Today, such earthiness might land a professor in front of a disciplinary tribunal, but Aschmann was from a vanishing time. Moreover, I remember him as a rather conservative intellectual, unafraid to challenge the more bombastic claims by the 1960s counterculture, albeit at a time when professors' political viewpoints ran the gamut from far left to far right. In these matters, Aschmann was more pragmatic than dogmatic; he had seen the way the private sector functioned in Latin America and found it more enlightening than government bureaucracy. As I would learn, however, politics goes beyond the partisan and can enter the realm of interpersonal relationships.

When I taught at the University of Minnesota in the early 1970s, a colleague in the Geography Department and I were discussing Aschmann's scholarship. To my surprise, this colleague contemptuously observed that "you can find a Homer Aschmann working at every gas station in Indiana," an elitist and nasty jab at Homer's earthiness and ability to mingle with locals. My response—that I wished more geographers like Homer Aschmann worked in geography departments, as that would raise those departments' collective IQs—was just one in a series of strongly stated opinions, and fiercely independent interdisciplinary observations, that earned me my walking papers.

The truth is, it took someone like Aschmann to understand the ways and values of common people who experience and shape places such as the

Atacama Desert. Aschmann respected the people he studied regardless of whether he personally agreed with their attitudes and beliefs. He instilled, or tried to instill, that respect in his students and also helped them understand that the role of a teacher is to expand rather than shrink horizons. Not all of Aschmann's students "got" this, and he could be a tough grader. Still, for those of us who hoped to teach someday, he was an inspiration. Homer, as we called him, could wax poetic about the people and landscapes that one would encounter in the Atacama and Sonoran Deserts; at first a bit intimidated, and then mesmerized, we'd often be motivated to go there ourselves. It was Aschmann whose repertoire demonstrated firsthand the importance of varied types of literature—the historically framed classics, the romantically themed novels and poems, and the recent scholarly literature. Taking courses in Spanish while enrolled in Aschmann's classes was a guarantee that you'd soon be dreaming in Spanish—not just any dreams, but dreams about places and the peoples who inhabit them. His fluency in German was also an inspiration, so much so that it became my third language in graduate school. Fifty years later, that skill is still helping me translate German reports on and maps of the Atacama Desert.

As noted by Professor José González of Antofagasta, deserts like the Atacama are above all places where the line between memory and forgetfulness is blurred. The literature hints at this, for as much or more has been forgotten as can ever be remembered about this remarkable place. Delightfully, others will come along to discover, and rediscover, that which was lost as well as that which is new. Readers should keep in mind that the time of forgetting can be remarkably short. In a recent message to me, historical geographer Daniel Arreola expressed his concern that a researcher had made a startling discovery about how frequently and passionately Texans use their state's map in advertising and promotion. Arreola thought I should write to the researcher and let him know that I had published a book about exactly that—*The Shape of Texas: Maps as Metaphors*—in 1995. In selecting that subtitle, I recalled how effectively geographer Roger Downs had discussed the connection between "Maps and Metaphors" in his seminal 1981 *Professional Geographer* article. In *The Shape of Texas*, I expanded on that theme to show how maps themselves become part of a deeply embedded folkloric and commercial narrative that evolves through time. That experience helped ground my understanding, and I applied some of the lessons learned to this book on the Atacama Desert, which further confirms that maps help people articulate their visions of a region. Scholars and others are now coming to realize that maps are not only indivisible from language but also part of a complex system of words and symbols that signify place. In a sense, maps not only work with metaphors but may *become* metaphors themselves—another reminder that the word and the image are inseparable. Appreciating that my colleague Arreola was looking out for me, and that he also hoped that my book would enlighten the researcher, I began my response. However, before getting very far, I remembered that about twenty-five years had passed since I wrote that book. Instinctively, I hit the "delete" key, recalling that no book is ever the last word, and that no subject is ever fully researched. After all, new discoveries are bound to await those who tackle the same subject a generation later.

With this connection in mind between geography and narratives lost and remembered, and my firm belief that the future of geography and discovery lies in postmodern interdisciplinary thinking, I would like to close this book with a poem that just came to mind. I've entitled it "In a Word" (En Una Palabra) and use it here because it is as much about language as it is about place, and it documents my biggest surprise in writing this book.

IN A WORD

Venturing into the Atacama
I searched for secrets obscured
by the amnesia of time
and the vastness of space,
never suspecting that its magical name
would become as meaningful to me
as the mysterious desert itself.

Notes

Preface

1. Isaiah Bowman, *Desert Trails of Atacama* (New York: American Geographical Society, 1924), v, 5.
2. Richard Francaviglia, "The Atacama Desert: A Five Hundred Year Journey of Discovery," *Terrae Incognitae* 48, no. 2 (September 2016): 105–38.

Chapter 1

1. This quote is from the Gospels (John 1:1) but refers to the earlier word(s) of God that are found in the Hebrew Bible, beginning with Genesis.
2. This passage is from Isaiah Bowman's field notes and is hard to date with certainty, but likely May 13–17, 1907, according to Robert Jaeger, who located it in the Bowman collection at the American Geographical Society Library, University of Wisconsin at Milwaukee, Wisconsin, and shared it with me via an email sent July 29, 2015.
3. Katherine G. Morrissey, *Mental Territories: Mapping the Inland Empire* (Ithaca: Cornell University Press, 1997).
4. *Bradshaw's Railway Manual Shareholders' Guide and Directory, 1869* (New York: Augustus M. Kelley, Publishers, 1969), s.v. "Arica and Tacna."
5. Jan Kozák and Vladimir Cermák, *The Illustrated History of Natural Disasters* (New York: Springer, 2010), 169–70.
6. "Climates of the Earth" map, by Glenn T. Trewartha, *Goode's World Atlas* (Chicago: Rand McNally, 1964), 12–13.
7. Donald Binns and Harold A. Middleton, *The Taltal Railway: A Chilean Mineral Line* (Bristol, England: Trackside Publications, 2010), 14.
8. For a detailed discussion of such weathering, see Ronald I. Dorn, *Rock Coatings: Developments in Earth Surface Processes*, vol. 6 (Amsterdam: Elsevier, 1998).
9. Guillermo Chong Díaz, Aníbal Gajardo Cubillos, Adrian J. Hartley, and Teresa Moreno, "Industrial Minerals and Rocks," chap. 7 in *The Geology of Chile*, ed. Teresa Moreno and Wes Gibbons (London: The Geological Society, 2007), 201, 204.
10. Annibale Mottana, Rudolfo Crespi and Guiseppe Liborio, *Simon & Schuster's Guide to Rocks and Minerals* (New York: Simon & Schuster, 1978), 305.
11. Charles R. Stern, Hugo Moreno, Leopoldo López-Escobar, et al., "Chilean Volcanoes," chap. 5 in *The Geology of Chile* (London: Geological Society, 2007), 147–78, 149 quoted.
12. "Natural Vegetation" map, by A. W. Küchler, *Goode's World Atlas*, 13th ed. (Chicago: Rand McNally, 1964), 20–21.
13. John N. Trager, "The Huntington Botanical Gardens presents the 2016 offerings of the International Succulent Introductions," *Cactus and Succulent Journal* 88 (May–June 2016): 111–12.

14. The song "Can't Help Falling in Love," written by Gary P. Skardina, Mati Sharron, Luigi Criatore, and Hugo Peretti, and recorded by Elvis Presley, 1960.

15. The song "Unchained Melody," written by H. Zaret and A. North, and recorded by Al Hibler (ca. 1955) and The Righteous Brothers (ca. 1962).

16. John Aarons and Claudio Vita-Finzi, *The Useless Land: A Winter in the Atacama Desert* (London: Robert Hale Limited, 1960), 15–16.

17. William E. Rudolph, *Vanishing Trails of Atacama*, American Geographical Society Research Series, no. 24 (New York: American Geographical Society, 1963), 8.

18. Jason Urbanus, "World Roundup" news section on Chile, *Archaeology: A Publication of the Archaeological Institute of America* 70 (May/June 2017): 24.

19. C. Latorre, Patricio Moreno, Gabriel Vargas, Antonio Maldonado, Rodrigo Villa-Martínez, Juan J. Armesto, Carolina Villagrán, Mario Pino, Lautaro Núñez, and Martin Grosjean, "Late Quaternary Environments and Paleoclimate," chap. 12 in *The Geology of Chile*, 309–28, 324 quoted.

20. Claudo Latorre, Calogero M. Santoro, Paula Ugalde, Eugenia M. Gayo, Daniela Osorio, Carolina Salas-Egaña, Ricardo De Pol-Holz, Delephine Joly, and Jason A. Rech, "Late Pleistocene Human Occupation of the Hyperarid Core in the Atacama Desert, Northern Chile," *Quaternary Science Review* 77 (October 2013): 19–30.

21. Isaiah Bowman, *South America: A Geography Reader*, Lands and Peoples Series (Chicago: Rand McNally, 1915), 89–90.

22. William E. Rudolph, *Vanishing Trails of Atacama* (New York: American Geographical Society, 1963), 7.

23. John Aarons and Claudio Vita-Finzi, *The Useless Land: A Winter in the Atacama Desert* (London: Robert Hale Limited,1960), 129.

24. Isaiah Bowman, *Desert Trails of Atacama* (New York: American Geographical Society, 1924), 155–61.

Chapter 2

1. See also Michael Welland, *The Desert: Lands of Lost Borders* (London: Reaktion Books, 2015), 13–15.

2. Felix Fabri, *The Book of the Wanderings of Felix Fabri*, as quoted in Michael Welland, *The Desert: Lands of Lost Borders*, 13–15.

3. William Harmless, *Desert Christians: An Introduction to the Literature of Early Monasticism* (New York: Oxford University Press, 2004), 60, 64, 70.

4. Since 1967, I have distributed informal questionnaires asking varied groups of people (students, residents) which of many types of physical environments (e.g., grassland, coast, tundra, deserts, jungle, forest, plains, mountains, etc.) they most associate with religion. Without fail, deserts, followed closely by mountains, are most often selected.

5. Carol Delaney, *Columbus and the Quest for Jerusalem* (New York: Free Press, 2011).

6. Stewart A. Weaver, *Exploration: A Very Short Introduction* (New York: Oxford University Press, 2015), 1, 3–6.

7. See Paul Gallez, *La cola del dragón: America del Sur en los mapas antiguos, medievales y renacentistas* (Bahia Blanca, Argentina: Instítuto Patagónico, 1990).

8. John Hébert, "The Map that Named America: Library Acquires 1507 Waldseemüller Map of the World," *Library of Congress Information Bulletin* (Washington, D.C., September 2003), accessed June 21, 2014, loc.gov/loc/lcib/0309/maps.html.

9. John Hébert, "America," in *Mapping Latin America: A Cartographic Reader*, ed. Jordana Dym and Karl Offen (Chicago: University of Chicago Press, 2011), 30.

10. Karen Olsen Bruhns, *Ancient South America* (Cambridge: Cambridge University Press, 1994), 42, 73.

11. Edwin L. Goodman, *The Explorers of South America* (Norman: University of Oklahoma Press, 1972), 38.

12. Terrence N. D'Altroy, *Provincial Power in the Inka Empire* (Washington, D.C.: The Smithsonian Institution, 1992), 115–16.

13. Weaver, *Exploration: A Very Short Introduction*, 59.

14. See Baker H. Morrow, *The South America Expeditions, 1540–1545: Alvar Nuñez Cabeza de Vaca* (Albuquerque: University of New Mexico Press, 2011).

15. Cynthia Radding, *Wandering Peoples: Colonialism, Ethnic Spaces, and Ecological Frontiers in Northwestern Mexico, 1700–1850* (Durham, NC: Duke University Press, 1997), xvii. See also J. Jorge Klor de Alva, "The Postcolonization of the (Latin) American Experience: A Reconsideration of 'Colonialism,' 'Postcolonialism,' and 'Mestizaje'," in *After Colonialism: Imperial Histories and Postcolonial Displacements*, ed. Gyan Prakash (Princeton: Princeton University Press, 1995), 241–75.

16. Claudio Esteva-Fabregat, *Mestizaje in Ibero-America* (Tucson: University of Arizona Press, 1987/1995), 33, 93–94.

17. "Studying farming in the driest desert in the world: Hayashida examines farm fields abandoned 500 years ago," Karen Wentworth, March 26, 2015, accessed July 2, 2015, news.unm.edu/news/studying-farming-in-the-driest-desert-in-the-world.

18. "Site Preservation Grants Awarded to Projects in Greece and Chile," *Archaeology* 68, no. 5 (September/October 2015): 64–65.

19. "Historia Chilena. Mapa de los pueblos originarios," also containing a map of these groups, May 29, 2007, accessed July 7, 2014, http://www.educarchile.cl/ech/pro/app/detalle?ID=132542.

20. William C. Bennett, "The Atacameño," in Julian H. Steward, ed., *Handbook of South American Indians: Vol. 2, The Andean Civilizations—Smithsonian Institution Bureau of American Ethnology, Bulletin 143* (Washington, D.C.: United States Government Printing Office, 1947), 599–601.

21. Ibid., 603–5.

22. John Hyslop and Mario Rivera, "An Expedition on the Inca Road in the Atacama Desert," *Archaeology* 37, no. 6 (November/December 1984): 33–39.

23. Goodman, *The Explorers of South America*, 54.

24. Ibid., 53–55.

25. Claudio Esteva-Fabregat, *Mestizaje in Ibero-America*, 84–87.

26. Pedro de Valdivia, "Cartas de Pedro de Valdivia, que tratan del descubrimiento y conquista de Chile," accessed August 8, 2009, http://www.cervantesvirtual.com/servlet/SirveObras/12593842001258285209068/p00000.

27. "Origen del Concepto Atacama: Su origen como Provincia y Región," on the blog *Historia, Política y Educación: Guillermo Cortes Lutz*, September 5, 2006, accessed January 14, 2009, guillermocorteslutz.blogspot.com/2006/09/origen-del-concepto-atacama.html.

28. Rebecca A. Carte, *Capturing the Landscape of New Spain: Baltasar Obregón and the 1564 Ibarra Expedition* (Tucson: University of Arizona Press, 2015), 54, 60–61.

29. See Jordana Dym and Karl Offen, "Introduction to Part 1, The Colonial Period: Explorations and Empires," in *Mapping Latin America: A Cartographic Reader*, ed. Dym and Offen (Chicago: University of Chicago Press, 2011), 20.

30. Lautaro Núñez Atencio, *Vida y Cultura en el Oasis de San Pedro de Atacama* (Santiago: Editorial Universitaria, 1991), 90–91.

31. Anita Carrasco, "One World, Many Ethics: The Politics of Mining and Indigenous Peoples in Chile," (doctoral dissertation, University of Arizona, 2011), 160; see also José Luis Martínez, *Pueblos del Chañar y el Algarrobo: Los Atacamas en el Siglo XVII* (Santiago: DIBAM, 1998), 20.

32. Radding, *Wandering Peoples*, 179.

33. Khairul Bashar Sajjadur Rasheed, "Man and the Desert: Northern Chile," (PhD dissertation, Columbia University, 1970), 169–72.

34. Ibid., 183–92, 210.

35. Adriana E. Hoffmann J., *Cactaceas En La Flora Silvestre de Chile* (Santiago: Ediciones Fundacion Claudio Gay, 1989), 53.

36. Gabriel Martínez, "Los Dioses de los Cerros en los Andes," *Journal de la Sociéte des Américanistes*, vol. 69, 3–4.

37. Carrasco, "One World, Many Ethics," 127–28.

38. David Narrett, *Adventurism and Empire: The Struggle for Mastery in the Louisiana-Florida Borderlands, 1762–1803* (Chapel Hill: University of North Carolina Press, 2015), 4.

39. Glyndwr Williams, *The Great South Sea: English Voyages and Encounters, 1570–1750* (New Haven: Yale University Press, 1997), 17–20.

40. Tom Conley, "Film and Exploration," entry in *The Oxford Companion to World Exploration*, David Buisseret, ed. (New York: Oxford University Press, 2007), vol. 1, A–L, 307.

41. Mirela Altic, "Jesuits at Sea: José Quiroga and José Cardiel—Two Complementary Views of Patagonia (1745–1746)," *Terrae Incognitae* 49, no. 2 (September 2017), 149–73; see also Lawrence C. Wroth, "Alonso de Ovalle's Large Map of Chile, 1646," *Imago Mundi* 14 (1959), 90–95.

42. Rasheed, "Man and the Desert: Northern Chile," 7.

43. "Nitrogen: 1—The World in its Place," *Fortune* 1, no. 9 (1930), 55.

44. James Griffith, *Beliefs and Holy Places: A Spiritual Geography of the Pimería Alta* (Tucson: University of Arizona Press, 1992), 100–101.

45. Bowman, *Desert Trails of Atacama*, 288–89.

46. "Myths and Mysteries Surround Chile's Desert Drawings," accessed April 4, 2016, www.nbcnews.com/science/weird-science/myths-mysteries-surround-chiles-desert-drawings-n27651.

47. William Dampier, *A New Voyage Round the World*, Introduction by Sir Albert Gray and a new Introduction by Percy G. Adams (New York: Dover, 1967), xxiv.

48. Dampier, *A New Voyage Round the World*, 8.

49. Diana and Michael Preston, *A Pirate of Exquisite Mind: Explorer, Naturalist, and Buccaneer—The Life of William Dampier* (New York: Walker & Company, 2004), 79.

50. Dampier, *A New Voyage Round the World*, 176.

51. Ibid., 71.

52. Ibid., 72.

53. Ibid., 101–2.

54. Diana and Michael Preston, *A Pirate of Exquisite Mind*, 74.

55. James Kelley, "Dampier, William," entry in *The Oxford Companion to World Exploration*, David Buisseret, ed. (Oxford and New York: Oxford University Press, 2007), vol. 1, A–L, 222–25.

Chapter 3

1. Edward A. Goodman, *The Explorers of South America* (Norman: University of Oklahoma Press, 1972), 352.

2. For a detailed examination of mapmaking in northern Europe at this time, see Mary Sponberg Pedley, *The Commerce of Cartography: The Patronage and Production of Maps in Eighteenth-Century France and England* (Chicago: University of Chicago Press, 2005).

3. Louis Feuillée, *Journal des observations physiques, mathématiques et botaniques: faites par l'ordre du Roi sur cotes orientales de l'Amérique Méridionale, & aux Indes Occidentales. Et dans un autre voyage fait par le méme ordre a la Nouvelle Espagne, & aux Isles de l'Amérique* (Paris: Pierre Giffart, 1714–1715).

4. Amedee Francois Frézier, *A Voyage to the South-Sea, and along the Coasts of Chili and Peru, in the Years 1712, 1713, and 1714* (London: Jonah Bowyer, 1735), 142.

5. Ibid., 140–41.

6. Ibid., 141–42.

7. Ibid., 143–45.

8. Ibid., 146–47.

9. Ibid., 148–51.

10. See Nicolas A. Robins, *Mercury, Mining, and Empire: The Human and Ecological Cost of Colonial Silver Mining in the Andes* (Bloomington: Indiana University Press, 2011).

11. Theodor de Bry, *Collection des Grands and Petits Voyages* (London: Molins Ltd, 1921), original illustration dated 1601, as shown in Manuel Fernández Canque, *Arica 1868: Un Tsunami y un Terremoto* (Santiago: Ediciones de la Dirección de Bibliotecas, Archivos y Museos, 2007), 45.

12. Williams, *The Great South Sea*, 10.

13. Goodman, *The Explorers of South America*, 192.

14. Covens et Mortier, *[Map of] Paraguay, Chili, Detroit de Magellan, 1742*, accessed June 11, 2014, www.davidrumsey.com/luna/servlet/detail/ Rumsey-8-1-31260-1150289.Carte-du-Paraguay,-du-chili,-du-Det.

15. *Hedendaagsche Historie of Tegenwoordige Staat van Amerika, Tweede Deel. Behelzende Het vervolg der befchryvinge van de Beziittingen der Kroon van Spanje in dat Werelsdeel; te weeten: Die van het Ryk van PERU en deszelfs Hoofstadt Lima; van TUCUMAN, PARAGUAY en CHILI; van 't Land van MAGELLAAN en Eilanden in de groote ZUIDZEE* (Amsterdam: Isaak Tirion, 1767), 98–99, 127.

16. See Thomas Smith, "Cruz Cano's Map of South America, Madrid 1775: Its Creation, Adversities, and Rehabilitation," *Imago Mundi* 20 (1966): 49–78; also Walter Ristow, "The Juan de la Cruz map of South America, 1775," reprinted from Festschrift, Clarence F. Jones, ed. Merle C. Prunty, Jr., *Northwestern University Studies in Geography, No. 6* (1962).

17. Thomas Jefferys, "Map of South America, northern and southern sections," (London: 1776), accessed June 11, 2014, www.davidrumsey. com/luna/servlet/detail/RUMSEY-8-1-1926-120032:South-America---Southern-Section-?qvq=W4s:/what/Atlas+Map/ when/1776/;lc:Rumsey-1&mi=33&trs=43.

18. David J. Weber, *Bárbaros: Spaniards and their Savages in the Age of Enlightenment* (New Haven: Yale University Press, 2005), 6–7, 41–42.

19. Iris H. W. Engstrand, *Spanish Scientists in the New World: The Eighteenth-Century Expeditions* (Seattle: University of Washington Press, 1981), 48–50.

20. Jorge Ortiz Soleto, "El Piloto Andrés Baleato y la cartografía peruana," accessed September 19, 2017, www.oannes.org.pe/ upload/201609221022181630002257.pdf.

21. See *Essay on the Geography of Plants: Alexander von Humboldt and Aimé Bonpland*, edited with an Introduction by Stephen T. Jackson and Sylvie Romanowski (Chicago: University of Chicago Press, 2009), 1–46.

22. "Deserts of Coastal Peru and Chile," accessed September 19, 2017, www. sacha.org/envir/deserts/intro.htm.

23. Goodman, *The Explorers of South America*, 301–3.

24. Alcide d'Orbigny, *La Relation du Voyage dans L'Amerique Méridionale… 1826–1833*, Vol. 4 (Paris: Chez. V. Levirault, 1859), 151–53.

25. Patience A. Schell, *The Sociable Sciences: Darwin and His Contemporaries in Chile* (London: Palgrave Macmillan US, 2013), 75–78.

26. David Rains Wallace, *Chuckwalla Land: The Riddle of California's Desert* (Berkeley: University of California Press, 2011), 22.

27. Victor Wolfgang von Hagen, *South America Called Them: Explorations of the Great Naturalists* (New York: Alfred Knopf, 1945), 214–15.

28. J. Ehleringer, H. A. Mooney, S. L. Gulmon, and P. Rundel, "Orientation and its consequences for Copiapoa (Cactaceae) in the Atacama Desert," *Oecologia* 46 (1980): 63–67.

29. Charles Darwin, *The Voyage of the Beagle* (1909 and 1937), 334–39, 342, 351; and Charles Darwin, *Charles Darwin's Beagle Diary* (1988), 323, 334–39.

30. Darwin's handwritten sketch map, accessed September 19, 2017,

darwin-online.org.uk/content/frameset?pageseq=1&itemID=CUL-DAR44.28&viewtype=image, as part of "The Complete Works of Darwin Online" (http://darwin-online.org.uk/).

31. Luz María Méndez Beltrán, *La Exportación Minera en Chile 1800–1840: Un estudio de la historia económica y social en la transición de la Colonia a la República* (Santiago: Editorial Unversitaria, 2004), 90.

32. Historian of transportation Steven Peck briefly discusses Darwin in light of mining and subsequent railroad developments in his unpublished "The Copiapó Locomotive: Historical and Technical Assessment" (master's thesis, Colorado State University, 2002), 57.

33. Isaiah Bowman, *Desert Trails of Atacama* (New York: American Geographical Society, 1924), 180–81.

34. Charles Darwin, *Journal of researches into the natural history and geology of the countries visited during the voyage of H.M.S. Beagle round the world*, 2nd ed. (London, 1860), 350.

35. Bowman, *Desert Trails of Atacama*, 177.

36. Personal email with William Culver, March 20, 2016.

37. Charles Darwin, as qtd. in William Howard Russell, *A Visit to Chile and the Nitrate Fields of Tarapacá* (London: J.S. Virtue & Co., 1890), 149.

38. Darwin, *The Voyage of the Beagle*, 365–68. For a more detailed analysis of this passage, see William Howard Russell, *A Visit to Chile*, 164–65.

39. Darwin, *Charles Darwin's Beagle Diary*, 345–47.

40. Charles Darwin, "Iquique" (1835), in John van Wyhe, ed., *The Complete Works of Charles Darwin Online*, accessed September 15, 2015, http://darwin-online.org.uk.

41. Ignacio Domeyko, *Mis Viajes: Memorias de un Exiliado*, 2 vols. (Santiago, 1978), Vol. 1, 403; also qtd. in José Joaquín Vallejo, Simon Collier, ed., *Sketches of Life in Chile, 1841–1851* (New York: Oxford University Press, 2002), xxiii.

42. Vallejo, *Sketches of Life in Chile*, 3, 8.

43. Ibid., 12–13, 27–29, 36–37.

44. Mark Twain, *Roughing It* (Hartford: American Publishing Company, 1872).

45. J. M. Gilliss, *The U.S. Naval Astronomical Expedition to the southern hemisphere, during the years 1849–'50–'51–'52* (Washington, D.C.: A.O.P. Nicholson, Printer, 1855), 450.

46. Ibid., 247.

47. Ibid., 245.

48. Ibid., 249.

49. Ibid., 250.

50. See Andrew C. A. Jampoler, *Sailors in the Holy Land: The 1848 American Expedition to the Dead Sea and the Search for Sodom and Gomorrah* (Annapolis: Naval Institute Press, 2005).

51. Gilliss, *The U.S. Naval Astronomical Expedition*, 244, 257.

52. Ibid., 246.

53. Oswald Hardey Evans, "Notes on the Raised Beaches of Taltal (Northern Chile)," *Quarterly Journal of the Geological Society* 63 (January 1907): 64–68.

54. Rudolph Philippi, *Reise durch die Wueste Atacama: auf Befehl der chilenischen Regierung im Sommer 1853–54* (Halle: Eduard Anton, 1860); and *Viage al Desierto de Atacama: hecho de orden del gobierno de Chile en el verano 1853–54* (Halle: Eduard Anton, 1860).

55. Hans-Rudolf Bork, Jürgen Bähr, Helga Bork, Michael Brombacher, Imre Josef Demhardt, Anja Habeck, Andreas Mieth, and Bernd Tschochner, "Die entwicklung der Oase San Pedro de Atacama, Chile," *Petermanns Geographische Mitteilungen* 46, no. 5 (2002): 56–63.

56. Anthony Rice, *Voyages of Discovery: Three Centuries of Natural History Exploration* (London and New York: The Natural History Museum, London, and Clarkson Potter Publishers, 1999), 90–120.

57. Philippi, *Viage al Desierto de Atacama*, 101–2.

58. Ibid., 5–6.

59. L. Fletcher, "On the meteorites, which have been found in the desert of Atacama and its neighborhood," *The Mineralogical Magazine* 8, no. 40 (1889): 243–49.

60. See "El Estudio de las Ciencias Naturales—Escrito en facsimile de don Rodulfo Amando Philippi," in Adriana E. Hoffmann J., *Cactaceas en la Flora Silvestre de Chile* (Santiago: Ediciones Fundacion Claudio Gay, 1989), 8.

61. Juanita Brooks, ed., *Journal of the Southern Indian Mission: The Diary of Thomas D. Brown* (Logan: Utah State University Press, 1972), 46–47.

62. For more on this subject, see Richard Francaviglia, *The Mapmakers of New Zion: A Cartographic History of Mormonism* (Salt Lake City: University of Utah Press, 2015), 162–68.

63. "Mormon's Abridgement Part XIV—We Did Find Upon the Land of Promise—Part II," November 3, 2014, accessed August 14, 2015, nephicode. blogspot.com/2014/11/mormons-abridgement-part-xiv-we-did.html.

64. See Brandon Plewe, ed., *Mapping Mormonism: An Atlas of Latter-day Saint History* (Provo: Brigham Young University Press, 2012), 172–73.

65. Herman Melville, *Moby-Dick; or, The Whale* (New York: Harper & Brothers, 1851), 214.

66. Herman Melville, "The Encantadas, or Enchanted Isles," *Putnam's Monthly Magazine*, March–May, 1854, and later published in *The Piazza Tales* (1856), 6. See also Hershel Parker, *Herman Melville, A Biography: Vol. 1, 1819–1851* (Baltimore: Johns Hopkins University Press, 1996), 200–202, 280.

67. Mrs. S. C. Hall, "A Railway Trip in Chili," *The St. James Magazine* (1862): 213–16.

68. Hoffmann, *Cactaceas en la Flora Silvestre de Chile*, 54.

69. Florencia Señoret Espinos and Juan Pablo Acosta Ramos, *Guía de Campo Cactáceas Nativas de Chile* (Concepción, Chile: CORMA—Corporacion Chilena de la Madura, 2013), 38–39, 102–3, 194–95, 220.

70. William Bollaert, *Antiquarian, ethnological, and other researches in New Granada, Equador, Peru and Chile* (London: Turner & Co., 1860), 117.

71. Wallace, *Chuckwalla Land*, 22–23.

72. Michael Welland, *The Desert* (London: Reaktion Press, 2014), 256–57.

73. Darwin, *Voyage of the Beagle*, 353.

74. Luis Aguirre, Francisco Hervé, and Mónica del Campo, "An Orbicular Tonalite from Caldera, Chile," *Journal of the Faculty of Science, Hokkaido University, Series 4, Geology and Mineralogy* (September 1976): 231—59.

75. Hall, "A Railway Trip in Chili," 214.

76. Méndez Beltrán, *La Exportación Minera en Chile*, 48, 147.

Chapter 4

1. Frederick H. Pough, *A Field Guide to Rocks and Minerals* (Cambridge, Mass: The Riverside Press, 1955), 155.

2. Rudolph Philippi, *Viage al Desierto*, 5, 107.

3. Simon Winchester, *The Map That Changed the World: William Smith and the Birth of Modern Geology* (New York: HarperCollins, 2001).

4. Carl A. Francis, "The Szenics Collection of Contemporary Chilean Minerals: A New Addition to the Harvard Mineralogical Museum," *Rocks & Minerals* 78 (January/February 2003): 46–51.

5. Louis Simonin, *La Vie Souterraine ou, Les Mines et les Mineurs* (Paris: Hachette, 1867).

6. Isaiah Bowman, *South America: A Reader* (Chicago: Rand McNally, 1915), 73–74.

7. See Anuario Administrativo de 1869, por Dr. Juan Francisco Velarde (La Paz de Ayacucho: Abril 1870), 335, and Refutación al Manifesto del Ministro de Relaciones Exteriores de Chile Sobre la Guerra con Bolivia (Lima: Imprenta Nacional, 1879), 69.

8. Aimé Pissis, *Geografía* Física de Chile (Paris: Instituto Geográfico de Paris, 1875). For references to the Atacama region, see 31, 150–51, 162, 179, 199, 206, 275–77.

9. Pedro José Amado Pissis, *Mapa mineralógico del desierto de Atacama, por A. Pissis* (Santiago: Lit. P. Cadot, Huérfanos, 1877), accessed January 30, 2017, www.memoriachilena.cl/602/w3-article-86699.html.

10. For a contextualization of Petermann's geographic information enterprise, see Imre Joseph Demhardt, *Der Erde ein Gesicht geben: Petermanns Geographische Mitteilungen und die Anfänge der modernen Geographie in Deutschland*, Katalog zur Ausstellung der Universitäts-und Forschungsbibliothek Erfurt/Gotha in Spiegelsaal auf Schloss Friedenstein in Gotha 23. Juni bis 9. Oktober 2005 (Gotha, 2006).

11. Jan Smits, *Petermann's Maps: Cartobibliography of the Maps in Petermanns Geographische Mitteilungen, 1855–1945* (Utrecht: HAS & DE GRAF, 2004), 167.

12. Karl Offen, "Minerals and War," in *Mapping Latin America: A Cartographic Reader*, Jordana Dym and Karl Offen, eds. (Chicago: University of Chicago Press, 2011), 139, 142.

13. The *Atacama* was an iron-hulled, 290-foot-long propeller-driven steamship built by John Elder & Company, Glasgow in 1870; see Ian Collard, *Pacific Steam Navigation Company: Fleet List & History* (Amberly: n.p., n.d.), 95; author's copy purchased online November 2016.

14. Jan Smits, *Petermann's Maps*, 177.

15. William F. Sater, *Andean Tragedy: Fighting the War of the Pacific, 1879–1884* (Lincoln: University of Nebraska Press, 2007), 177.

16. Bruce W. Farcau, *The Ten Cents War: Chile, Peru, and Bolivia in the War of the Pacific, 1879–1884* (Westport, CT: Praeger Publishers, 2000), 45; for more information on the War of the Pacific, see also William F. Sater, *Andean Tragedy*; and Charles Victor Grosnier de Varigny, *La guerra del Pacífico* (Santiago de Chile: Imprenta Cervantes, 1922).

17. Daniel Arthur McCray, "Eternal Ramifications of the War of the Pacific," (master's thesis, University of Florida, 2005).

18. See "Boundary tensions between Peru & Chile continue," Boundary News: Durham University, October 22, 2015, accessed March 20, 2016, https://www.dur.ac.uk/ibru/news/boundary_news/?itemno=25974.

19. See César Vásquez Bazan, "Robo chileno del desierto de Atacama," accessed August 24, 2014, https://cavb.blogspot.com/2013/05/desierto-de-atacama-no-pertenece-chile.html.

20. See William Edmundson, *The Nitrate King: A Biography of "Colonel" John Thomas North* (New York: St. Martin's Press, 2011).

21. William Howard Russell, *A Visit to Chile and the Nitrate Fields of Tarapacá, Etc.* (London: J. S. Virtue & Co., 1890), 300, 301–2, 308–9.

22. Russell, *A Visit to Chile*, 303.

23. *Serena's* sister ship was the *Mendoza*. Both were 320 feet long and had two decks, according to Ian Collard, *Pacific Steam Navigation Company*, 108–9.

24. Russell, *A Visit to Chile*, 138–39, 144.

25. Ibid., 160–61, 164–65.

26. Ibid., 160–61, 166–67.

27. Richard Francaviglia, *Go East, Young Man: Imagining the American West as the Orient* (Logan: Utah State University Press, 2011).

28. For a detailed survey of this railway, its equipment, and operations, see Donald Binns, *The Nitrate Railways Company Limited* (Skipton, England: Trackside Publications, n.d., ca. 2000).

29. Russell, *A Visit to Chile*, 175–76, 367.

30. Ibid., 174, 179.

31. Ibid., 191–92.

32. L. Fletcher, "On the meteorites, which have been found in the desert of Atacama and its neighborhood," *The Mineralogical Magazine* 8, no. 40 (October 1889): 243, 264.

33. "Francisco J. San Román," accessed January 19, 2015, https://es.wikipedia.org/wiki/Francisco_J_San_Román.

34. Francisco J. San Román, *Reseña industrial e histórica de la minería i metalurjia de Chile* (Santiago: Imprenta Nacional, 1894), 39–40.

35. Francisco J. San Román, *Desierto i cordilleras de Atacama por Francisco J. San Román, Vol. 1, Tomo Primero*, Itinerario de las exploraciones y Tomo II, Mision a Los Estados Unidos (Santiago de Chile, 1896), 185.

36. Francisco J. San Román, *Desierto i cordilleras de Atacama, Vol. 2, Tomo Tercero: Hidrologia* (Santiago de Chile, 1902), 21.

37. William W. Culver and Cornel J. Reinhart, "Capitalist Dreams: Chile's Response to Nineteenth-Century World Copper Competition," *Comparative Studies in Society and History* 31, no. 4 (October 1989): 738, 739, 742–43.

38. Isaiah Bowman, *Field Notebook, Antofagasta to La Paz, Titicaca and Desaguadero* (1913), American Geographical Society collection, Special Collections, Sheridan Libraries, Johns Hopkins University, 5, 7.

39. See Donald Binns and Harold A. Middleton, *The Taltal Railway: A Chilean Mineral Line* (Bristol, England: Trackside Publications, 2010).

40. Bowman, *Desert Trails of Atacama*, 74.

41. Ibid., 178–79.

42. Benjamin L. Miller and Joseph T. Singewald, *The Mineral Deposits of South America* (New York: McGraw Hill Book Company, 1919), 248, 254.

43. Ibid., 238.

44. Ibid., 244, 246, 285.

45. Paul Maar, "Technology, Labor, and the Collapse of Chile's Nitrate Industry," *Middle States Geographer* 46 (2013): 19–26; see also Jaime Wisniak and Ingrid Garcés, "The Rise and Fall of the salitre (sodium nitrate) industry," *Indian Journal of Chemical Technology* 8 (September 2001): 427–38.

46. Gabriela Mistral, *Desolation: A Bilingual Edition of Desolación, 1923*, translation, Introduction, Afterword by Michael P. Predmore and Liliana Baltra (Pittsburgh, PA: Latin American Literary Review Press, 2013), 121, 131, 151, 175, 349–51.

47. Catherine Phillips, "The Lost Cactus Garden of 'Quien Sabe'," *Cactus and Succulent Journal* 87, no. 2 (March/April 2015): 48, 58.

48. Catherine Phillips, "Plants known to Ysabel Wright and grown in her garden at Quien Sabe," an unpublished list provided to the author in an informative email, October 4, 2015, identifying Fric as a possible collector of these plants.

49. Harry F. Keller, "Notes on Some Chilean Copper Minerals," *Proceedings of the American Philosophical Society* 47 (January–April 1908): 79–85.

50. Carl A. Francis, "The Szenics Collection of Contemporary Chilean Minerals," 47.

51. Peter C. Keller, ed., "Collector's diary: Iquique, Copiapo, and Chañarcillo: Mark Chance Bandy, 1935," *The Mineralogical Record* 15, no. 2 (March–April 1984): 67, 68, 70.

52. Personal email communication between the author and mineralogist Wendell Wilson of Tucson, Arizona, February 26, 2016.

53. Although proustite (a silver sulfarsenide) is well known for its ruby-red color, the mineral pyrargyrite (a silver-antimony sulfosalt) is also called by the vernacular name "ruby silver," though its color is often a darker red than proustite.

54. Sarah Lea, "Joseph Cornell: Wanderlust," in *Joseph Cornell: Wanderlust* (London: Royal Academy of Arts, 2015), 31.

55. Lynda Roscoe Hartigan, "Musings on Joseph Cornell's Alchemy of the Mind," in *Joseph Cornell: Wanderlust* (London: Royal Academy of Arts, 2015), 66–67.

56. "Santa Maria School massacre," accessed February 3, 2016, https://en.wikipedia.org/wiki/Santa_Maria_School_massacre.

57. Michael Monteón, "The British in the Atacama Desert: The Cultural Bases of Economic Imperialism," *The Journal of Economic History* 35, no. 1 (March 1975): 133.

58. Miller and Singewald, *The Mineral Deposits of South America*, 252.

59. Elisabeth Glaser-Schmidt, "The Guggenheims and the Coming of the Great Depression in Chile, 1923–1934," *Business and Economic History* 24, no. 1 (Fall 1995): 184.

60. Homer Aschmann, "The Natural History of a Mine," *Economic Geography* 46, no. 2 (April 1970): 185.

61. See Paul Marr, "Ghosts of the Atacama: The Abandonment of Nitrate Mining in the Tarapacá Region of Chile," *Middle States Geographer* 40 (2007): 22–31.

62. José Antonio González Pizarro, "Capitales, empresas y trabajo en la industria salitrera," in *Región de Antofagasta: pasado, presente y futuro* (Antofagasta: Universidad Católica del Norte), 125–39; see also his article "La minería como factor en la demarcación de limites político-administrativos de la Provincia de Antofagasta, 1880–1925," (n.p., n.d.).

Chapter 5

1. Anthony M. Kronbauer, ed., *The International Standard Atlas of the World* (Chicago: Library Publishers, Inc., 1947, with various editions to 1961), 142, 307.

2. Alexis Everett Frye, *Frye's Grammar School Geography* (Boston: Ginn & Co., 1902), was also available in subsequent years and published in various state editions.

3. *Goode's World Atlas* (Chicago: Rand McNally and Company), published since 1932 and currently available as the 22nd edition, 2015.

4. *Grand atlas gallimard pour le XXIe siècle* (Paris: Guides Gallimard, 1997), 49, 50, 51, 53.

5. Peveril Meigs, *Geography of Coastal Deserts* (Paris: UNESCO, 1966), "Sector 36: Atacama Desert," 110–14.

6. Rudolph, *Vanishing Trails of Atacama*, 75.

7. Rasheed, "Man and the Desert: Northern Chile," 1, 34–36.

8. Hans Rudolf Bork et al., "Die Entwicklung der Oase San Pedro de Atacama, Chile," *Petermanns Geographische Mitteilungen* 146, no. 5 (2002): 56.

9. Anita Carrasco, "Jobs and Kindness: W. E. Rudolph's role in the shaping of perceptions of mining company–indigenous community relations in the Atacama Desert, Chile," *The Extractive Industries and Society* 2 (2015): 353, 358.

10. Ibid., 357.

11. Personal communication to author from William Culver, January 2, 2017. See also Pamela Constable and Arturo Valenzuela, *A Nation of Enemies: Chile Under Pinochet* (New York: W. W. Norton & Co., 1993).

12. Janet Finn, *Tracing the Veins: Of Copper, Culture, and Community from Butte to Chuquicamata* (Berkeley: University of California Press, 1998), 119.

13. Carrasco, "One World, Many Ethics," 72, 124, 157.

14. Ibid., 177, 208, 215–16, 277, 286.

15. John N. Trager, "The Huntington Botanical Gardens presents the 2016 offering of International Succulent Introductions," *Cactus and Succulent Journal* 88, no. 3 (May–June 2016): 110–12.

16. Terry C. Wallace and Michelle K. Hall-Wallace, "Minerals of the Andes: Emeralds, Gold, and Silver from the Sky," *Rocks & Minerals* 78, no. 1 (January/February 2003): 34–37.

17. Rudolph, *Vanishing Trails of Atacama*, 68–69.

18. Rasheed, "Man and the Desert: Northern Chile," 165.

19. Ryan Seelau and Laura M. Seelau, *The Pukará of Quitor: An Indigenous Self-Determination Case Study*, Bauu Institute and Press, 2013), 8, 19–20.

20. Julian Mellet, *Viajes por el interior de la América meridional 1808–1820* (Santiago: Editorial del Pacífico, 1959), 106.

21. See Chris Wilson, *The Myth of Santa Fe: Creating a Modern Regional Tradition* (Albuquerque: University of New Mexico Press, 1997).

22. *Lonely Planet Chile & Easter Island* (n.p., n.d.), 174, 176, 177, 197.

23. Russell, *A Visit to Chile*, 231.

24. Donald Binns, *The Anglo-Chilean Nitrate & Railway Company (Ferrocarril de Tocopilla al Toco): A History of the Company and its Locomotives* (Skipton, North Yorkshire, England: Trackside Publications, 1995), 33.

25. See Daniel Weiss, "Colors of the Priesthood: An intriguing source of power is revealed in an ancient Andean tomb," *Archaeology* 69, no. 2 (March/April 2016): 36–37.

26. Bowman, *Desert Trails of Atacama*, 40.

27. "Atacama Desert," in New World Encyclopedia, accessed May 26, 2014, http://www.newworldencyclopedia.org/entry/Atacama_Desert.

28. Chris Moss, "Atacama Desert: Trip of a Lifetime," *Telegraph*, March 10, 2015, accessed February 3, 2017, www.telegraph.co.uk/travel/destinations/south-america/chile/articles/Atacama-Desert-Trip-of-a-Lifetime/.

29. Richard Francaviglia, "The Cemetery as an Evolving Cultural Landscape," *Annals, Association of American Geographers* 61, no. 3 (September 1971): 501–9.

30. Rudolph, *Vanishing Trails of Atacama*, 4.

31. "Northern Chile Floods, March 2015—Facts, Figures, and Photos," April 8, 2015, accessed May 25, 2015, http://floodlist.com/america/northern-chile-floods-march-2015-facts-figures-and-photos.

32. Jonathan D. A. Clarke, "Antiquity of aridity in the Chilean Atacama Desert," *Geomorphology* 73 (2006): 101–14.

33. John Aarons and Claudio Vita-Finzi, *The Useless Land: A Winter in the Atacama Desert* (London: Robert Hale Limited, 1960), 114–15, 126–27.

34. Rudolph, *Vanishing Trails of Atacama*, 51.

35. Bernardo T. Arriaza, *Beyond Death: The Chinchorro Mummies of Ancient Chile* (Washington, D.C.: Smithsonian Institution Press, 1995), xiii, 30.

36. Jarrett A. Lobell, "Atacama's Decaying Mummies," in *Archaeology: A publication of the Archaeological Institute of America* 68, no. 5 (September/October 2015): 12.

37. Aschmann, "The Natural History of a Mine," 178.

38. For more on this subject, see Richard Francaviglia, *Hard Places: Reading the Landscape of America's Historic Mining Districts* (Iowa City: University of Iowa Press, 1991), 166–84.

39. John H. White, Jr., *American Locomotives: An Engineering History, 1830–1880* (Baltimore: The Johns Hopkins University Press, 1997), 311.

40. "Topographic maps of the Americas," accessed April 15, 2014, www.mcgill.ca/library/find/maps/topoamerica.

41. "San Pedro de Atacama," accessed August 1, 2015, www.geocaching.com/geocache/GCKNFJ_san-pedro-de-atacama?guid=7dc942c5-d220-4ebc-9249-83e033562877.

42. John H. White, Jr., *A History of the American Locomotive: Its Development, 1830–1880* (Baltimore, MD: The Johns Hopkins University Press, 1968), 311–19.

43. *Eyewitness Travel Chile & Easter Island* (London: DK, 2013), 168.

44. *Nelles Map of Chile [and] Patagonia*, scale 1:250,000, [with] Special Maps: Around Santiago, Easter Island, Peninsula Valdés, Torres del Paine [and] City maps: Central Santiago, Arica, Iquique, Antofagasta, La Serena, Punta Arenas (Munich, 2009).

45. W. Rauh, "The Peruvian-Chilean Deserts," in *Ecosystems of the World*, vol. 12A *Hot Deserts and Arid Shrublands*, Michael Evenari, Imanuel Noy-Meir, and David W. Goodall, eds. (Amsterdam: Elsevier, 1985), 259–60.

46. Raquel Pinto Bahamonde and Arturo Kirberg Benavides, *Cactus del Extremo Norte de Chile* (Santiago, Chile: Raquel Pinto Bahamonde, 2009).

47. *Chile*, 1:1,750,000 (Richmond, B.C., Canada: ITMB Publishing Ltd, n.d.).

48. Lake Sagaris, *Bone and Dream: Into the World's Driest Desert* (Toronto: Alfred A. Knopf, 2000), 2, 49, 52.

49. Ericka Kim Verba, "Sewing Resistance," in *Mapping Latin America: A Cartographic Reader*, Jordana Dym and Karl Offen, eds. (Chicago: University of Chicago Press, 2011), 259.

50. *Nostalgia for the Light*, a film written and directed by Patricio Guzmán (Icarus Films Home Video, 2010).

51. Paula Allen, "Grief in the Atacama Desert," accessed May 28, 2014, https://notesontheamericas.wordpress.com/2013/05/30/grief-in-the-atacama-desert. It should be noted that Isabel Allende Llona's father was first cousin to Salvador Allende. Moreover, she should not be confused with Isabel Allende Bussi, a socialist politician who is Salvador Allende's daughter.

52. Paula Allen, "Flowers in The Desert: The Search for Chile's Disappeared," accessed May 30, 2014, https://www.kickstarter.com/projects/1756888157/flowers-in-the-desert-the-search-for-chiles-disapp.

53. Isabel Allende, *My Invented Country*, accessed May 28, 2014, http://isabelallende.com/en/book/invented/excerpt.

54. Letra, "Cantata Santa Maria de Iquique," accessed January 6, 2017, https://www.musica.com/letras.asp?letra=1739354.

55. The Smiling Buddhas, *Atacama* (distributed by CDBABY, n.p., n.d.), music CD.

56. Jean Luc Ponty & His Band, *The Acatama* [*sic*] *Experience* (JLP Productions, Inc. and KOCH Entertainment LP, 2007), music CD.

57. Roberto Ampuero, *El alemán de Atacama* (Santiago: Editorial Planeta, 1997), 74, 117, 214, 220.

58. Hernán Rivera Letelier, *La Reina Isabel cantaba rancheras* (Santiago: Alfaguara, 1994), 64, 91, 117, 153.

59. Hernán Rivera Letelier, *El vendedor de pájaros* (Santiago: Alfaguara, 2014), 15, 48, 50, 65, 75, 79, 216–217.

60. Héctor Tobar, *Deep Down Dark: The Untold Stories of 33 Men Buried in a Chilean Mine, and the Miracle that Set Them Free* (New York: Farrar, Straus and Giroux, 2014), 3, 7, 14.

61. Ibid., 15, 38–39, 67, 75.

62. Ibid., 103, 114, 138, 187, 194.

63. *Discover the World: Chile* (distributed by Worldwide Academic Media, n.p., 2011), DVD.

64. "Atacama Desert (NASA Ames participation)," May 9, 2007, accessed May 30, 2014, www.nasa.gov/centers/ames/research/expeditions/atacama.html.

65. Pablo Sobron et al., "Geochemical profile of a layered outcrop in the Atacama analogue using laser-induced breakdown spectroscopy: Implications for Curiosity investigations in Gale," *Geophysical Research Letters* 40, no. 10 (May 28, 2013).

66. Francaviglia, *Go East, Young Man*, 28–43.

67. Rice, *Voyages of Discovery*, 323.

68. Andrea DenHoed, "An Agoraphobic Photographer's Virtual Travels, on Google Street View," accessed June 30, 2017, https://www.newyorker.com/culture/photo-booth/an-agoraphobic-photographers-virtual-travels-on-google-street-view.

69. "Agoraphobic Traveller," accessed June 30, 2017, https://www.instagram.com/streetview.portraits/?hl=en.

70. Personal conversation between the author and Ecuadorian writer-scholar Ana Julia Rugel Hollis, July 25, 2016.

71. Checking in October 2017 as this book was about to go to press, I discovered that this puzzle map could be purchased online in Chile, but there was no provision for shipping it out of the country.

Conclusion

1. Carl A. Francis, "The Szenics Collection of Contemporary Chilean Minerals: A New Addition to the Harvard Mineralogical Museum," *Rocks & Minerals* 78 (January/February 2003): 47, 48.

2. Jérome Gattacceca et al., "The densest meteorite collection area in hot deserts: The San Juan meteorite field (Atacama Desert, Chile)," *Meteoritics & Planetary Science* 46, no. 9 (2011): 1281, 1284, 1285.

3. Glenn T. Trewartha, *The Earth's Problem Climates*, 2nd ed. (Madison: University of Wisconsin Press, 1981), 29, 34–35.

4. See Patricia E. Vidiella, Juan J. Armesto, and Julio R. Gutiérrez, "Vegetation changes and sequential flowering after rain in the southern Atacama Desert," *Journal of Arid Environments* 43 (1999): 449–58.

5. "52 Places to Go in 2017," *The New York Times.com*, January 4, 2017, accessed January 25, 2017, https://www.nytimes.com/interactive/2017/travel/places-to-visit.html?.

6. Death Valley Natural History Association online newsletter, February 2017, accessed February 12, 2017, https://dvnha.org/59--sp-134?start=5.

7. See Lea Dilworth, *Imagining Indians in the Southwest: Persistent Images of a Primitive Past* (Washington, D.C.: Smithsonian Institution Press, 1996), 5–6.

8. For a discussion of cartography in northern New Spain, see Dennis Reinhartz and Gerald Saxon, eds., *The Mapping of the Entradas into the Greater Southwest* (Norman: University of Oklahoma Press, 1998).

9. Mirela Altic, "Baja California in 1739: An Early Exploration by Ferdinand Konscak," *Terrae Incognitae* 47, no. 2 (September 2015): 106–26.

10. Raymond B. Craib, *Cartographic Mexico: A History of State Fixation and Fugitive Landscapes* (Durham, NC: Duke University Press, 2004), 9, 81.

11. See J. Bird, "The 'Copper Man': A Prehistoric Miner and His Tools from Northern Chile," in E. Benson, ed., *Pre-Columbian Metallurgy of South America* (Washington, D.C.: Dumbarton Oaks, 1979), 105–32.

12. William Rudolph, *Vanishing Trails of Atacama*, 80.

13. See "Hydroponic Growing in Chile," accessed May 7, 2016, www.hydroponics.com.au/hydroponic-growing-in-chile.

14. Richard Gray, "Ancient Spirals in Peruvian Desert Used as 'Sophisticated' Irrigation System," *Archaeology News Network*, April 14, 2016, accessed May 19, 2016, https://archaeologynewsnetwork.blogspot.com/2016/04/ancient-spirals-in-peruvian-desert-used.html#rhopqDFOvHQPP7bV.97.

15. D. Graham Burnett, *Masters of All They Surveyed: Exploration, Geography, and a British El Dorado* (Chicago: University of Chicago Press, 2000).

16. José Antonio Gonzáles Pizarro, "El conocimiento del territorio por viajeros y exploradores," in *Región de Antofagasta: Pasado, presente, y futuro* (Antofagasta: Universidad Católica del Norte, 2013), 83–93.

17. Carl J. Bauer, *Siren Song: Chilean Water Law as a Model for International Reform* (Washington, D.C.: Resources for the Future, 2004), 135.

18. Manuel Prieto, "Privatizing Water and Articulating Indigeneity: The Chilean Water Reforms and the Atacameño People (Likan Antai)," (PhD dissertation, University of Arizona, 2014).

19. Leland Pederson, *The mining industry of the Norte Chico, Chile* (Evanston, Ill: Northwestern University Studies in Geography, 1966).

20. Homer Aschmann, "The Natural History of a Mine," *Economic Geography* 46, no. 2 (1970): 185.

21. William L. Thomas, ed., *Man's Role in Changing the Face of the Earth* (Chicago: University of Chicago Press, 1956).

22. "Chile's court confirms 2nd highest fine in history for Caserones mine," accessed June 14, 2016, www.mining.com/chiles-court-confirms-2nd-highest-fine-in-history-for-caserones-mine.

23. "Mining Industry Plans Massive Use of Seawater in Arid Northern Chile," Inter Press Service, Terramérica, accessed February 6, 2017, www.ipsnews.net/2013/08/mining-industry-plans-massive-use-of-seawater-in-arid-northern-chile/.

24. Raquel Pinto Bahamonde and Arturo Kirberg Benavides, *Cactus del extremo Norte de Chile* (Santiago: Raquel Pinto Bahamonde, 2009), Prólogo.

25. United Nations Environment Programme, Environment for Development, Global Deserts Outlook, Table 3.1: Changes in Desert Temperatures and Rainfall, accessed January 1, 2016, https://www.telesurtv.net/english/analysis/Is-Climate-Change-Leading-to-Floods-Droughts-in-Chile-20150331-0004.html.

26. Bowman, *Desert Trails of Atacama*, 236–38.

27. Carl Sauer, "The Personality of Mexico," in John Leighly, ed., *Land & Life: A Selection from the Writings of Carl Ortwin Sauer* (Berkeley: University of California Press, 1963), 104–18.

28. José Antonio González Pizarro, "La formación de la nortinidad," in *Región de Antofagasta: Pasado, presente, y futuro* (Antofagasta: Universidad Católica del Norte, 2013), 233.

29. José Antonio Goazález Pizarro, *La épica del salitre en el desierto de Atacama: 1880–1967: Trabajo, technologías, vida cotidiana, conflicto y cultura* (Saarbrucken, Germany: Editorial Académica Española, 2016).

30. Martin Pasqualetti, "An Iconoclast on the Loose," in Pasqualetti, ed., *The Evolving Landscape: Homer Aschmann's Geography* (Baltimore: Johns Hopkins University Press, 1997), 5.

BIBLIOGRAPHY

Aarons, John, and Claudio Vita-Finzi. *The Useless Land: A Winter in the Atacama Desert*. London: Robert Hale Limited, 1960.

The Acatama [sic] Experience. A music album by Jean Luc Ponty & His Band. JLP Productions Inc. and KOCH Entertainment LP, 2007.

The Agoraphobic Traveller. An Instagram account by Jacqui Kenny. https://www.instagram.com/streetview.portraits/?hl=en

Aguirre, Luis, Francisco Hervé, and Mónica del Campo. "An Orbicular Tonalite from Caldera, Chile." *Journal of the Faculty of Science, Hokkaido University, Series 4: Geology and Mineralogy* (September 1976): 231–59.

Allende, Isabel. *My Invented Country*. http://isabelallende.com/en/book/invented/excerpt.

Altic, Mirela. "Baja California in 1739: An Early Exploration by Ferdinand Konscak." *Terrae Incognitae* 47, no. 2 (September 2015): 106–26.

———. "Jesuits at Sea: José Quiroga and Jóse Cardiel—Two Complementary Views of Patagonia (1745–1746)." *Terrae Incognitae* 49, no. 2 (September 2017): 149–73.

Americae Sive Novi Orbis Nova Descriptio. A map from the atlas *Theatrum Orbis Terrarum*. Antwerp: Abraham Ortelius, 1570. David Rumsey Map Collection, no. 10000.013.

Ampuero, Roberto. *El Alemán de Atacama*. Santiago: Editorial Planeta, 1997.

Antofagasta (Chili) and Bolivia Railway. A map, scale 1:2,500,000. London: Waterlow & Sons, ca. 1916. American Geographical Society Library, no. 254-c.A57 D-1915?.

Anuario Administrativo de 1869, por Dr. Juan Francisco Velarde. An annual report. La Paz, Bolivia: April 1870.

"A Railway Trip in Chili." *The St. James Magazine* (May 1862): 213–16. Anonymous, but sometimes attributed to Mrs. S. C. Hall.

Arriaza, Bernardo T. *Beyond Death: The Chinchorro Mummies of Ancient Chile*. Washington, D.C.: Smithsonian Institution Press, 1995.

Aschmann, Homer. "The Natural History of a Mine." *Economic Geography* 46, no. 2 (April 1970): 171–90.

ATACAMA. A map, scale 1:1,000,000. Reprinted with revisions to June 1942; hypsometric and bathometric tints revised in 1947. New York: American Geographical Society Library, no. 050-b or online no. am004245.

Atacama. A music album by *The Smiling Buddhas*. CD distributed by CDBABY, n.d.

"Atacama Desert." Entry in *NewWorldEncyclopedia.org*. http://www.newworldencyclopedia.org/entry/Atacama_Desert.

"Atacama Desert (NASA Ames participation)." May 9, 2007. www.nasa.gov/centers/ames/research/expeditions/atacama.html.

Bauer, Carl J. *Siren Song: Chilean Water Law as a Model for International Reform*. Washington, D.C.: Resources for the Future, 2004.

Bauer, Ralph. *The Cultural Geography of Colonial Literatures: Empire, Travel, Modernity*. Cambridge University Press, 2003.

Bennett, William C. "The Atacameño." In *Handbook of South American Indians: Vol. 2, The Andean Civilizations—Smithsonian Institution Bureau of American Ethnology, Bulletin 143*. Edited by Julian H. Steward. Washington, D.C.: United States Government Printing Office, 1947.

Binns, Donald. *The Anglo-Chilean Nitrate & Railway Company (Ferrocarril de Tocopilla al Toco): A History of the Company and its Locomotives*. Skipton, North Yorkshire, England: Trackside Publications, 1995.

———. *The Nitrate Railways Company Limited*. Skipton, North Yorkshire, England: Trackside Publications, n.d., ca. 2000.

Binns, Donald, and Harold A. Middleton. *The Taltal Railway: A Chilean Mineral Line*. Henbury, Bristol, England: Trackside Publications, 2010.

Bird, J. "'The Copper Man': A Prehistoric Miner and His Tools from Northern Chile." In *Pre-Columbian Metallurgy of South America*, edited by E. Benson, 105–32. Washington, D.C.: Dumbarton Oaks, 1979.

Blakemore, Harold. *From the Pacific to La Paz: The Antofagasta (Chili) and Bolivia Railway Company, 1888–1988*. London: Lester Crook Academic Publishing, 1990.

Bollaert, William. *Antiquarian, ethnological, and other researches in New Granada, Equador, Peru and Chile, with observations on the pre-Incarial, Incarial, and other monuments of Peruvian nations*. London: Turner & Co., 1860.

Bork, Hans-Rudolf, Jürgen Bähr, Helga Bork, Michael Brombacher, Imre Josef Demhardt, Anja Habeck, Andreas Mieth, and Bernd Tschochner. „Die Entwicklung der Oase San Pedro de Atacama, Chile." *Petermanns Geographiche Mitteilungen* 46, no. 5 (2002): 56–63.

Bork, Hans-Rudolf et al. "Die entwicklung der Oase San Pedro de Atacama, Chile." *Petermanns Geographische Mitteilungen* 146, no. 5 (2002): 56–63.

The Boundaries of Chili in Atacama Settled by History. From the Peruvian Review. Lima: Liberal Printing Office, 1879.

"Boundary tensions between Peru and Chile continue." In *Boundary news—Durham University*, October 22, 2015. https://www.dur.ac.uk/ibru/news/ boundary_news/?itemno=25974.

Bowman, Isaiah. *Desert Trails of Atacama*. New York: American Geographical Society, 1924.

———. *Field Notebook, Antofagasta to La Paz, Titicaca and Desaguadero* (1913). American Geographical Society Collection, Sheridan Libraries, Johns Hopkins University.

———. *General Location map of the Desert and Puna of Atacama, in northern Chile, northwestern Argentina, and southwestern Bolivia*. In *Desert Trails of Atacama*, 10. New York: American Geographical Society, 1924.

———. *South America: A Geography Reader*. Chicago and New York: Rand McNally & Company, 1915.

Brevis exactaq[ue] totivs novi orbis eivsq[ue] insvlarum descriptio recens a Joan Bellro edita. A map by Pedro de Cieza de León. From *La Primera Parte Dell Historie*

Del Peru. Venitia: Appresso Giordano Ziletti, 1560. Beinecke Rare Book and Manuscript Library, Yale University, no. Peru Cwpr j553ci.

Brooks, Juanita, ed. *Journal of the Southern Indian Mission: The Diary of Thomas D. Brown.* Logan: Utah State University Press, 1972.

Bruhns, Karen Olsen. *Ancient South America.* Cambridge: Cambridge University Press, 1994.

Burnett, D. Graham. *Masters of All They Surveyed: Exploration, Geography, and a British El Dorado.* Chicago: University of Chicago Press, 2000.

Cameron, Eion M., Matthew I. Leybourne, and Carlos Palacios. "Atacamite in the oxide zone of copper deposits in northern Chile: Involvement of deep formation waters?" *Miner Deposita* 42 (2007): 201–18.

"Cantata Santa Maria de Iquique." A folksong by Letra. https://www.musica.com/letras.asp?letra=1739354.

Carrasco, Anita. "Jobs and Kindness: W. E. Rudolph's role in the shaping of perceptions of mining company–indigenous community relations in the Atacama Desert, Chile." *The Extractive Industries and Society* 2 (2015): 352–59.

———. "One World, Many Ethics: The Politics of Mining and Indigenous Peoples in Chile." PhD diss., University of Arizona, 2011.

Carta Jeográfica del Desierto i Cordilleras de Atacama. Levantada por la Comision Esploradora de Atacama. A map by Francisco J. San Román. Santiago de Chile: Direccion de Obras Publicas, Seccion de Jeografia y Minas, 1892.

Carta Jeográfica y Minera de los Departamentos de Vallenar y Freirina de la Provincia de Atacama. A map prepared under the direction of José Del C. Fuenzalida by engineer N. Eschagaral. Santiago: Sociedad Imprenta y Litografía, 1914. American Geographical Society Library, no. 254-c.A82 A-1914.

Carta Minera y Geográfica de Provincia de Antofagasta, Creada por Ley 12 Julio 1888 Sobre el Territorio del Antiguo Desierto de Atacama. A blueprint map, scale 1:500,000. n.p., n.d. Revised to include mining statistics and data by decade to 1920, with additional information to 1923. American Geographical Society Library, no. 254-c.A57 A-[1923].

Carta Topográfica [de] Departamentos de Antofagasta i Tocopilla: Contiene la Ubicacion de Salitreras, Minas, Aguadadas, etc., Ademas Los Límites de Subdelegaciaones y Otros Datos Importantes, por Manuel A. Rojas. A map, scale 1:500,000. Santiago: Humo y Walker, 1918. American Geographical Society Library, no. 254-c.A57 A[1918].

Carte, Rebecca A. *Capturing the Landscape of New Spain: Baltasar Obregón and the 1564 Ibarra Expedition.* Tucson: University of Arizona Press, 2015.

Carte des Terrains Métallifères du CHILI, dressée par M. Simonin dáprès Gay et Domeyko. A map. Paris: Librarie de L. Hachette, 1867.

Carte Du Paraguay, Du Chili, Du Detroit De Magellan & C. des P. P. Alfonse d'Ovalle et Nicolas Techo et sur les Relations et memoires de Brouwer, Narbouroug, Mt. de Beauchesne & Par GUILLAUME DE L'ISLE. A map by Guillaume Delisle. Amsterdam: Jean Covens et Corneille Mortier, 1742. David Rumsey Map Collection, no. 4638.102.

Carte Du Paraguay, Du Chili, Du Detroit De Magellan & C. Dressée sur les Descriptions des PP. Alfonse d'Ovalle, et Nicolas Techo, et sur les Relations et Memoires de Brouwer,

Narbouroug, Mt. de Beauchesne &c. Par GUILLAUME De L'ISLE. A map by Guillaume Delisle. Paris: Liebaux le fils, 1703 or 1708. David Rumsey Map Collection, no. 4764.103.

Chile. 2nd ed. A map, scale 1:1,750,000. Richmond, B.C., Canada: ITMB Publishing Ltd, 2012.

Chile. 5th ed. A map, scale 1:2,000,000. Includes Northern Chile, Central Chile, Southern Chile/Patagonia, Arica, Iquique, Antofagasta, Viña del Mar, La Serena, Valparaíso, Santiago, Robinson Crusoe Island, Rapa Nui (Easter Island), Torres del Paine National Park, Punta Arenas. Hauzenberg, Germany: BORCH, 2011.

Chile Adventure Travel Map (scale 1:1,750,000). Evergreen, Colorado: National Geographic Maps, 2010.

Chile [&] Patagonia. A map, scale 1:2,500,000. Munich: Nelles, 2009.

"Chile's court confirms 2nd highest fine in history for Caserones mine." www.mining.com/ chiles-court-confirms-2nd-highest-fine-in-history-for-caserones-mine.

Chili. A map by Aaron Arrowsmith. Boston: Thomas & Andrews, 1812. David Rumsey Map Collection, no. 0028056.

Chili. A map by Fielding Lucas, Jr. Baltimore: Fielding Lucas, 1817. David Rumsey Map Collection, no. 4866.067.

Chili: Carte Géographique, Statistique et Historique du Chili. Paris: J. Carez, 1825. David Rumsey Map Collection, no. 0102.050.

Chili: Geographical, Statistical, and Historical Map of Chile. Engraved by J. Finlayson and attributed to H. C. Carey and I. Lea. Philadelphia: H. C. Carey & Lea, 1822. David Rumsey Map Collection, no. 0122.046.

Chong Díaz, Guillermo, Aníbal Gajardo Cubillos, Adrian J. Hartley, and Teresa Moreno. "Industrial Minerals and Rocks." In *The Geology of Chile*, edited by Teresa Moreno and Wes Gibbons, 201–14. London: The Geological Society, 2007.

Clarke, Jonathan D. A. "Antiquity of Aridity in the Chilean Atacama Desert." *Geomorphology* 73 (2006): 101–14.

Climates of the Earth. A map by Glenn T. Trewartha. In *Goode's World Atlas*, 12–13. Chicago: Rand McNally, 1964.

Collard, Ian. *Pacific Steam Navigation Company: Fleet List & History.* Amberly: n.p., n.d. Author's soft-bound copy purchased online November 2016.

Conley, Tom. "Film and Exploration." An entry in *The Oxford Companion to World Exploration*, Vol. 1, A–L, edited by David Buisseret, 307. New York: Oxford University Press, 2007.

Constable, Pamela, and Arturo Valenzuela. *A Nation of Enemies: Chile Under Pinochet.* New York: W. W. Norton & Co., 1993.

Craib, Raymond B. *Cartographic Mexico: A History of State Fixation and Fugitive Landscapes.* Durham, NC: Duke University Press, 2004.

Culver, William W., and Cornel J. Reinhart. "Capitalist Dreams: Chile's Response to Nineteenth-Century World Copper Competition." *Comparative Studies in Society and History* 31, no. 4 (October 1989): 722–44.

D'Altroy, Terrence N. *Provincial Power in the Inka Empire.* Washington, D.C.: The Smithsonian Institution, 1992.

Dampier, William. *A New Voyage Round the World by William Dampier.* With an Introduction by Sir Albert Gray and a new Introduction by Percy G. Adams. New York: Dover, 1967.

"Dampier, William." Entry by James Kelley in *The Oxford Companion to World Exploration*, Vol. 1, A–L, edited by David Buisseret, 222–25. Oxford and New York: Oxford University Press, 2007.

Darwin, Charles. *Charles Darwin's Beagle Diary.* Edited by Richard Darwin Keynes. Cambridge: Cambridge University Press, 1988.

———. *Journal of researches into the natural history and geology of the countries visited during the voyage of the H.M.S. Beagle round the world.* 2nd ed. London, 1860.

———. Map, undated, attributed to. Featured in a podcast with James Moore on *OnBeing.org.* "Evolution and Wonder: Understanding Charles Darwin." February 5, 2009. https://onbeing.org/programs/james-moore-evolution-and-wonder-understanding-charles-darwin/. Map source: "Reprinted with permission of Syndics of Cambridge Library, DAR 44:28."

———. Sketch map of a portion of northern Chile from Coquimbo to Copiapó (1835). *The Complete Works of Darwin Online.* darwin-online.org.uk/content/frameset?pageseq=1&itemID=CUL-DAR44.28&viewtype=image.

———. *The Voyage of the Beagle.* In *The Harvard Classics*, Vol. 29, edited by Charles W. Eliot. New York: P. F. Collier & Son, 1909/1937.

Death Valley Natural History Association online newsletter, February 2017. https://dvnha.org/59--sp-134?start=5.

de Bry, Theodor. *Collection des Grands and Petits Voyages.* London: Molins Ltd, 1921.

Delaney, Carol. *Columbus and the Quest for Jerusalem.* New York: Free Press, 2011.

Demhardt, Imre Joseph. *Der Erde ein Gesicht geben: Petermanns Geographische Mitteilungen und die Anfänge der modernen Geographie in Deutschland.* Katalog zur Ausstellung der Univsersitäts-und Forschugsbibliothek Erfurt / Gotha in Spiegelsaal auf Schloss Friedenstein in Gotha 23. Juni bis 9. Oktober 2005. Gotha, 2006.

DenHoed, Andrea. "An Agoraphobic Photographer's Virtual Travels, on Google Street View." *TheNewYorker.com.* June 29, 2017. https://www.newyorker.com/culture/photo-booth/an-agoraphobic-photographers-virtual-travels-on-google-street-view.

Desert and Puna of Atacama. A map by Isaiah Bowman. In *Desert Trails of Atacama*, 25. New York: American Geographical Society, 1924.

Desierto de Atacama. A map by Rudolph A. Philippi. In *Reise durch die Wueste Atacama: auf Befehl der chilenischen Regierung im Sommer.* Halle: Eduard Anton, 1860.

Dillon, M. O., A. E. Hoffmann J. "Lomas formations of the Atacama Desert, northern Chile." January 1997. https://www.researchgate.net/publication/313293549_Lomas_formations_of_the_Atacama_Desert_northern_Chile.

Dilworth, Lea. *Imagining Indians in the Southwest: Persistent Images of a Primitive Past.* Washington, D.C.: Smithsonian Institution Press, 1996.

Discover the World: Chile. A DVD distributed by Worldwide Academic Media. n.p., 2011.

Domeyko, Ignacio. *Ensayo sobre los depósitos metaliferos de Chile, con relacion a su jeologia i configuración esterior*. Santiago de Chile: Imprenta Nacional, 1876.
———. *Mis Viajes: Memorias de un exiliado*. 2 vols. Santiago: 1978.

d'Orbigny, Alcide. *La Relation du Voyage dans l'Amerique* Méridionale... *1826–1833*. Vol. 4. Paris: Chez. V. Levirault, 1859.

Dorfman, Ariel. *Desert Memories: Journeys Through the Chilean North*. Washington, D.C. : National Geographic, 2004.

Dorn, Ronald I. *Rock Coatings: Developments in Earth Surface Processes*. Amsterdam: Elsevier, 1998.

Dr. R. A. Philippi's Erforschung der sogenannten Wüste Atacama Nov[ember] 1853– Februar[y] 1854. A map published by August Petermann. Gotha, Germany: Justus Perthes, 1856.

Dym, Jordana, and Karl Offen, eds. *Mapping Latin America: A Cartographic Reader*. Chicago: University of Chicago Press, 2011.

Edmundson, William. *The Nitrate King: A Biography of "Colonel" John Thomas North*. New York: St. Martin's Press, 2011.

Ehleringer, J., H. A. Mooney, S. L. Gulmon, and P. Rundel. "Orientation and its consequences for Copiapoa (Cactaceae) in the Atacama Desert." *Oecologia* 46, (1980): 63–67.

Engstrand, Iris H. W. *Spanish Scientists in the New World: The Eighteenth-Century Expeditions*. Seattle: University of Washington Press, 1981.

Espinos, Florencia Señoret, and Juan Pablo Acosta Ramos. *Guía de Campo Cactáceas Nativas de Chile*. Concepción, Chile: CORMA—Corporacion Chilena de la Madura, 2013.

Essay on the Geography of Plants: Alexander von Humboldt and Aimé Bonpland. Edited by Stephen T. Jackson and Sylvie Romanowski. Chicago: University of Chicago Press, 2009.

Esteva-Fabregat, Claudio. *Mestizaje in Ibero-America*. Tucson: University of Arizona Press, 1987/1995.

Evans, Oswald Hardey. "Notes on the Raised Beaches of Taltal (Northern Chile)." *Quarterly Journal of the Geological Society* 63 (January 1907): 64–68.

Eyewitness Travel Chile & Easter Island. London: DK, 2013.

Farcau, Bruce W. *The Ten Cents War: Chile, Peru, and Bolivia in the War of the Pacific, 1879–1884*. Westport, CT: Praeger Publishers, 2000.

Fernández Canque, Manuel. *Arica 1868: Un Tsunami y un Terremoto*. Santiago: Ediciones de la Dirección de Bibliotecas, Archivos y Museos, 2007.

Feuillée, Louis. *Journal des observations physiques, mathématiques et botaniques: faites par ordre du roi sur cotes orientales de l'Amérique méridionale, & aux Indes occidentales. Et dans un autre voyage fait par le méme ordre a la Nouvelle Espagne, & aux Isles de l'Amérique*. Paris: Pierre Giffart, 1714–1715.

Fifer, J. Valerie. *William Wheelwrigh—Steamship and Railroad Pioneer: Early Yankee Enterprise in the Development of South America*. Newburyport, MA: Historical Society of Old Newbury, 1998.

"52 Places to Go in 2017." *The New York Times.com*. January 4, 2017. https://www.nytimes.com/interactive/2017/travel/places-to-visit.html?_r=d.

Finn, Janet. *Tracing the Veins: Of Copper, Culture, and Community from Butte to Chuquicamata*. Berkeley: University of California Press, 1998.

Fletcher, L. "On the meteorites, which have been found in the desert of Atacama and its neighborhood." *The Mineralogical Magazine* 8, no. 40 (October 1889): 223–64.

"Flowers in The Desert: The Search for Chile's Disappeared." A project by Paula Allen. https://www.kickstarter.com/projects/1756888157/flowers-in-the-desert-the-search-for-chiles-disapp.

Francaviglia, Richard. "The Atacama Desert: A Five Hundred Year Journey of Discovery." *Terrae Incognitae* 48, no. 2 (September 2016): 105–38.

———. "The Cemetery as an Evolving Cultural Landscape." *Annals, Association of American Geographers* 61, no. 3 (September 1971): 501–9.

———. *Go East, Young Man: Imagining the American West as the Orient*. Logan: Utah State University Press, 2011.

———. *Hard Places: Reading the Landscape of America's Historic Mining Districts*. Iowa City: University of Iowa Press, 1991.

———. *The Mapmakers of New Zion: A Cartographic History of Mormonism*. Salt Lake City: University of Utah Press, 2015.

———. "The Treasures of Aladdin: Orientalizing the Mining Frontier." *The Mining History Journal* annual issue (2011): 87–97.

Francis, Carl A. "The Szenics Collection of Contemporary Chilean Minerals: A New Addition to the Harvard Mineralogical Museum." *Rocks & Minerals* 78 (January/February 2003): 46–51.

"Francisco J. San Román." An entry on *Wikipedia.org*. https://es.wikipedia.org/wiki/Francisco_J_San_Román.

Frézier, Amédée François. *A voyage to the South-Sea, and along the coasts of Chili and Peru, in the years 1712, 1713, and 1714*. London: Jonah Bowyer, 1717.

Frye, Alexis Everett. *Frye's Grammar School Geography*. Boston: Ginn & Co., 1902.

Gallez, Paul. *La cola del dragón: America del Sur en los mapas antiguos, medievales y renacentistas*. Bahia Blanca, Argentina: Institúto Patagónico, 1990.

Gattacceca, Jérome, Millarca Valenzuela, Minoru Uehara, A. J. Timothy Jull, Marlene Giscard, Pierre Rochette, Regis Braucher, Clement Suavet, Matthieu Gounelle, Diego Morata, Pablo Munayco, Michele Bourot-Denise, Didier Bourles, and François Demory. "The densest meteorite collection area in hot deserts: The San Juan meteorite field (Atacama Desert, Chile)." *Meteoritics & Planetary Science* 49, no. 9 (2011): 1276–87.

Gilliss, J. M. *The U.S. Naval Astronomical Expedition to the southern hemisphere, during the years 1849–'50–'51–'52. Vol. 1: Chile*. Washington, D.C.: A.O.P. Nicholson, Printer, 1855.

Glaser-Schmidt, Elisabeth. "The Guggenheims and the Coming of the Great Depression in Chile, 1923–1934." *Business and Economic History* 24, no. 1 (Fall 1995): 176–85.

"Global Deserts Outlook, Table 3.1: Changes in Desert Temperatures and Rainfall." United Nations Environment Programme, Environment for

Development. https://www.telesurtv.net/english/analysis/Is-Climate-Change-Leading-to-Floods-Droughts-in-Chile-20150331-0004.html.

González Pizarro, José Antonio. "Capitales, empresas y trabajo en la industria salitrera." In *Región de Antofagasta: pasado, presente y futuro*, 125–39. Antofagasta: Universidad Católica del Norte, 2013.

———. "El conocimiento del territorio por viajeros y exploradores." In *Región de Antofagasta: Pasado, presente, y futuro*, 83–93. Antofagasta: Universidad Católica del Norte, 2013.

———. *La épica del salitre en el desierto de Atacama: 1880–1967: Trabajo, technologías, vida cotidiana, conflicto y cultura*. Saarbrucken, Germany: Editorial Académica Española, 2016.

———. "La formación de la nortinidad." In *Región de Antofagasta: Pasado, presente, y futuro*, 223–37. Antofagasta: Universidad Católica del Norte, 2013.

———. "La minería como factor en la demarcación de limites político-administrativos de la Provincia de Antofagasta, 1880–1925." n.p., n.d.

Goodman, Edwin L. *The Explorers of South America*. Norman: University of Oklahoma Press, 1972.

Grand atlas gallimard pour le XXIe siècle. Paris: Guides Galimard, 1997.

Gray, Richard. "Ancient Spirals in Peruvian Desert Used as 'Sophisticated' Irrigation System." A blogpost on *Archaeology News Network*. April 14, 2016. https://archaeologynewsnetwork.blogspot.com/2016/04/ancient-spirals-in-peruvian-desert-used.html#rhopqDFOvHQPP7bV.97.

"Grief in the Atacama Desert." A blog entry by Paula Allen. https://notesontheamericas.wordpress.com/2013/05/30/grief-in-the-atacama-desert.

Griffith, James. *Beliefs and Holy Places: A Spiritual Geography of the Pimería Alta*. Tucson: University of Arizona Press, 1992.

Guia General de Chile, Chile Industrial, Comercial y Social. Vol. 1, Northern Chile. Santiago: Empresa Inter-America, 1923/1924.

Harding, Josiah. "The Desert of Atacama (Bolivia)." *Journal of the Royal Geographical Society of London* (1877): 249–53.

Harmless, William, S. J. *Desert Christians: An Introduction to the Literature of Early Monasticism*. New York: Oxford University Press, 2004.

Hartigan, Lynda Roscoe. "Musings on Joseph Cornell's Alchemy of the Mind." In *Joseph Cornell: Wanderlust*, 56–69. London: Royal Academy of Arts, 2015.

Hébert, John R. "America." In *Mapping Latin America: A Cartographic Reader*, edited by Jordana Dym and Karl Offen, 29–32. Chicago: University of Chicago Press, 2011.

———. "The Map that Named America: Library Acquires 1507 Waldseemüller Map of the World." *Library of Congress Information Bulletin*. Washington, D.C., September 2003. loc.gov/loc/lcib/0309/maps.html.

Hedendaagsche Historie of Tegenwoordige Staat van Amerika, Tweede Deel. Behelzende Het vervolg der befchryvinge van de Beziittingen der Kroon van Spanje in dat Werelsdeel; te weeten: Die van het Ryk van PERU en deszelfs Hoofdstadt Lima; van TUCUMAN, PARAGUAY en CHILI; van't Land van MAGELLAAN en Eilanden in de groote ZUIDZEE. Amsterdam: Isaak Tirion, 1767.

"Historia Chilena. Mapa de los pueblos originarios." A modern map showing the locations of original Indigenous populations. May 29, 2007. http://www.educarchile.cl/ech/pro/app/detalle?ID=132542.

Hoffmann J., Adriana E. *Cactaceas en la Flora Silvestre de Chile*. Santiago: Ediciones Fundacion Claudio Gay, 1989.

"Hydroponic Growing in Chile." A post on Hydroponics.com. www.hydroponics.com.au/hydroponic-growing-in-chile.

Hyslop John, and Mario Rivera. "An Expedition on the Inca Road in the Atacama Desert." *Archaeology* 37, no. 6 (November/December 1984): 33–39.

Izquierdo, Philippi J., ed. *Vistas de Chile (por Rodulfo Amando Philippi)*. Santiago: Editorial Universitaria, 1973, reprinted 2015.

Jacobs, Michael. *Ghost Train Through the Andes: On My Grandfather's Trail in Chile and Bolivia*. London: John Murray, 2006.

Jampoler, Andrew C. A. *Sailors in the Holy Land: The 1848 American Expedition to the Dead Sea and the Search for Sodom and Gomorrah*. Annapolis: Naval Institute Press, 2005.

Jefferys, Thomas. *A Map of South America containing Terra-Firma, Guayana, New Granada, Amazonia, Brasil, Peru, Paraguay, Chaco, Tucuman, Chili and Patagonia, from Mr. d'Anville with several Improvements and Additions and the Newest Discoveries*. London: Sayer & Bennett, 1794. David Rumsey Map Collection, no. 0346.034.

Kabat, Alan R., and Eugene V. Coan. "The Life and Work of Rudolph Amandus Philippi (1808–1904)." *Malacologia* 60, no. 1–2 (2017): 1–30.

Karte der Salzwüste Atacama und des Grenzgebiets zwischen Chile, Bolivia & Peru. A lithographic original color map drafted by Bruno Domann for A. Petermann. Gotha, Germany: Justus Perthes, 1879.

Kattermann, Fred. "Molecular analysis of the genus *Eriosyce: Eriosyce* Section Neoporteria subsection *Neoporteria*, Part 1, *Corryocactus* and *Eriosyce* section *Eriosyce* subsection *Islaya*," *Cactus and Succulent Journal*, 89, no. 5, September–October 2017, 204–213.

———. Molecular analysis of the genus *Eriosyce: Eriosyce* Section *Neoporteria* subsection *Neoporteria*, Part II, section *Eriosyce* subsection *Eriosyce*," *Cactus and Succulent Journal*, 89, no. 6, November–December 2017, 266–275.

Keller, Harry F. "Notes on Some Chilean Copper Minerals." *Proceedings of the American Philosophical Society* 47 (January–April 1908): 79–85.

Keller, Peter C., ed. "Collector's Diary—Iquique, Copiapo, and Chañarcillo: Mark Chance Bandy, 1935." *The Mineralogical Record* 15, no. 2 (March–April 1984): 67–74.

Klor de Alva, J. Jorge. "The Postcolonization of the (Latin) American Experience: A Reconsideration of 'Colonialism,' 'Postcolonialism,' and 'Mestizaje.'" In *After Colonialism: Imperial Histories and Postcolonial Displacements*, edited by Gyan Prakash, 241–75. Princeton: Princeton University Press, 1995.

Kozák, Jan, and Vladimir Cermák. *The Illustrated History of Natural Disasters*. New York: Springer, 2010.

Kronbauer, Anthony M., ed. *The International Standard Atlas of the World*. Chicago: Library Publishers, Inc., 1947, with various editions to 1961.

La Serena [plan of] Scituée la Cote du Chily par 29d 55' de lat Australe au bas de la Valleé de Coquimbo, Planche XIX, 128. A map by Amédée Frézier. Paris: 1713.

Latorre, Claudio, Calogero M. Santoro, Paula Ugalde, Eugenia M. Gayo, Daniela Osorio, Carolina Salas-Egaña, Ricardo De Pol-Holz, Delephine Joly, and Jason A. Rech. "Late Pleistocene Human Occupation of the Hyperarid Core in the Atacama Desert, Northern Chile." *Quaternary Science Review* 77 (October 2013): 19–30.

Latorre, Claudio, Patricio Moreno, Gabriel Vargas, Antonio Maldonado, Rodrigo Villa-Martínez, Juan J. Armesto, Carolina Villagrán, Mario Pino, Lautaro Núñez, and Martin Grosjean. "Late Quaternary Environments and Paleoclimate." In *The Geology of Chile*, 309–28. London: The Geological Society, 2007.

Lea, Sarah. "Joseph Cornell: Wanderlust." In *Joseph Cornell: Wanderlust*, 18–39. London: Royal Academy of Art, 2015.

Lobell, Jarrett A. "Atacama's Decaying Mummies." *Archaeology* 68, no. 5 (September/October 2015): 12.

Lonely Planet Chile & Easter Island. A guidebook. n.p., n.d.

Maar, Paul. "Technology, Labor, and the Collapse of Chile's Nitrate Industry." *Middle States Geographer* 46 (2013): 19–26.

Mapa Geográfico de America Meridional, Dispuesto y Gravado por D. Juan de la Cruz Cano y Olmedilla. A map first published in Madrid, 1775, and subsequently, including by William Faden. London: 1799.

Mapa Mineralógico del Desierto de Atacama, por A. Pissis. A map by Pedro José Amado Pissis. Santiago: Lit. P. Cadot, Huérfanos, 1877.

Map of South America, northern and southern sections. By Thomas Jefferys. London: 1776. www.davidrumsey.com/luna/servlet/detail/RUMSEY-8-1-1926-120032:South-America---Southern-Section-?qvq=W4s:/what/Atlas+Map/when/1776/;lc:Rumsey-1&mi=33&trs=43.

Map of the Republic of Chile (Sheet 1, northern Chile). Compiled by the U.S. Naval Astronomical Expedition, from the surveys of Messrs. Pissis & Allan Campbell, the maps of Claude Gay, and unpublished originals from the archives of Santiago and from Astronomical Determinations by the Expedition. In J. M. Gilliss. *The U.S. Naval Astronomical Expedition* (Washington, D.C.: A.O.P. Nicholson, Printer, 1855).

Martínez, Gabriel. "Los dioses de los cerros en los Andes." *Journal de la Sociéte des Américanistes* 69, no. 1 (1983): 85–115.

Martínez, José Luis. *Pueblos del Chañar y el Algarrobo: Los Atacamas en el Siglo XVII* (Santiago: DIBAM, 1998).

McCray, Daniel Arthur. "Eternal Ramifications of the War of the Pacific." Master's thesis, University of Florida, 2005.

Meigs, Peveril. *Geography of Coastal Deserts.* Paris: UNESCO, 1966.

Mellet, Julian. *Viajes por el interior de la América Meridional 1808–1820.* Santiago: Editorial del Pacífico, 1959.

Melville, Herman. "The Encantadas, or Enchanted Isles." *Putnam's Monthly Magazine* (March–May 1854). Later published in *The Piazza Tales*, 1856.

———. *Moby-Dick; or, The Whale.* New York: Harper & Brothers, 1851.

Méndez Beltrán, Luz María. *La Exportación Minera en Chile 1800–1840: Un estudio de la historia económica y social en la transición de la Colonia a la República.* Santiago: Editorial Universitaria, 2004.

Miers, John. *Travels in Chile and La Plata, including Accounts Respecting the Geography, Geology, Statistics, Government, Finances, Agriculture, Manners and Customs, and the Mining Operations in Chile.* 2 Vols. London: Baldwin, Cradock, and Joy, 1826.

Miller, Benjamin L., and Joseph T. Singewald. *The Mineral Deposits of South America.* New York: McGraw Hill Book Company, 1919.

"Mining Industry Plans Massive Use of Seawater in Arid Northern Chile." A post by Marianela Jarroud for the Inter Press Service, Terramérica. August 8, 2013. www.ipsnews.net/2013/08/mining-industry-plans-massive-use-of-seawater-in-arid-northern-chile/.

Mistral, Gabriela. *Desolation: A Bilingual Edition of Desolación, 1923.* Translation, Introduction, and Afterword by Michael P. Predmore and Liliana Baltra. Pittsburgh, PA: Latin American Literary Review Press, 2013.

Monteón, Michael. "The British in the Atacama Desert: The Cultural Bases of Economic Imperialism." *The Journal of Economic History* 35, no. 1 (March 1975): 117–33.

"Mormon's Abridgement Part XIV—We Did Find Upon the Land of Promise—Part II." A post on the blog *NephiCode.* November 3, 2014. nephicode.blogspot.com/2014/11/mormons-abridgement-part-xiv-we-did.html.

Morrissey, Katherine G. *Mental Territories: Mapping the Inland Empire.* Ithaca: Cornell University Press, 1997.

Morrow, Baker H. *The South America Expeditions, 1540–1545: Alvar Nuñez Cabeza de Vaca.* Albuquerque: University of New Mexico Press, 2011.

Moss, Chris. "Atacama Desert: Trip of a Lifetime." *Telegraph.co.* March 10, 2015. www.telegraph.co.uk/travel/destinations/south-america/chile/articles/Atacama-Desert-Trip-of-a-Lifetime.

Mottana, Annibale, Rudolfo Crespi, and Guiseppe Liborio. *Simon & Schuster's Guide to Rocks and Minerals.* New York: Simon & Schuster, 1978.

Muñoz, Santiago. *Jeografía descriptiva de las provincias de Atacama i Antofagasta.* Santiago: Imprenta Gutenberg, 1894.

"Myths and Mysteries Surround Chile's Desert Drawings." NBCnews.com. February 11, 2014. www.nbcnews.com/science/weird-science/myths-mysteries-surround-chiles-desert-drawings-n27651.

Narrett, David. *Adventurism and Empire: The Struggle for Mastery in the Louisiana-Florida Borderlands, 1762–1803.* Chapel Hill: University of North Carolina Press, 2015.

Natural Vegetation. 13th ed. A world map in *Goode's World Atlas* by A. W. Küchler, 20–21. Chicago: Rand McNally, 1964.

Nelles Map of Chile [and] Patagonia. Scale 1:250,000. With Special Maps: Around Santiago, Easter Island, Peninsula Valdés, Torres del Paine [and] City maps: Central Santiago, Arica, Iquique, Antofagasta, La Serena, Punta Arenas. Munich: Nelles, 2009.

Nitrate Operations in Chile as projected by the Cosach [Chilean Nitrate Company]. Second of two-map set. *Fortune* (October 1930): 57.

Nitrate Operations in Chile before rationalization. First of two-map set. *Fortune* (October 1930): 56.

"Nitrogen: I—The World in its Place." *Fortune* (October 1930): 55–56.

"Nitrogen: II—Chile." *Fortune* (October 1930): 56–59.

"Northern Chile Floods, March 2015—Facts, Figures, and Photos." April 14, 2016. floodlist.com/america/northern-chile- floods-march-2015-facts-figures-photos.

Nostalgia for the Light. A film written and directed by Patricio Guzmán. Icarus Films Home Video, 2010.

Núñez Atencio, Lautaro. *Vida y Cultura en el Oasis de San Pedro de Atacama.* Santiago: Editorial Universitaria, 1991.

Offen, Karl. "Minerals and War." In *Mapping Latin America: A Cartographic Reader*, edited by Jordana Dym and Karl Offen, 139–43. Chicago: University of Chicago Press, 2011.

"Origen del Concepto Atacama: Su origen como Provincia y Región." A post by Guillermo Cortes Lutz in *Historia, Política y Educación* on his blog. September 5, 2006. guillermocorteslutz.blogspot.com/2006/09/origen-del-concepto-atacama.html.

Ortiz Soleto, Jorge. "El Piloto Andrés Baleato y la cartografía peruana." www.oannes.org.pe/upload/20160922102218163000257.pdf.

Ovalle, Alonso de. *Histórica relación del Reyno de Chile, y las missiones, y ministérios que exercita en la Compañía de Jesus.* Rome: 1646.

Parker, Hershel. *Herman Melville, A Biography: Vol. 1, 1819–1851.* Baltimore: Johns Hopkins University Press, 1996.

Pasqualetti, Martin. "An Iconoclast on the Loose." In *The Evolving Landscape: Homer Aschmann's Geography*, edited by Pasqualetti, 5. Baltimore: Johns Hopkins University Press, 1997.

Peck, Steven. "The Copiapó Locomotive: Historical and Technical Assessment." Master's thesis, Colorado State University, 2002.

Pederson, Leland. *The Mining Industry of the Norte Chico, Chile.* Evanston, Ill: Northwestern University Studies in Geography, 1966.

Pedley, Mary Sponberg. *The Commerce of Cartography: The Patronage and Production of Maps in Eighteenth-Century France and England.* Chicago: University of Chicago Press, 2005.

Penck, Walther. *Puna de Atacama: Bergfahren und Jagen in der Cordillere von Südamerika.* Stuttgart: J. Englehorns Nachf, 1933.

Peru. A map in *Atlas Major Sive Cosmographia Blaviana, Qua Solvm, Salvm, Coelvm, Accvratissime Describvntvr.* Amsterdam: Joan Blaeu, 1665. David Rumsey Map Collection, no. 10017.000.

Philippi, Rudolph Amandus. "El Estudio de las Ciencias Naturales—Escrito en Facsimil de don Rodulfo Amando Philippi." In *Cactaceas en la Flora Silvestre de Chile*, edited by Adriana E. Hoffmann J., 8. Santiago: Ediciones Fundacion Claudio Gay, 1990.

————. *Reise durch die Wueste Atacama: auf Befehl der chilenischen Regierung im Sommer 1853–54.* Halle: Eduard Anton, 1860.

————. *Viage al Desierto de Atacama: hecho de orden del gobierno de Chile en el verano 1853–54.* Halle: Eduard Anton, 1860.

Phillips, Catherine. "The Lost Cactus Garden of 'Quien Sabe.'" *Cactus and Succulent Journal* 87, no. 2 (March/April 2015): 48–60.

————. "Plants Known to Ysabel Wright and grown in her garden at Quien Sabe." An unpublished list provided to author.

Pinto Bahamonde, Raquel, and Arturo Kirberg Benavides. *Cactus del extremo Norte de Chile.* Santiago, Chile: Raquel Pinto Bahamonde, 2009.

Pissis, Aimé. *Geografía Física de Chile.* Paris: Instituto Geográfico de Paris, 1875.

Plano de Antofagasta. A map, scale 1:10,000, by Leoncio Guerra. Antofagasta: Lemare & Co., 1923. American Geographical Society Library, no. 254-d.A57 A-1923.

Plewe, Brandon, ed. *Mapping Mormonism: An Atlas of Latter-day Saint History.* Provo: Brigham Young University Press, 2012.

Pough, Frederick H. *A Field Guide to Rocks and Minerals.* Cambridge, Mass: The Riverside Press, 1955.

Preston, Diana and Michael. *A Pirate of Exquisite Mind: Explorer, Naturalist, and Buccaneer—The Life of William Dampier.* New York: Walker & Company, 2004.

Prieto, Manuel. "Privatizing Water and Articulating Indigeneity: The Chilean Water Reforms and the Atacameño People." PhD diss., University of Arizona, 2014.

Radding, Cynthia. *Wandering Peoples: Colonialism, Ethnic Spaces, and Ecological Frontiers in Northwestern Mexico, 1700–1850.* Duke University Press, 1997.

Rasheed, Bashar Sajjadur Khairul. "Man and the Desert: Northern Chile." PhD diss., Columbia University, 1970.

Rauh, W. "The Peruvian-Chilean Deserts." In *Ecosystems of the World 12A Hot Deserts and Arid Shrublands,* edited by Michael Evenari, Imanuel Noy-Meir, and David W. Goodall, 239–67. Amsterdam: Elsevier, 1985.

Reich, Martin, Carlos Palacios, Miguel Parada, Udo Fehn, Elion M. Cameron, Matthew I. Leybourne, and Alejandro Zúñiga. "Atacamite formation by deep saline waters in copper deposits from the Atacama Desert, Chile: evidence from fluid inclusions, groundwater geochemistry, TEM, and 36Cl data." *Mineralium Deposita* 43 (2008): 663–75.

Reinhartz, Dennis, and Gerald Saxon, eds. *The Mapping of the Entradas into the Greater Southwest.* Norman: University of Oklahoma Press, 1998.

Reyes Ortiz, Serapio. *Refutación al manifiesto del Ministro de Relaciones Exteriores de Chile, sobre la guerra con Bolivia.* Lima: Imprenta Nacional, 1879.

Rice, Tony. *Voyages of Discovery: Three Centuries of Natural History Exploration.* London and New York: The Natural History Museum, London, and Clarkson Potter Publishers, 1999.

Ristow, Walter. *The Juan de la Cruz map of South America, 1775.* Reprinted from Festschrift: edited by Clarence F. Jones. By Merle C. Prunty, Jr. *Northwestern University Studies in Geography* 6 (1962).

Rivera, Hernán Letelier. *El Vendedor de Pájaros.* Santiago: Alfaguara, 2014.

————. *La Reina Isabel cantaba rancheras.* Santiago: Alfaguara, 1994.

————. *Mi Nombre es Malarrosa.* Santiago: Alfaguara, 2008.

Robins, Nicolas A. *Mercury, Mining, and Empire: The Human and Ecological Cost of Colonial Silver Mining in the Andes.* Bloomington: Indiana University Press, 2011.

Rudolph, William E. *Vanishing Trails of Atacama—American Geographical Society Research Series No. 24.* New York: American Geographical Society, 1963.

Russell, William Howard. *A Visit to Chile and the Nitrate Fields of Tarapacá, Etc.* London: J. S. Virtue & Co., 1890.

Sagaris, Lake. *Bone and Dream: Into the World's Driest Desert.* Toronto: Alfred A. Knopf, 2000.

"San Pedro de Atacama." An entry in *Geocaching.com.* www.geocaching.com/geocache/GCKNFJ_san-pedro-de-atacama?guid=7dc942c5-d220-4ebc-9249-83e033562877.

San Román, Francisco J. *Desierto i cordilleras de Atacama, Vol. 2, Tomo Tercero: Hidrologia.* Santiago de Chile: Imprenta Nacional, 1902.

————. *Desierto i cordilleras de Atacama por Francisco J. San Román, Vol. 1, Tomo Primero. Itinerario de las exploraciones y Tomo II, Mision a Los Estados Unidos.* Santiago de Chile: Imprenta Nacional, 1896.

————. *Estudios Jeolójicos i Mineralójicos del Desierto i Cordilleras de Atacama. Vol. II.* Santiago: Sociedad Nacional de Minería, 1911.

————. *Reseña Industrial e Histórica de la Minería I Metalurjia de Chile.* Santiago de Chile: Imprenta Nacional, 1894.

Sater, William F. *Andean Tragedy: Fighting the War of the Pacific, 1879–1884.* Lincoln: University of Nebraska Press, 2007.

Sauer, Carl. "The Personality of Mexico." In *Land & Life: A Selection from the Writings of Carl Ortwin Sauer.* Edited by John Leighly, 104–18. Berkeley: University of California Press, 1963).

Schell, Patience A. *The Sociable Sciences: Darwin and His Contemporaries in Chile.* New York: Palgrave Macmillan, 2013.

Seelau, Ryan, and Laura M. Seelau, *The Pukará of Quitor: An Indigenous Self-Determination Case Study.* Bauu Institute and Press, 2013.

Simonin, Louis. *La Vie Souterarrine ou, Les Mines et les Mineurs.* Paris: Hachette, 1867.

"Site Preservation Grants Awarded to Projects in Greece and Chile." *Archaeology* 68, no. 5 (September/October 2015): 64–65.

Sketch map of northern Chile showing approximate location of nitrate lands. Based on a map in "The Nitrate Fields of Chile," by Walter Tower, 209–30, (1913), and in Miller and Singewald, *Mineral Deposits of South America*, 285, (1919).

Sketch of the Mines & Estates belonging to the Copiapo Mining Company, in the Province of Copiapo in Chili. A map by Isaiah Bowman in *Desert Trails of Atacama.* New York: American Geographical Society, 1924.

Skizze des Litorals von Bolivia. A map by Hermann Wagner, drafted for *Petermanns Geographische Mittheilungen* by Otto Koffmahn. Gotha, Germany: Justus Perthes, 1876.

Smith, Thomas. "Cruz Cano's Map of South America, Madrid 1775: Its Creation, Adversities, and Rehabilitation." *Imago Mundi* 20 (1966): 49–78.

Smits, Jan. *Petermann's Maps: Cartobibliography of the Maps in Petermanns Geographische Mitteilungen, 1855–1945*. Utrecht: HAS & DE GRAF, 2004.

Sobron, Pablo et al. "Geochemical profile of a layered outcrop in the Atacama analogue using laser-induced breakdown spectroscopy: Implications for Curiosity investigations in Gale." *Geophysical Research Letters* 40, no. 10 (May 28, 2013).

Stern, Charles R. et al. "Chilean Volcanoes." In *The Geology of Chile*, 147–78. London: The Geological Society, 2007.

"Studying farming in the driest desert in the world. Hayashida examines farm fields abandoned 500 years ago." A post by Karen Wentworth. March 26, 2015. news.unm.edu/news/studying-farming-in-the-driest-desert-in-the-world.

Südamerika. A map by F. A. Brockhaus. Leipzig: F. A. Brockhaus, 1863. David Rumsey Map Collection, no. 6819.055.

Sundt, Lorenzo. *Estudios jeolójicos i topográficos del desierto i puna de Atacama*, Vol. 1. Santiago: Sociedad Nacional de Minería, 1909.

Thomas, William L., ed. *Man's Role in Changing the Face of the Earth*. Chicago: University of Chicago Press, 1956.

Tobar, Héctor. *Deep Down Dark: The Untold Stories of 33 Men Buried in a Chilean Mine, and the Miracle that Set Them Free*. New York: Farrar, Straus and Giroux, 2014.

"Topographic maps of the Americas." An entry in *McGill.ca*. www.mcgill.ca/library/find/maps/topoamerica.

Tower, Walter S. "The Nitrate Fields of Chile." *Popular Science Monthly* 83 (September 1913): 209–30.

Trager, John N. "The Huntington Botanical Gardens presents the 2016 offerings of International Succulent Introductions." *Cactus and Succulent Journal* 88, no. 3 (May–June 2016): 111–12.

Trewartha, Glenn T. *The Earth's Problem Climates*. 2nd ed. Madison: University of Wisconsin Press, 1981.

Tschudi, J. J. Von. *Reisen durch Südamerika*. Vol. 5, Chile. Leipzig: F. A. Brockhaus, 1866–1869, reprinted 1971.

Twain, Mark. *Roughing It*. Hartford, Conn: American Publishing Company, 1872.

United Provinces of La Plata, Banda Oriental, Chile. A map by John Arrowsmith. London: John Arrowsmith, 1834. David Rumsey Map Collection, no. 0036.050.

Urbanus, Jason. "World Roundup" news section on Chile. *Archaeology* 70, no. 3 (May/June 2017): 24.

Valdivia, Pedro de. *Cartas de Pedro de Valdivia, que tratan del descubrimiento y conquista de Chile*. Santiago de Chile: Fondo Histórico y Bibliográfico José Toribio Medina, 1953.

Vallejo, José Joaquín. *Sketches of Life in Chile*. Edited by Simon Collier. Oxford University Press, 2002.

van Wyhe, John, ed. *The Complete Work of Charles Darwin Online.* http://darwin-online.org.uk.

Varigny, Charles Victor Grosnier de. *La guerra del Pacífico.* Santiago de Chile: Imprenta Cervantes, 1922.

Vásquez Bazan, César. "Robo chileno del desierto de Atacama—Testimónios cartográficos y constitucionáles y documentos de la epoca prueban que limite norte de Chile es el paralelo 25." November 23, 2016. https://cavb.blogspot.com/2013/05/desierto-de-atacama-no-pertenece-chile.html.

Verba, Ericka Kim. "Sewing Resistance." In *Mapping Latin America: A Cartographic Reader*, edited by Jordana Dym and Karl Offen, 258–62. Chicago: University of Chicago Press, 2011.

Vidiella, Patricia E., Juan J. Armesto, and Julio R. Gutiérrez. "Vegetation changes and sequential flowering after rain in the southern Atacama Desert." *Journal of Arid Environments* 43 (1999): 449–58.

von Hagen, Victor Wolfgang. *South America Called Them: Explorations of the Great Naturalists.* New York: Alfred Knopf, 1945.

Waldseemüller Map. Also known as *Universalis cosmographia secundum Ptolomaei traditionem et Americi Vespucci Alioru[m]que illustrationes.* Twelve sheets comprising one map. St. Dié, France: 1507. U.S. Library of Congress.

Wallace, David Rains. *Chuckwalla Land: The Riddle of California's Desert.* Berkeley: University of California Press, 2011.

Wallace, Terry C., and Michelle K. Hall-Wallace. "Minerals of the Andes: Emeralds, Gold, and Silver from the Sky." *Rocks & Minerals* 78, no. 1 (January/February 2003): 12–38.

Weaver, Stewart A. *Exploration: A Very Short Introduction.* New York: Oxford University Press, 2015.

Weber, David J. *Bárbaros: Spaniards and Their Savages in the Age of Enlightenment.* New Haven: Yale University Press, 2005.

Weiss, Daniel. "Colors of the Priesthood: An intriguing source of power is revealed in an ancient Andean tomb." *Archaeology* 69, no. 2 (March/April 2016): 36–37.

Welland, Michael. *The Desert: Lands of Lost Borders.* London: Reaktion Books, 2015.

White, John H., Jr. *American Locomotives: An Engineering History, 1830–1880.* Baltimore, MD: The Johns Hopkins University Press, 1997.

———. *A History of the American Locomotive: Its Development, 1830–1880.* Baltimore, MD: The Johns Hopkins University Press, 1968.

Williams, Glyndwr. *The Great South Sea: English Voyages and Encounters, 1570–1750.* New Haven: Yale University Press, 1997.

Wilson, Chris. *The Myth of Santa Fe: Creating a Modern Regional Tradition.* Albuquerque: University of New Mexico Press, 1997.

Winchester, Simon. *The Map That Changed the World: William Smith and the Birth of Modern Geology.* New York: HarperCollins, 2001.

Wisniak, Jaime, and Ingrid Garcés. "The rise and fall of the salitre (sodium nitrate) industry." *Indian Journal of Chemical Technology* 8 (September 2001): 427–38.

Wroth, Lawrence C. "Alonso de Ovalle's large map of Chile, 1642." *Imago Mundi* (1959): 90–95.

INDEX

Note: locators in italics refer to images.